AF361679

Advances in Diversity, Conservation, and Taxonomy of Madagascar's Amphibians and Reptiles—Feature Issue of the Section Herpetology

Advances in Diversity, Conservation, and Taxonomy of Madagascar's Amphibians and Reptiles—Feature Issue of the Section Herpetology

Editors

Franco Andreone
Angelica Crottini
Andolalao Rakotoarison
Fandresena Rakotoarimalala

Basel • Beijing • Wuhan • Barcelona • Belgrade • Novi Sad • Cluj • Manchester

Editors

Franco Andreone
Museo Regionale di
Scienze Naturali
Torino
Italy

Angelica Crottini
Universidade do Porto
Vairão
Portugal

Andolalao Rakotoarison
Université de l'Itasy
Soavinandriana Itasy
Madagascar

Fandresena Rakotoarimalala
Université d'Antananarivo
Antananarivo
Madagascar

Editorial Office
MDPI AG
Grosspeteranlage 5
4052 Basel, Switzerland

This is a reprint of articles from the Special Issue published online in the open access journal *Animals* (ISSN 2076-2615) (available at: https://www.mdpi.com/journal/animals/special_issues/513737ZPKY).

For citation purposes, cite each article independently as indicated on the article page online and as indicated below:

Lastname, A.A.; Lastname, B.B. Article Title. *Journal Name* **Year**, *Volume Number*, Page Range.

ISBN 978-3-7258-1787-0 (Hbk)
ISBN 978-3-7258-1788-7 (PDF)
doi.org/10.3390/books978-3-7258-1788-7

Cover image: *Mantidactylus pauliani*, a Critically Endangered frog species from the Ankaratra Massif (photograph by Franco Andreone).

Contents

Editorial

Conserving Madagascar's Amphibians and Reptiles Requires Collaboration between Scientists

Franco Andreone [1,*], Angelica Crottini [2,3,4], Andolalao Rakotoarison [5,6] and Fandresena Rakotoarimalala [7]

[1] Museo Regionale di Scienze Naturali, Via G. Giolitti, 36, I-10123 Torino, Italy
[2] CIBIO, Centro de Investigação em Biodiversidade e Recursos Genéticos, InBIO Laboratório Associado, Campus de Vairão, Universidade do Porto, 4485-661 Vairão, Portugal; tiliquait@yahoo.it
[3] Departamento de Biologia, Faculdade de Ciências, Universidade do Porto, 4169-007 Porto, Portugal
[4] BIOPOLIS Program in Genomics, Biodiversity and Land Planning, CIBIO, Campus de Vairão, 4485-661 Vairão, Portugal
[5] Mention Environnement, Université de l'Itasy, Faliarivo Ambohidanerana, Soavinandriana Itasy 118, Madagascar; andomailaka@gmail.com
[6] School for International Training, Antananarivo 101, Madagascar
[7] Mention Zoologie et Biodiversité Animale, Faculté des Sciences, Université d'Antananarivo, Antananarivo 101, Madagascar; fandresenarak@gmail.com
* Correspondence: franco.andreone@regione.piemonte.it or franco.andreone@gmail.com

Citation: Andreone, F.; Crottini, A.; Rakotoarison, A.; Rakotoarimalala, F. Conserving Madagascar's Amphibians and Reptiles Requires Collaboration between Scientists. *Animals* **2024**, *14*, 2091. https://doi.org/10.3390/ani14142091

Received: 8 June 2024
Revised: 2 July 2024
Accepted: 11 July 2024
Published: 17 July 2024

Madagascar is well known for its exceptional biodiversity and striking endemicity levels, which are accompanied by high rates of deforestation and habitat alteration [1,2]. In recent decades, it has been possible to witness a progressive shrinking of Malagasy forests, which makes it more urgent than ever to improve our knowledge of this unique biota [3].

Among its peculiar fauna, amphibians and reptiles undoubtedly play an important ecological role and represent very speciose vertebrate groups: at the time of writing (May 2024), 414 species of amphibians (100% endemic [4]) and 437 species of reptiles (98% endemic) are known, and each year numerous species are described [5]. This situation highlights the urgent need to continue to document and describe the incredible biota of Madagascar and to promote its conservation. To discuss these threats and develop a strategy to protect the amphibians of Madagascar, two "A Conservation Strategy for the Amphibians of Madagascar" (ACSAM) meetings, an initiative coordinated by the regional IUCN/SSC Amphibian Specialist Group—Madagascar, were organized (the first in 2006 in Antananarivo and the second in 2014 in Ranomafana) [6,7]. These initiatives were mostly dedicated to the knowledge and protection of the highly diverse amphibian communities of Madagascar and the safeguard of their ecosystems, providing suggestions for needed actions to contribute to their conservation. More recently, an International Chameleon Day [8] was also launched and highlighted the peculiar role played by these animals as reptilian ambassadors [9].

Conservation in Madagascar, a country with clear economic difficulties, can realistically be achieved through a continuative synergy between knowledge, education, and people's awareness. The development of a tight collaboration between local and foreign researchers and institutions has contributed to the development of a rich production of scientific contributions on the amphibians and reptiles of Madagascar [10,11]. Several publications are now produced by a variegated mix of local and foreign contributors, with national researchers frequently taking the lead, e.g., [12]. This is a much welcome development, followed by an increase in the number of scientific papers that include only Malagasy researchers. These outputs are often the result of fundamental research but play a significant role in informing decision-makers and stakeholders to facilitate the identification of more efficient conservation actions.

For these reasons, we are delighted to present the current Special Issue of Animals dedicated to advances in diversity, conservation, and taxonomy of Madagascar's amphibians and reptiles. We hope that this issue, which has benefitted from a total waiver of publication fees and is published in open access, will represent a valuable contribution to the development of scientific knowledge about Madagascar's biodiversity. In this issue, two papers have been produced by a team composed of only Malagasy researchers [13,14] and three contributions by a mix of local and foreign researchers [15–17].

The topics treated in this collection are sufficiently varied; more specifically, four papers deal with taxonomy and species distribution [13,16–18], and six with the ecology and conservation of Malagasy herpetofauna [14,15,19–22]. This repartition and the high number of discovered and described species that accumulated over the last three decades highlight how studies of geographic and taxonomic diversity remain a priority and should be considered a "must" to continually improve the comprehension of Madagascar's herpetofauna [5].

We are glad to have succeeded in putting together this collection of papers where taxonomy, ecology, and biogeography approaches will hopefully be able to contribute to inform amphibian and reptile conservation. Studies on the ecology and behavior of Malagasy amphibians and reptiles are becoming increasingly available [23–25] and represent a significant contribution to the understanding of the life histories of these animals. The increase in attention toward these facets of the biology of amphibians and reptiles is particularly beneficial and remains an important aspect worth further investigation.

Lastly, the impact of amphibian and reptile trade, as shown by Carpenter and Andreone [26], is a relevant contribution to an aspect that cannot be neglected, i.e., the necessary bridge between ecology and economy in a developing country, as also evidenced by a recent editorial [27]. An approach that considers conservation, economy, and local development remains crucial for the sustainable management of Malagasy natural resources.

Conflicts of Interest: The authors declare no conflict of interest.

References

1. Andreone, F.; Carpenter, A.I.; Crottini, A.; D'Cruze, N.; Dubos, N.; Edmonds, D.; Garcia, G.; Luedtke, J.; Megson, S.; Rabemananjara, F.C.E.; et al. Amphibian conservation in Madagascar: Old and novel threats for a peculiar fauna. In *Status and Threats of Afrotropical Amphibians. Sub-Saharian Africa, Madagascar, Western Indian Ocean Islands. Vol. II*; Heatwole, H., Roedel, M., Eds.; Chimaira: Frankfurt am Main, Germany, 2021; Chapter 4; pp. 147–186, ISBN 9783899373103.
2. Ralimanana, H.; Perrigo, A.L.; Smith, R.J.; Borrell, J.S.; Faurby, S.; Rajaonah, M.T.; Randriamboavonjy, T.; Vorontsova, M.S.; Cooke, R.S.C.; Phelps, L.N.; et al. Madagascar's extraordinary biodiversity: Threats and opportunities. *Science* **2022**, *378*, eadf1466. [CrossRef] [PubMed]
3. Vieilledent, G.; Grinand, C.; Rakotomalala, F.A.; Ranaivosoa, R.; Rakotoarijaona, J.-R.; Allnutt, T.F.; Achard, F. Combining global tree cover loss data with historical national forest-cover maps to look at six decades of deforestation and forest fragmentation in Madagascar. *Biol. Conserv.* **2018**, *222*, 189–197. [CrossRef]
4. AmphibiaWeb. University of California, Berkeley, CA, USA. 2024. Available online: https://amphibiaweb.org. (accessed on 19 May 2024).
5. Antonelli, A.; Smith, R.J.; Perrigo, A.L.; Crottini, A.; Hackel, J.; Testo, W.; Farooq, H.; Torres Jiménez, M.F.; Andela, N.; Andermann, T.; et al. Madagascar's extraordinary biodiversity: Evolution, distribution, and use. *Science* **2022**, *378*, abf0869. [CrossRef] [PubMed]
6. Andreone, F.; Carpenter, A.I.; Cox, N.; du Preez, L.; Freeman, K.; Furrer, S.; Garcia, G.; Glaw, F.; Glos, J.; Knox, D.; et al. The challenge of conserving amphibian megadiversity in Madagascar. *PLoS Biol.* **2008**, *6*, e118. [CrossRef] [PubMed]
7. Rakotoarison, A.; Ndriantsoa, S.H.; Rabemananjara, F.C.E.; Rabibisoa, N.H.C.; Rakotonanahary, T.F.; Randriamahazo, J.A.R.H.; Andreone, F. More than 15 years of amphibian conservation in Madagascar under the flag of IUCN SSC Amphibian Specialist Group. *Nat. Sicil.* **2022**, *46*, 305–312. [CrossRef]
8. Samoelah, L.F. First International Chameleon Day: Madagasikara Voakajy Focuses Its Efforts on the Conservation of *Calumma tarzan*. 2024. Available online: https://www.madagasikara-voakajy.org/blog/first-international-chameleon-day-madagasikara-voakajy-focuses-its-efforts-on-the-conservation-of-calumma-tarzan (accessed on 10 May 2024).
9. Jenkins, R.K.B.; Tognelli, M.F.; Bowles, P.; Cox, N.; Brown, J.L.; Chan, L.; Andreone, F.; Andriamazava, A.; Andriantsimanarilafy, R.R.; Anjeriniaina, M.; et al. Extinction Risks and the Conservation of Madagascar's Reptiles. *PLoS ONE* **2014**, *9*, e100173. [CrossRef]
10. Andreone, F. Frogging Madagascar: A free chat on frogs and frog conservationists across the Red Island. In *A Conservation Strategy for the Amphibians of Madagascar*; Andreone, F., Ed.; Museo Regionale di Scienze Naturali di Torino: Torino, Italy, 2008; Volume 45, pp. 21–46, ISBN 9788886041720.

11. Vences, M.; Jovanovic, O.; Glaw, F. Historical analysis of amphibian studies in Madagascar: An example for increasing research intensity and international collaboration. In *A Conservation Strategy for the Amphibians of Madagascar*; Andreone, F., Ed.; Museo Regionale di Scienze Naturali di Torino: Torino, Italy, 2008; Volume 45, pp. 47–58.

12. Rakotondrina, A.J.V.; Andriantsimanarilafy, R.R.; Razafimanahaka, H.J.; Raselimanana, A.P.; Gumbs, R.; Ofori-Boateng, C.; Taft, J.M.; Ratsoavina, F.M. New distribution records and population density of the Critically Endangered Tarzan chameleon (*Calumma tarzan*), eastern Madagascar. *Afr. J. Herpetol.* **2024**, *73*, 36–47. [CrossRef]

13. Randriamialisoa; Andriantsimanarilafy, R.R.; Rakotondrina, A.J.V.; Rakotoarisoa, J.A.; Nasaina, R.T.; Rabearivony, J.; Raselimanana, A.P. Overview of reptile diversity from Bobaomby Complex, northern tip of Madagascar. *Animals* **2023**, *13*, 3396. [CrossRef]

14. Radonirina, H.O.; Randriamahatantsoa, B.; Rabibisoa, N.H.C. Population status and vulnerability of *Mantidactylus pauliani* from Ankaratra Protected Area, Madagascar. *Animals* **2023**, *13*, 2706. [CrossRef] [PubMed]

15. Guarino, F.M.; Andreone, F.; Mezzasalma, M.; Licata, F.; Puoti, S.; Santos, B.; Cocca, W.; Fidy, J.F.S.; Ndriantsoa, S.H.; Noel, J.; et al. Life history traits and longevity of the invasive Asian common toad *Duttaphrynus melanostictus* (Schneider, 1799) in Madagascar. *Animals* **2023**, *13*, 2099. [CrossRef] [PubMed]

16. Rabibisoa, N.H.C.; Welt, R.S.; Raxworthy, C.J. Revision of the subgenus *Ochthomantis* frogs from Madagascar (Amphibia: Mantellidae) with the description of four species and resurrection of *Mantidactylus catalai* and *M. poissoni*. *Animals* **2023**, *13*, 2800. [CrossRef]

17. Radonirina, H.O.; Randriamahatantsoa, B.; Rabibisoa, N.H.C.; Raxworthy, C.J. Amphibians and reptiles of the Montagne des Français: An update of the distribution and regional endemicity. *Animals* **2023**, *13*, 3361. [CrossRef]

18. Mezzasalma, M.; Brunelli, E.; Odierna, G.; Guarino, F.M. Chromosome diversity and evolution of the endemic Malagasy velvet geckos of the genus *Blaesodactylus* (Reptilia, Gekkonidae). *Animals* **2023**, *13*, 2068. [CrossRef]

19. Mausberg, N.; Dausmann, K.H.; Glos, J. In search of suitable breeding sites: Habitat heterogeneity and environmental filters determine anuran diversity of western Madagascar. *Animals* **2023**, *13*, 3744. [CrossRef]

20. Licata, F.; Eusebio Bergò, P.; Edmonds, D.; Andreone, F.; Rosa, G.M. Spatial ecology of an arboreal iguana (*Oplurus cyclurus*) in a treeless landscape. *Animals* **2023**, *13*, 3198. [CrossRef]

21. Carpenter, A.I.; Andreone, F. Malagasy amphibian wildlife trade revisited: Improving management knowledge of the trade. *Animals* **2023**, *13*, 2324. [CrossRef]

22. Lehtinen, R.M. Empirical evidence for the rescue effect from a natural microcosm. *Animals* **2023**, *13*, 1907. [CrossRef]

23. Mori, A.; Randriamboavonjy, T.M. Field observation of maternal attendance of eggs in a Madagascan snake, *Leioheterodon madagascariensis*. *Curr. Herpetol.* **2010**, *29*, 91–95. [CrossRef]

24. Dubos, N.; Morel, L.; Crottini, A.; Freeman, K.; Honoré, J.; Lava, H.; Nöel, J.; Porton, I.; Rendrirendry, G.; Rosa, G.M.; et al. High interannual variability of a climate-driven amphibian community in a seasonal rainforest. *Biodivers. Conserv.* **2019**, *29*, 893–912. [CrossRef]

25. Porcel, X.; Dubos, N.; Nöel, J.; Lava, H.; Velo, J.H.; Georges; Melo, M.; Rosa, G.M.; Andreone, F.; Crottini, A. Male parental care in Malagasy stream-dwelling frogs of the *Mantidactylus femoralis* group (Anura: Mantellidae: *Ochthomantis*). *Herpetol. Notes* **2022**, *15*, 55–61.

26. Carpenter, A.I.; Andreone, F. Valorisation of Madagascar's wildlife trade and wildlife tourism: What are the conservation benefits? *Conservation* **2023**, *3*, 509–522. [CrossRef]

27. Luiselli, L. Africa's biodiversity in-between ecology and economy. *Afr. J. Ecol.* **2024**, *62*, e13268. [CrossRef]

Article

Amphibians and Reptiles of the Montagne des Français: An Update of the Distribution and Regional Endemicity

Herizo Oninjatovo Radonirina [1,*], Bernard Randriamahatantsoa [2], Nirhy H. C. Rabibisoa [1,2,*] and Christopher John Raxworthy [3]

1   Doctoral School of Natural Ecosystems, University of Mahajanga, Mahajanga 401, Madagascar
2   Sciences de la Vie et de l'Environnement, Faculté des Sciences, de Technologies et de l'Environnement, University of Mahajanga, BP 652, Mahajanga 401, Madagascar; bernardzoo01@gmail.com
3   American Museum of Natural History, Department of Herpetology, New York, NY 10024, USA; rax@amnh.org
*   Correspondence: radonirinaherizo@gmail.com (H.O.R.); nhcrabibisoa@gmail.com (N.H.C.R.); Tel.: +261-340471088 (H.O.R.)

Simple Summary: In 2014 and 2020, we conducted research in the Montagne des Français Protected Area. This area is recognized for its remarkable fauna and high rate of endemism. The objective of this study was to prioritize areas and determine the diversity and distribution of herpetofauna species, as well as their conservation status, endemism, and threats. In this study, we compared species richness between two surveys conducted in 2014 and 2020. We found that three species of amphibians and one reptile species had a new distribution. The 2020 survey also discovered *Langaha pseudoalluaudi*, a snake that has not been observed since 2007. The information gathered from this survey could be of use to site managers in the revision of conservation management plans.

Abstract: The harmonious landscape of Montagne des Français is a protected area in the far north of Madagascar. Our herpetofauna surveys were conducted on the eastern and western slopes according to habitat variations within the massif for 2014 and 2020. Our research updates the herpetofauna species richness, spatial distribution ranges, and ecological guilds within the protected area. We used direct opportunistic observations, systematic refuge examinations, and pitfall traps with drift fences at three sites to sample animals. Nineteen amphibian and fifty reptile species were recorded during this study. Three amphibian species and one reptile species, in addition to the snake *Langaha pseudoalluaudi*, last recorded in 2007, were discovered at Montagne des Français. Here, we present a database update for the herpetofauna species from Montagne des Français and provide a specific morphological description of the morphospecies considered as a new extension or localized distribution. This new database can help site managers develop new strategic conservation plans in response to habitat modification.

Keywords: amphibians; reptiles; Montagne des Français; update; distribution; regional endemicity

Citation: Oninjatovo Radonirina, H.; Randriamahatantsoa, B.; Rabibisoa, N.H.C.; Raxworthy, C.J. Amphibians and Reptiles of the Montagne des Français: An Update of the Distribution and Regional Endemicity. *Animals* **2023**, *13*, 3361. https://doi.org/10.3390/ani13213361

Academic Editors: Franco Andreone, Angelica Crottini, Andolalao Rakotoarison and Fandresena Rakotoarimalala

Received: 20 June 2023
Revised: 24 October 2023
Accepted: 27 October 2023
Published: 29 October 2023

1. Introduction

Madagascar is recognized globally as a biodiversity hotspot [1]. High levels of endemism and diversity are observed due to its isolation [2–4]. The creation of protected areas is underway to conserve biodiversity and ecosystem services [5]. The island also has a highly endemic herpetofauna, with 90% of the species found nowhere else [6]. Montagne des Français, which is one of Madagascar's protected areas, is situated in the northern region. The site is a limestone massif [7] and is home to an exceptional community of herpetofauna [8], including several endemic species [9].

Unfortunately, the biodiversity of the flora and fauna in this protected area has been affected by human activities. These anthropogenic pressures have caused habitat loss and significant alterations to the ecosystem [7,10], resulting in landscape erosion. Currently, dry forests are one of the most threatened ecosystems on the island as they are subjected to frequent fires [11,12]. Among vertebrates, herpetofauna are especially sensitive to habitat alterations and disturbances [13,14]. In addition, numerous studies have shown that species within the herpetofauna category are among the most endangered vertebrates on Earth [15–17]. This is typically observed in tropical regions [18]. Although research efforts have been ongoing for decades, the inventory of Madagascar's herpetofauna remains incomplete [19,20].

Research in Montagne des Français has provided incomplete data on the distribution and status of species [7,20,21]; however, ongoing field studies continue to collect new information. A large body of literature indicates that Montagne des Français is home to many reptile and amphibian species. Among them are *Heteroliodon fohy* [22] and *Thamnosophis martae* [23] snakes, along with *Brookesia tristis* [24]. The gecko *Paroedura lohatsara* is also present in this locality, along with various frog species, including *Stumpffia staffordi* [22,25] and *Mantella viridis* [26,27]. The majority of these species are microendemic or regionally endemic, meaning that they are exceptionally or primarily found in the Montagne des Français and its neighboring regions.

Many of the species under consideration have recently undergone assessments revealing that they are "Critically Endangered" or "Endangered", highlighting their precarious position regarding extinction. Human activities, such as agriculture and logging, have severely reduced their habitats and caused fragmentation and degradation, all of which pose significant threats to their continued survival. The distribution maps of several species suggest that a biogeographical border exists between Montagne des Français/Orangea and limestone massifs situated to the south, underscoring the importance of safeguarding and managing Montagne des Français as a critical habitat for these reptiles [28]. Furthermore, species inventories have only been conducted in a few areas of Montagne des Français. In this study, we investigated two sites that are much less visited, Ampitiliantsambo and Sahabedara.

This article presents recent findings on the biodiversity, biogeography, and endemicity of herpetofauna residing in Montagne des Français. Field surveys were conducted on both the eastern and western slopes of the mountain in 2014 and 2020, incorporating ecological parameters, such as habitat type, stream flow, and degree of degradation. The database on biodiversity resulting from the two studies of amphibians and reptiles serves not only to prioritize species but also to ease pressure on the protected core area and aid the site manager in executing the new management plan. These new data can also contribute to the implementation of a new conservation strategy, given the many threats that surround this protected area [29,30]. Optimal management of protected areas requires access to scientific data and comprehensive knowledge of biodiversity, which should be incorporated into the design of conservation and management strategies [31]. Biological inventories are highly effective means of acquiring such information [32].

2. Materials and Methods

2.1. Study Site

Montagne des Français is a limestone plateau intersected by narrow canyons, located in northern Madagascar, between 12°18 and 12°27′ latitudes south and 49°21 and 49°23′ longitudes east, with a maximum elevation of 425 m asl. The forest massif is very close to the city of Antsiranana, measuring 6049 ha (Figure 1). This site has had a fully protected status as a Category V new protected area since 28 April 2015 (Decree N°. 2015-780) and is managed by the NGO "Service d'Appui pour la Gestion de l'Environment" which is located in the city of Antsiranana.

- Sahabedara is located at an altitude of 159 m between 049°21′41.7″ E and 12°23′05.8″ S. The area consists of a gallery forest with sandy and rocky terrains. The forest is moderately degraded, with an open canopy, and is situated outside the core area of Montagne des Français.

- Ampitiliantsambo lies between 049°23′05.5″ E and 12°23′16.6″ S at an altitude of 204 m. It supports a natural semi-deciduous forest and an open-canopy fragmented forest. This site is situated in the peripheral area of Montagne des Français.
- Andavakoera is located between 049°20′58.1″ E and 12°19′49.6″ S at an altitude of 173 m. It consists of a Sambirano forest or a relict gallery forest with sand and rocks. The site represents the core area of Montagne des Français PA.

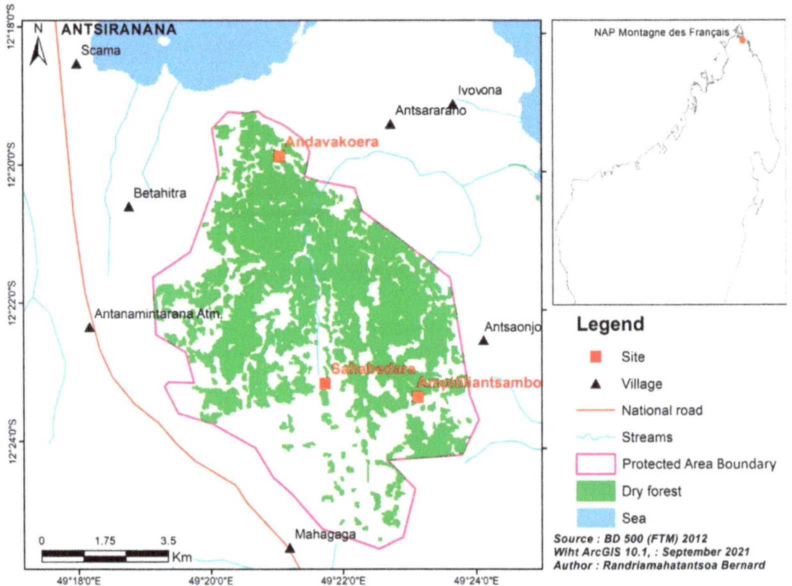

Figure 1. Map of the study sites in Montagne des Français.

2.2. Data Collection

We conducted a biological inventory at the Andavakoera site from January 13 to 24 in 2014, and from May 3 to 24 in 2020. The first survey focused only on the Andavakoera site and the second on Sahabedara, Ampitiliantsambo, and Andavakoera (Figure 1). The Andavakoera site is situated in the core area, while Sahabedara and Ampitiliantsambo are in the peripheral areas. In 2014, a group of four individuals conducted ten days of fieldwork, whereas in 2020, a group of seven individuals conducted six days of fieldwork per site. We established four transects to ensure that the different habitat types at each site were surveyed. The transect length ranged from 1500 m to 3000 m. We determined each transect based on the type of environment, plant and geological formations, topography (valley bottoms, flanks, canyons, and summits), and degree of degradation. The field guide developed by [8] was used to identify species.

During these expeditions, three standard field techniques were used to sample the herpetofauna community:

(1) Pitfall trapping: We used pitfall traps made up of buckets (275 mm deep, 290 mm top internal diameter, and 220 mm bottom internal diameter). To allow water to drain, we removed the handles and made a few tiny holes at the bottom. We buried the buckets with their upper rims flushed with the earth at intervals of 10 m along a drift fence. We used 100 m long by 0.5 m high black plastic sheeting to create the drift fence. We positioned the fence to cut through the center of each pitfall trap and fastened it vertically to short wooden pegs. To prevent individuals from passing beneath the barrier, we dug the bottom of the plastic approximately 50 mm deep into the ground. Each day, in the early morning and late afternoon, we checked the pitfall traps and removed all the captives. For eight nights, we employed three drift fence pitfall arrays at each location.

(2) Refuge examinations: We conducted active searches both during the day and night in all possible habitats and altitudes. The majority of the searches, which lasted between 30 min and three hours, were concentrated in the interior of the forest near pathways and trails for cryptic reptile species and arboreal frogs, as well as in streams and the related riparian vegetation for amphibians. Examining refuges are suitable for both active daytime and nocturnal resting animals.

(3) Opportunistic searching [33]: This method complemented the refuge examinations and was usually carried out at the same time and along the same transect. This involved creating an inventory of amphibians and reptiles perceived along the transect. The transect was visited once per day to mitigate disturbance to the animals. As certain groups of species are nocturnal, observations were also conducted at night.

2.3. Data Analysis

Population structure was determined by examining several parameters, including the number of species present at each site, their abundance, status, ecological distribution, and habitat frequency. Concerning the habitat description, we used two forest types according to habitat loss, which was also used in [34]. The first one represents a forest with a weak pressure (noted F) where there are more big trees ($\geq$25 m), and the second one has numerous observed anthropogenic activities as well as fewer big trees (noted A). We used a classification system similar to that developed by [35] to categorize species based on their abundance. This system is as follows: abundant (encountered regularly in large numbers throughout the year), common (encountered regularly year-round), infrequent (encountered unpredictably and in small numbers), and rare (rarely encountered). All identifications of the species are only based on morphological analysis. For frogs, we used stable morphological features like mensuration, iris peripheral coloration, and webbing formula; and for Squamata, we used mensuration, characteristics of rostral appendix, and scale and spine features.

3. Results

3.1. Species Richness

We recorded a total of 19 amphibian and 50 reptile species at Montagne des Français during the two surveys of 2014 and 2020, which provided an overall herpetofaunal diversity of 69 species. Some species are presented within the Supplementary Materials. Of these species, 94% are endemic to Madagascar, 10% are locally endemic, and 21% are regionally endemic. From this overall result, in 2014, we identified 10 amphibians of three families and 35 reptiles of six families. In 2020, we found 17 amphibians from two families and 44 reptiles from six families (Table 1). In 2014, we only investigated the Andavakoera site. However, in 2020, we recorded amphibians and reptiles from Sahabedara (14 amphibians/37 reptiles), Ampitiliantsambo (15 amphibians/30 reptiles), and Andavakoera (8 amphibians/24 reptiles) (Figure 2). We noticed that Andavakoera had fewer species recorded in 2020 than in 2014.

3.2. Abundance and Ecological Preference

We found that 9 species (13%) were abundant, 21 species (30%) were common, 17 species (24%) were infrequent, and 22 species (31%) were rare (Table 1). Of the recorded amphibians, *Mantella viridis* was dominant (34% of the total) in Andavakoera, whereas other sites were dominated by *Boophis tephraeomystax*, *Mantidactylus bellyi*, and *Mantidactylus ulcerosus*. We also observed that 28 species were strictly arboreal (37%), 32 were in terrestrial environments (43%), 10 were semi-aquatic (13%), and 5 were found in rocky environments (7%). Half of the observed species (50%) were found within forested areas, 10% were recorded in anthropogenically disturbed areas, and 40% were in both forested and anthropogenically disturbed areas.

Table 1. Species composition of the herpetofaunal community recorded in the NAP Montagne des Français during the years 2014 and 2020. Site: AND: Andavakoera, AMP: Ampitiliantsambo, SAH: Sahabedara. End: endemicity, IUCN: CR = Critically Endangered, EN = Endangered, VU = vulnerable, LC = Least Concern, NE = Not Evaluated. Relative abundance (RA): A = abundant, C = common, I = infrequent, R = rare. Ecological distribution (ED): Ab = arboreal, Tr = terrestrial, S = semiaquatic, Ro = rocky. Habitat: F = forest relatively intact, A= anthropogenically disturbed area. Endemicity (END): E = endemic, RE = regionally endemic, LE = locally endemic, NE = not endemic.

TAXON	2014 AND	AND	2020 AMP	SAH	IUCN	RA	ED	Habitat	END
AMPHIBIANS									
MANTELLIDAE									
Aglyptodactylus securifer	+	−	−	−	LC	I	Tr	F	E
Boophis marojezensis	−	−	+	−	NE	R	Ab	F	E
Boophis occidentalis	−	−	+	+	NE	R	Ab	F, A	E
Boophis sp.	−	−	−	+	NE	R	Ab	F, A	E
Boophis tephraeomystax	+	+	+	+	LC	A	Ab, S	F, A	E
Gephyromantis pseudoasper	−	−	+	+	LC	R	S	F	E
Laliostoma labrosum	+	−	+	−	LC	I	Tr	F, A	RE
Mantella viridis	+	+	+	+	EN	A	Tr	F, A	LE
Mantidactylus bellyi	+	+	+	+	LC	A	Tr	F, A	E
Mantidactylus betsileanus	+	−	−	−	LC	I	Tr	F	E
Mantidactylus ulcerosus	+	+	+	+	LC	A	Tr	F, A	E
PTYCHADENIDAE									
Ptychadena mascareniensis	+	+	+	+	LC	C	Tr, S	F, A	C
MICROHYLIDAE									
Stumpffia analamaina	−	−	+	+	LC	I	S	F	E
Stumpffia angeluci		+	+	+	LC	C	S	F	RE
Stumpffia gimmeli	+	−	−	+	LC	C	S	F	RE
Stumpffia cf. *madagascariensis*	−	−	+	+	LC	R	S	F	RE
Stumpffia roseifemoralis	−	+	+	+	EN	C	S	F	E
Stumpffia staffordi	−	+	+	+	VU	R	S	F	E
DICROGLOSSIDAE									
Hoplobatrachus tigerinus	+	−	−	−	LC		Tr	F, A	NE
REPTILES									
BOIDAE									
Acrantophis madagascariensis	−	−	+	+	LC	R	Tr	F, A	E
Sanzinia volontany	+	−	−	+	LC	C	Ab	F	E
CHAMAELEONIDAE									
Brookesia ebenaui	+	−	−	+	VU	I	Tr	F	E
Brookesia stumpffi	+	+	−	−	VU	C	Tr	F	E
Brookesia tristis	+	−	−	−	NE	R	Tr	F	E
Furcifer oustaleti	+	−	+	−	LC	A	Ab	F, A	E
Furcifer pardalis	+	−	+	−	LC	C	Ab	F, A	RE
Furcifer petteri	+	+	+	+	VU	I	Ab	F, A	E
GEKKONIDAE									
Blaesodactylus boivini	+	+	+	+	VU	C	Ab	F, A	RE
Geckolepis maculata	+	−	+	−	NE	C	Ab	F, A	E
Geckolepis typica	+	+	−	−	LC	C	Ab	F, A	E
Hemidactylus frenatus	+	+	−	+	LC	A (	Ab, Ro	A	NE
Hemidactylus mercatorius	+	−	−	−	LC	A	Ab, Ro	A	NE
Lygodactylus heterurus	−	−	+	+	LC	R	Ab	F	RE
Paroedura hordiesi	+	−	−	+	CR	R	Ro	F, A	LE
Paroedura lohatsara	+	+	+	+	CR	R	Ro	F, A	LE
Paroedura stumpffi	+	−	−	+	LC	C	Ro	F, A	RE
Phelsuma abbotti	+	+	+	+	LC	I	Ab	F	E

Table 1. *Cont.*

TAXON	2014 AND	AND	2020 AMP	SAH	IUCN	RA	ED	Habitat	END
Phelsuma grandis	+	−	+	−	LC	A	Ab	F	RE
Uroplatus fetsy	−	−	+	+	NE	R	Ab	F	RE
Uroplatus giganteus	−	−	−	+	VU	R	Ab	F	E
Uroplatus henkeli	+	−	+	−	VU	C	Ab	F	E
Uroplatus sameiti	+	+	−	−	LC	C	Ab	F	E
LAMPROPHIDAE									
Alluaudina bellyi	−	+	−	−	LC	R	Tr	F	E
Dromicodryas bernieri	−	−	+	+	LC	C	Tr	A	E
Dromicodryas quadrilineatus	+	−	+	+	LC	C	Tr	A	E
Heteroliodon fohy	+	−	−	−	EN	R	Tr	F	LE
Ithycyphus miniatus	+	−	+	+	LC	I	Ab	F	E
Langaha madagascariensis	+	−	−	−	LC	I	Ab	F, A	E
Langaha pseudoalluaudi	−	−	−	+	LC	R	Ab	F, A	E
Leioheterodon modestus	+	−	−	+	LC	C	Tr	F, A	E
Leioheterodon madagascariensis	+	+	+	+	LC	C	Tr	F, A	E
Liophidium therezieni	+	−	−	−	EN	R	Tr	F	LE
Liophidium torquatum	+	−	+	+	LC	I	Tr	F	E
Lycodryas inopinae	+	−	+	+	LC	R	Ab	F, A	RE
Lycodryas granuliceps	−	−	+	−	LC	I	Ab	F, A	E
Lycodryas pseudogranuliceps	−	−	−	+	LC	C	Ab	F, A	E
Madagascarophis colubrinus	+	+	+	+	LC	C	Tr	F, A	E
Madagascarophis fuchsi	+	+	−	−	CR	I	Tr	F	LE
Mimophis occultus	−	+	+	+	LC	C	Tr	F, A	E
Phisalixella cf. *arctifasciata*	+	−	+	−	NE	R	Ab	F	LE
Phisalixella arctifasciata	+	−	+	+	LC	I	Ab	F	RE
Pseudoxyrhopus quinquelineatus	+	−	−	+	LC	I	Tr	F	RE
Thamnosophis martae	−	+	+	−	EN	I	Tr	F	RE
Thamnosophis stumpffi	−	−	+	−	VU	I	Tr	F	E
SCINCIDAE									
Madascincus miafina	−	−	−	+	LC	R	Tr	F	E
Trachylepis elegans	+	+	+	+	LC	A	Tr	A	E
Trachylepis tavaratra	−	−	−	+	VU	C	Tr	A	RE
Voeltzkowia sp.	+	−	−	−		R	Tr	A	
TYPHLOPIDAE									
Indotyphlops braminus	+	−	+	+	NE	R	Tr	F	E
Total of Amphibians	10	8	14	14					
Total of Reptiles	36	24	30	37					

3.3. Newly Recorded Species at Montagne des Français

We identified three frogs (*Boophis* cf. *occidentalis*, *B.* cf. *marojezensis*, and *B.* sp.) and one snake (*Phisalixella* cf. *arctifasciata*) previously unknown to the area. Additionally, we found *Langaha pseudoalluaudi*, which previously had not been seen in the area since 2007. Below, we present a description of these observations:

3.3.1. *Boophis* cf. *occidentalis*

This is a medium-sized deciduous forest species (SVL = 44.3 mm). The seven individuals observed had a green dorsum and yellow lateral line running between the snout tip and inguinal region. All the ventral surfaces were pink (Figure 2). The webbing formula is 1 (0.25) 2i (0.5) 2e (0.25) 3i (0.25) 3e (0.25) 4i (1) 4e (1) 5 (0). Males have evident tubercles, especially on the dorsum, which are absent in females. We observed this species on a branch overhanging the stream and above the riverbank from 6–9 p.m.

Figure 2. Photographic documentation of the morphospecies found and rediscovered species in the Montagne des Français area. (**A**): *Boophis* cf. *occidentalis*; (**B**): *Boophis* sp.; (**C**): *Boophis* cf. *marojezensis*; (**D**): *Phisalixella* cf. *arctifasciata*; (**E**): *Langaha pseudoalluaudi* (male); (**F**): *Langaha pseudoalluaudi* (female).

3.3.2. *Boophis* cf. *marojezensis*

This is a small forest species (SVL = 23.5 mm) found close to a small stream. One individual had a cream color on the dorsum and a blue peripheral ring on its iris. The fingertips and toe tips were pigmented with a golden color. The webbing formula is 1 (0.5) 2i (0) 2e (0.25) 3i (1) 3e (0.5) 4i (1) 4e (1) 5 (0) and this characteristic shows some differentiation according to the webbing formula by Glaw et Vences (2007). This species was resting on a leaf of a shrub overhanging a small stream at 9:15 p.m.

3.3.3. *Boophis* sp.

This is a large forest species (SVL = 65.5 mm). This species could belong to the *Boophis goudotii* group based on the presence of a turquoise-blue outer iris periphery and a large SVL. The webbing formula is 1 (0.25) 2i (0) 2e (0) 3i (0.5) 3e (0.25) 4i (0.75) 4e (0.75) 5 (0). It was resting on a branch of a shrub overhanging a small stream at 10:15 p.m.

3.3.4. *Phisalixella* cf. *arctifasciata*

This arboreal snake was found at Montagne des Français during our 2014 and 2020 investigations. It is a large species (TL: 864 mm; SVL: 288 mm) but differs from *P. arctifasciatus* by the presence of 135 dark transverse bands between the neck and the tail tip. It was active during the night, around 11:00 p.m., on a branch at a height of 3.5 m next to a small stream.

3.3.5. *Langaha pseudoalluaudi*

This is a large arboreal forest species (TL = 1275 mm for males; 1260 m for females). These male and female individuals were found in the same environment. Both specimens were discovered under identical conditions, on shrubby plants measuring 2 m and 2.5 m at the periphery of the relict forest, situated in an open environment at an altitude of 80 m.

4. Discussion

Our study provides new data on the distribution and endemism of amphibians and reptiles within Montagne des Français PA. The results build on previous studies [7–9,36] of the unique herpetofaunal community of Montagne des Français. All these studies have already highlighted the level of endemicity and habitat loss due to various threats, including the proximity of this zone to human settlements and their activities.

4.1. Species Richness

Incredible faunal diversity was found in the extreme northern karst formations of Madagascar. Sixty-nine species were encountered during this study. Of these, we recorded 10 amphibians from three families and 35 reptiles from six families in 2014 in Andavakoera. In addition, the survey we conducted in 2020 allowed us to encounter 16 amphibians from two families and 44 reptiles from six families. The endemicity level was 94%, which is similar to that in [7,9,34]. Despite the relatively short fieldwork, we were able to increase sampling effort by having a large research team composed of eight people. Moreover, we investigated other sites, such as Sahabedara and Ampitiliantsambo, located in the peripheral areas of Montagne des Français PA, in addition to Andavakoera, to further assess the herpetofaunal diversity of the area. Although these additional sites were not frequently surveyed, we found a higher species diversity of amphibians and reptiles than that of Andavakoera during the survey of 2020 (Table 1), including the highly cryptic snake *Langaha pseudoalluaudi* rediscovered in Sahabedara, which was last recorded by [7] in 2007. Furthermore, most of the newly recorded species were discovered in these sites, namely, *Boophis* cf. *marojezensis*, *B.* cf. *occidentalis*, *B.* sp., and the snake *Physalixella* cf. *arctifasciata*, which morphologically present a big challenge for classification and Montagne des Francais appears as a new biogeographical distribution for them. Importantly, we emphasize the need for bioacoustic and molecular studies on the newly recorded species we present here, as we cannot confirm their status based on morphology and appearance alone.

Comparing previous research data with those gathered during the current inventory, it can be shown that *Zonosaurus aenus*, *Thamnosophis lateralis*, and *T. stumpffi* were not identified by [7,9,25]. Relatedly, the recent study conducted by [25] did not confirm the occurrence of *Dromicodryas bernieri* and *Langaha pseudoalluaudi*, while the previous studies in 2007 [7,9] did. The absence of these animals could be attributed to seasonal factors. If surveys were conducted when they are inactive, it may make observing them challenging or impossible. Additionally, the lack of detection in some years might be due to survey differences between the western slope, Andavakoera, and the eastern slope, Sahabedara and Ampitiliantsambo, or due to the presence of bushfire, charcoal production, or other habitat factors affecting species presence during surveys.

4.2. Importance of Conservation

While the Montagne des Français is located near the urban area of Antsiranana, anthropogenic activities such as charcoal production [7] and timber harvesting provoke threats to species habitat and negatively affect herpetofauna [36]. Moreover, the sites of Sahabedara and Ampitiliantsambo were more disturbed by land use leading to a fragmented and degraded forest. Some species of herpetofauna in tropical dry forests have been shown to be highly resilient [37]. Indeed, certain species are found in forests of varying quality. For example, *Langaha pseudoalluaudi*, *Lycodryas* sp., *Sanzinia volontany*, and *Madagascarophis colubrinus* may be able to adapt their behavior to disturbed habitats. We observed these species frequently in areas with closed canopies and in more open habitats. At the same time, other species are more sensitive to minor habitat changes; for example, *Mantella viridis*, *Boophis* cf *majori*, *B.* cf *occidentalis*, and *B.* sp., which are typically restricted to intact humid forest. We suggest that the frequent and continued exploitation of forest will drastically reduce the habitat of amphibians and reptiles, meaning that they will be more prone to extinction [7,13,15]. The most threatened species in Montagne des Français are three amphibian species, *Mantella viridis* (EN), *Stumpffia roseifemoralis* (EN), and *Stumpffia staffordi* (VU), and nineteen reptile species, including three critically endangered, five endangered, and eleven vulnerable (cf. Table 1). To ensure the viability of the site, permanent monitoring and conservation planning is needed for all of them [21,38–40]. Particularly, we recommend the conservation of species that have a restricted distribution throughout the northern half of Madagascar, namely, *Mantella viridis*, *Paroedura lohatsara*, *Zonosaurus tsingy*, *Thamnosophis martae*, and *Heteroliodon fohy*.

5. Conclusions

The fieldwork conducted in 2014 and 2020 contributes to the existing literature regarding the diversity and occurrence of the herpetofaunal community of Montagne des Français, especially in Ampitiliantsambo, where investigation has not been undertaken before. Overall, this study highlighted the occurrence and the update of the distribution of 19 amphibian and 50 reptile species living within the Montagne des Français PA. In the face of a rapidly changing landscape, it is pivotal to consider an effective management and conservation plan for Montagne des Français regarding its proximity to Antsiranana city and frequent land use. Human activity in the boundary of this zone is one of the factors that drives habitat loss and consequently species decline. Intensive long-term monitoring has to be planned to assess herpetofaunal trends in the northern area of Madagascar, informing effective policies.

Supplementary Materials: The following supporting information can be downloaded at: https://www.mdpi.com/article/10.3390/ani13213361/s1, Figure S1: Photo of amphibians; Figure S2: Photo of reptiles.

Author Contributions: Conceptualization, H.O.R., B.R. and C.J.R.; methodology, B.R., H.O.R. and C.J.R.; validation, N.H.C.R. and B.R.; investigation, H.O.R., B.R. and C.J.R.; writing—original draft preparation, H.O.R.; writing—review and editing, H.O.R. and N.H.C.R.; supervision, N.H.C.R. All authors have read and agreed to the published version of the manuscript.

Funding: This research received no external funding.

Institutional Review Board Statement: Field surveys were approved by the Direction of Protected Area Systems (permit number: 238/18/MEEF/SG/DGF/DSAP/SCB.Re).

Informed Consent Statement: Not applicable.

Data Availability Statement: The data presented in this study are available on request from the corresponding author. The data are not publicly available due to privacy or ethical restrictions.

Acknowledgments: We thank the team from the Sciences Department of the University of Mahajanga for data collections in Montagne des Français surveys.

Conflicts of Interest: The authors declare no conflict of interest.

References

1. Myers, N.; Mittermeier, R.A.; Mittermeier, C.G.; Da Fonseca, G.A.; Kent, J. Biodiversity Hotspots for Conservation Priorities. *Nature* **2000**, *403*, 853–858. [CrossRef] [PubMed]
2. Raxworthy, C.J.; Pearson, R.G.; Zimkus, B.M.; Reddy, S.; Deo, A.J.; Nussbaum, R.A.; Ingram, C.M. Continental Speciation in the Tropics: Contrasting Biogeographic Patterns of Divergence in the Uroplatus Leaf-tailed Gecko Radiation of Madagascar. *J. Zool.* **2008**, *275*, 423–440. [CrossRef]
3. Vences, M.; Wollenberg, K.C.; Vieites, D.R.; Lees, D.C. Madagascar as a Model Region of Species Diversification. *Trends Ecol. Evol.* **2009**, *24*, 456–465. [CrossRef]
4. Goodman, S.M. Updated Estimates of Biotic Diversity and Endemism for Madagascar—Revisited after 20 Years. *Oryx* **2023**, *57*, 561–565. [CrossRef]
5. Heinen, J. International Trends in Protected Areas Policy and Management. *Prot. Area Manag.* **2012**, 1–18. [CrossRef]
6. Goodman, S.M.; Benstead, J.P. *Natural History of Madagascar*; University of Chicago Press: Chicago, IL, USA, 2003; ISBN 0-226-30306-3. [CrossRef]
7. D'Cruze, N.; Sabel, J.; Green, K.; Dawson, J.; Gardner, C.; Robinson, J.; Starkie, G.; Vences, M.; Glaw, F. The First Comprehensive Survey of Amphibians and Reptiles at Montagne Des Français, Madagascar. *Herpetol. Conserv. Biol.* **2007**, *2*, 87–99.
8. Glaw, F.; Vences, M. *A Field Guide to the Amphibians and Reptiles of Madagascar*, 3rd ed.; Vences & Glaw: Köln, Germany, 2007; ISBN 978-3-929449-03-7.
9. Frontier Madagascar. *Montagne des Français Biodiversity and Conservation Evaluation*; Frontier-Madagascar Environmental Research; Society for Environmental Exploration: London, UK; Institut Halieutique et des Sciences Marines: Antsiranana, Madagascar, 2007; p. 33.
10. Sabel, J.; Green, K.; Dawson, J.; Robinson, J.; Gardner, C.; Starkie, G.; D'Cruze, N. The Conservation Status of Mammals and Avifauna in the Montagne Des Français Massif, Madagascar. *Madag. Conserv. Dev.* **2009**, 4. [CrossRef]
11. Harper, G.J.; Steininger, M.K.; Tucker, C.J.; Juhn, D.; Hawkins, F. Fifty Years of Deforestation and Forest Fragmentation in Madagascar. *Environ. Conserv.* **2007**, *34*, 325–333. [CrossRef]
12. Soarimalala, V.; Raherilalao, M. Pression et Menace Dans La Région Forestière Sèche Malgache. *For. Sèches Madag. SM Goodman Wilmé Eds Malagasy Nat.* **2008**, *1*, 1–57.
13. Vallan, D. Effects of Anthropogenic Environmental Changes on Amphibian Diversity in the Rain Forests of Eastern Madagascar. *J. Trop. Ecol.* **2002**, *18*, 725–742. [CrossRef]
14. Vallan, D.; Vences, M.; Glaw, F. Two New Species of the Boophis Mandraka Complex (Anura, Mantellidae) from the Andasibe Region in Eastern Madagascar. *Amphib. Reptil.* **2003**, *24*, 305–319.
15. Cordier, J.M.; Aguilar, R.; Lescano, J.N.; Leynaud, G.C.; Bonino, A.; Miloch, D.; Loyola, R.; Nori, J. A Global Assessment of Amphibian and Reptile Responses to Land-Use Changes. *Biol. Conserv.* **2021**, *253*, 108863. [CrossRef]
16. Heigl, F.; Horvath, K.; Laaha, G.; Zaller, J.G. Amphibian and Reptile Road-Kills on Tertiary Roads in Relation to Landscape Structure: Using a Citizen Science Approach with Open-Access Land Cover Data. *BMC Ecol.* **2017**, *17*, 1–11. [CrossRef] [PubMed]
17. Clulow, S.; Clulow, J.; Marcec-Greaves, R.; Della Togna, G.; Calatayud, N.E.; Yuan, Y. Common Goals, Different Stages: The State of the ARTs for Reptile and Amphibian Conservation. *Reprod. Fertil. Dev.* **2022**, *34*, i–ix. [CrossRef] [PubMed]
18. Mittermeier, R.A.; Carr, J.L.; Swingland, I.R.; Werner, T.B.; Mast, R.B. Conservation of Amphibians and Reptiles. In *Herpetology: Current Research on the Biology of Amphibians and Reptiles*; Adler, K., Ed.; Society for the Study of Amphibians and Reptiles: St. Louis, MO, USA, 1992; pp. 59–80.
19. Pabijan, M.; Gehring, P.-S.; Koehler, J.; Glaw, F.; Vences, M. A New Microendemic Frog Species of the Genus Blommersia (Anura: Mantellidae) from the East Coast of Madagascar. *Zootaxa* **2011**, *2978*, 34–50. [CrossRef]
20. D'Cruze, N.; Henson, D.; Olsson, A.; Emmett, D. The Importance of Herpetological Survey Work in Conserving Malagasy Biodiversity: Are We Doing Enough? *Herpetol. Rev.* **2009**, *40*, 19.
21. D'Cruze, N. Conserving Amphibian and Reptile Diversity in North Madagascar: Contributions from Baseline Herpetological Survey Work. Ph.D. Thesis, Oxford Brookes University, Oxford, UK, 2011.
22. Megson, S.; Mitchell, P.; D'Cruze, N. Reptilia, Serpentes, Colubridae, Heteroliodon Fohy: Distribution Extension. *Check List* **2009**, *5*, 692–694. [CrossRef]
23. Glaw, F.; Nagy, Z.T.; Köhler, J.; Franzen, M.; Vences, M. Phylogenetic Relationships of a New Species of Pseudoxyrhophiine Snake (Reptilia: Lamprophiidae: Thamnosophis) Suggest a Biogeographical Link between Western and Northern Madagascar. *Org. Divers. Evol.* **2009**, *9*, 13–22. [CrossRef]
24. Glaw, F.; Köhler, J.; Townsend, T.M.; Vences, M. Rivaling the World's Smallest Reptiles: Discovery of Miniaturized and Microendemic New Species of Leaf Chameleons (Brookesia) from Northern Madagascar. *PLoS ONE* **2012**, *7*, e31314. [CrossRef]
25. Megson, S.; Mitchell, P.; Köhler, J.; Marsh, C.; Franzen, M.; Glaw, F.; D'Cruze, N. A Comprehensive Survey of Amphibians and Reptiles in the Extreme North of Madagascar. *Herpetol. Notes* **2009**, *2*, 31–44.
26. Mercurio, V.; Andreone, F. New Distribution Data of the Green Mantella, Mantella Viridis, from Northern Madagascar (Anura: Mantellidae). *Herpetol. Notes* **2008**, *1*, 3–7.
27. Pintak, T.; Bohme, W. *Mantella viridis* sp. n.(Anura: Ranidae: Mantellinae) Aus Nord-Madagaskar. *Salamandra* **1988**, *24*, 119–124.
28. Andriantsimanarilafy, R.R.; Rakotondrina, A.J.; Glaw, F.; München, S.S. A New Locality and New Colour Variant of the Giant Stick Insect Achrioptera Manga from the Northern Tip of Madagascar. *SPIXIANA* **2019**, *42*, 283–284.

29. Vallan, D. Influence of Forest Fragmentation on Amphibian Diversity in the Nature Reserve of Ambohitantely, Highland Madagascar. *Biol. Conserv.* **2000**, *96*, 31–43. [CrossRef]
30. Andreone, F.; Cadle, J.E.; Cox, N.; Glaw, F.; Nussbaum, R.A.; Raxworthy, C.J.; Stuart, S.N.; Vallan, D.; Vences, M. Species Review of Amphibian Extinction Risks in Madagascar: Conclusions from the Global Amphibian Assessment. *Conserv. Biol.* **2005**, *19*, 1790–1802. [CrossRef]
31. Kremen, C.; Razafimahatratra, V.; Guillery, R.P.; Rakotomalala, J.; Weiss, A.; Ratsisompatrarivo, J. Designing the Masoala National Park in Madagascar Based on Biological and Socioeconomic Data. *Conserv. Biol.* **1999**, *13*, 1055–1068. [CrossRef]
32. Nussbaum, R.; Raxworthy, C.; Raselimanana, A.; Ramanamanjato, J.-B. Amphibians and Reptiles of the Reserve Naturelle Integrale d'Andohahela, Madagascar. *Fieldiana Zool.* **1999**, *94*, 155–174.
33. Raxworthy, C.J. Reptiles, Rainforest and Conservation in Madagascar. *Biol. Conserv.* **1988**, *43*, 181–211. [CrossRef]
34. Durkin, L.; Steer, M.; Belle, E. Herpetological Surveys of Forest Fragments between Montagne d'Ambre National Park and Ankarana Special Reserve, Northern Madagascar. *Herpetol. Conserv. Biol.* **2011**, *6*, 114–126.
35. Wilson, L.D.; McCranie, J.R. The Herpetofauna of the Cloud Forests of Honduras. *Amphib. Reptile Conserv.* **2004**, *3*, 34.
36. D'Cruze, N.; Kumar, S. Effects of Anthropogenic Activities on Lizard Communities in Northern Madagascar: Anthropogenic Activities and Lizard Communities in Northern Madagascar. *Anim. Conserv.* **2011**, *14*, 542–552. [CrossRef]
37. Suazo-Ortuño, I.; Alvarado-Díaz, J.; Mendoza, E.; López-Toledo, L.; Lara-Uribe, N.; Márquez-Camargo, C.; Gil Paz-Gutiérrez, J.; David Rangel-Orozco, J. High Resilience of Herpetofaunal Communities in a Human-Modified Tropical Dry Forest Landscape in Western Mexico. *Trop. Conserv. Sci.* **2015**, *8*, 396–423. [CrossRef]
38. D Cruze, N.C.; Green, K.E.; Robinson, J.E.; Gardner, C.J. A Rapid Assessment of the Amphibians and Reptiles of an Unprotected Area of Dry Deciduous Forest in North Madagascar. *Herpetol. Bull.* **2006**, *96*, 17.
39. Jenkins, R.K.; Tognelli, M.F.; Bowles, P.; Cox, N.; Brown, J.L.; Chan, L.; Andreone, F.; Andriamazava, A.; Andriantsimanarilafy, R.R.; Anjeriniaina, M. Extinction Risks and the Conservation of Madagascar's Reptiles. *PLoS ONE* **2014**, *9*, e100173. [CrossRef] [PubMed]
40. Vences, M.; Andreone, F.; Glaw, F.; Raminosoa, N.; Randrianirina, J.E.; Vieites, D.R. Amphibians and Reptiles of the Ankaratra Massif: Reproductive Diversity, Biogeography and Conservation of a Montane Fauna in Madagascar. *Ital. J. Zool.* **2002**, *69*, 263–284. [CrossRef]

Article

Overview of Reptile Diversity from Bobaomby Complex, Northern Tip of Madagascar

Randriamialisoa [1,*], Raphali R. Andriantsimanarilafy [2], Alain J. V. Rakotondrina [2,*], Josué A. Rakotoarisoa [3], Ranaivoson T. Nasaina [4], Jeanneney Rabearivony [5] and Achille P. Raselimanana [6]

[1] Association Groupe d'Etudes et de Recherche sur les Primates de Madagascar, Lot II M 78 Bis Antsakaviro, BP 779, Antananarivo 101, Madagascar

[2] Madagasikara Voakajy, BP 5181, Antananarivo 101, Madagascar; arraphali@voakajy.mg

[3] Fondation Tany Meva, Lot I A I 1bis Ambatobe, BP 4300, Antananarivo 103, Madagascar; josue.rakotoarisoa@gmail.com

[4] Mention Zoologie et Biodiversité Animale, Facultés des Sciences, Université d'Antananarivo, BP 906, Antananarivo 101, Madagascar; tambyranaivoson@gmail.com

[5] Facultés des Sciences, Université d'Antsiranana, BPO, Antsiranana 201, Madagascar; r.jeanneney@gmail.com

[6] Association Vahatra, BP 3972, Antananarivo 101, Madagascar; raselimananaachille@gmail.com

* Correspondence: randriamialisoa29@gmail.com (R.); alain@voakajy.mg (A.J.V.R.); Tel.: +261-34-16-32727 (R.); +261-34-41-79082 (A.J.V.R.)

Citation: Randriamialisoa; Andriantsimanarilafy, R.R.; Rakotondrina, A.J.V.; Rakotoarisoa, J.A.; Nasaina, R.T.; Rabearivony, J.; Raselimanana, A.P. Overview of Reptile Diversity from Bobaomby Complex, Northern Tip of Madagascar. *Animals* 2023, 13, 3396. https://doi.org/10.3390/ani13213396

Academic Editors: Franco Andreone, Angelica Crottini, Andolalao Rakotoarison and Fandresena Rakotoarimalala

Received: 30 June 2023
Revised: 10 October 2023
Accepted: 27 October 2023
Published: 1 November 2023

Simple Summary: Madagascar is home to diverse ecosystems with many endemic reptiles, many of which are threatened by the increasing loss of their habitat. Unfortunately, while some areas have been relatively well studied, there is a paucity of data in most of them. This knowledge gap hinders efforts to collect scientific information in order to conserve the remaining habitat. We conducted surveys in the Bobaomby Complex, in the northern tip of Madagascar to assess its potential for the creation of a new protected area. We found 42 reptile species of which 39 are endemic and the discovery of many threatened species with restricted range distributions. The findings fill the knowledge gaps on the herpetofauna of the Bobaomby Complex. We recommend the inclusion of the Bobaomby Complex into the network of protected areas of Madagascar.

Abstract: Many studies on reptiles have been conducted across Madagascar but some areas are poorly known in terms of the diversity of reptiles such as the Bobaomby Complex in the northern tip of Madagascar. In February and March 2018, we conducted a biodiversity survey within five sites. This biological survey is to collect scientific information for helping new protected creations. Three main methods were used including pitfall trap, visual and acoustic searching along the transect and refuge examination. In total, we recorded 42 species including 5 chameleons, 8 skinks, 11 geckos, 16 snakes and 2 blinds snake species. All recorded species are endemic to Madagascar except *Hemidactylus frenatus*, *Ebenavia inunguis* and *Phelsuma abbotti*. Rare species known only from a few specimens have been recorded in the Bobaomby Complex: *Heteroliodon fohy*, *Pseudoxyrhopus ambreensis*, and *Madascincus arenicola*. Thirteen species are classified as threatened on the IUCN Red List, of which three are Critically Endangered: *Paracontias minimus*, *Madascincus arenicola*, and *Paroedura lohatsara*; three are Endangered: *Heteroliodon fohy*, *Lycodryas inopinae*, and *Phisalixella variabilis*; and seven are Vulnerable: *Brookesia ebenaui*, *Furcifer petteri*, *Blaesodactylus boivini*, *Uroplatus ebenaui*, *Uroplatus henkeli*, *Liophidium therezieni* and *Flexiseps ardouini*. Our results reveal the importance of the Bobaomby Complex for conserving reptile diversity and highlight the need to protect it.

Keywords: reptile survey; deciduous forest; micro-endemic; protected area; northern Madagascar

---

1. Introduction

Madagascar is remarkable for its diversified ecosystems and a high rate of species endemism [1]. The island is home to 436 native non-marine reptile species [2] and with more than

99% of endemicity [3]. However, many species are threatened by habitat loss and fragmentation caused by expanding agriculture as well as logging and wood harvesting [4]. Based on the IUCN Red List assessment, at least 29.9% of Malagasy reptile species are classified as threatened [5].

Protected areas (PAs) are crucial worldwide to ensure the conservation of biodiversity in the face of loss and fragmentation of ecosystems [6]. According to the Durban Vision in 2003, the Malagasy government committed to tripling the surface area of PAs in Madagascar in order to guarantee the long-term conservation of its rich and unique biodiversity. Since this declaration, many new PAs have been created, mainly encompassing humid rainforests of the eastern part of the island. However, other ecosystems are far less well-represented in the Malagasy PA system. For example, many dry forests in Madagascar remain unprotected although they are among the most threatened ecosystems on the island [7]. Moreover, Malagasy dry forests are known for their rich herpetofauna communities, but many sites have not received extensive attention from researchers and their herpetofauna diversity is poorly known [8]. Protected areas have been subject to more survey efforts than unprotected areas [9]. Further exploration of these understudied sites is crucial to provide valuable information on biodiversity to inform future conservation efforts. This study aims to provide scientific information on a reptile community to establish a management and conservation plan for sites without current protection. Indeed, the biological exploration of forests outside PAs is essential [10] to assess its importance for conservation, as in the case of the Bobaomby Complex in the extreme north of Madagascar. Previous research here in the form of a biological survey of the Ampombofofo and Anjiabe forest sites has resulted in the finding of 6 endemic amphibians and 37 endemic reptiles [11]. These findings prompted the designation of these sites as Key Biodiversity Areas due to the presence of three Critically Endangered, two Endangered and one Vulnerable species [10]. However, some sites within the Bobaomby Complex such as Beantely, Antsisikala and Ambanililabe are poorly studied. Here, we report findings from surveys we carried out to assess the importance of the Bobaomby Complex in terms of the diversity of reptiles to support the proposal of a new protected area.

2. Materials and Methods

2.1. Study Location

The Bobaomby Complex is located in the extreme north of Madagascar (Figure 1), District Antsiranana II, Diana Region. The reptile survey was carried out in the remaining forest of the Bobaomby Complex from February to March 2018, a hot and rainy season coinciding with optimal biological activities for most species and favorable to their census [12,13]. Five sites, Beantely, Antsisikala, Ambanililabe, Anjiabe and Ampombofofo (Figure 1), were inspected. Each site was visited for five successive days.

The study area belongs to the western dry deciduous forest ecoregion [14] and includes the northern part of the island. The vegetation has a canopy of 10 to 15 m, sometimes reaching 20 m. Rocky or karst formations backed by dense deciduous vegetation and tree savannahs characterize the landscape in the Bobaomby Complex (Figure 2). The vegetation and habitat characteristics of the study sites were assessed from field observations (Table 1).

Table 1. Specific characteristics of the vegetation in the study sites.

Sites	Geographic Coordinates	Altitude (m)	Vegetation Characteristics
Beantely (Site 1)	12°16′33.2″ S 49°10′05.7″ E	119–200	Disturbed forest, closed canopy (10–20 m height), diffuse undergrowth, thick litter, clay and rocky soils, temporary streams
Antsisikala (Site 2)	12°10′31.8″ S 49°12′56.5″ E	53–150	Disturbed forest, open canopy (5–8 m height), light undergrowth, thin or even absent leaf litter, clay and rocky soils, temporary streams
Ambanililabe (Site 3)	12°11′21.4″ S 49°17′33.7″ E	11–50	Disturbed forest located near coastal zone, open canopy (5–8 m high), light undergrowth, thin or even absent leaf litter, clay and rocky soils, temporary streams
Anjiabe (Site 4)	12°07′02.6″ S 49°20′06.8″ E	11–91	Intact forest, semi-open canopy (5–15 m high), abundant undergrowth, very thick litter, clay or sandy soils, temporary stream
Ampombofofo (Site 5)	12°05′38.76″ S 49°20′23.89″ E	24–70	Relatively intact forest, closed canopy (5–15 m high), well-stocked undergrowth, very thick litter, sandy soils, permanent watercourse, presence of marshes and ponds

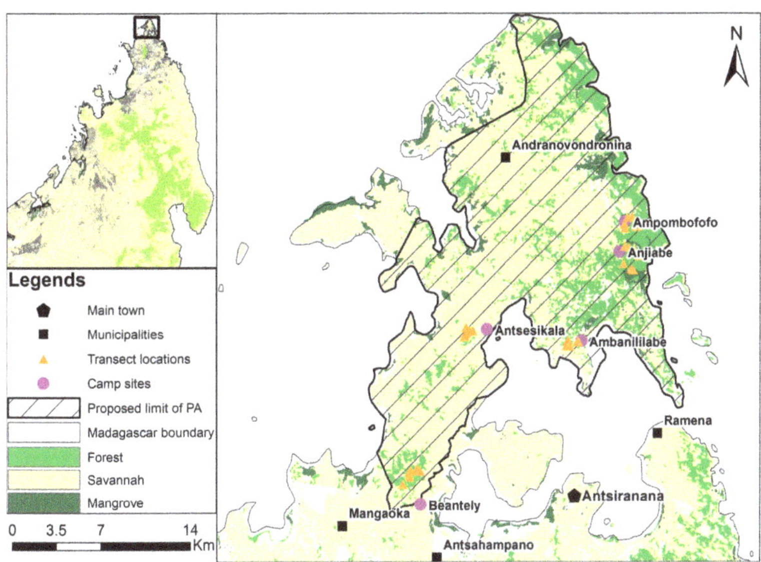

Figure 1. Location of the study sites.

Figure 2. Different types of habitat within the five sites. (**A**) Beantely, a primary forest dominated by savannah; (**B**) Ambanililabe, a degraded forest surrounded by a savannah and mangrove forest; (**C**): Antsisikala, a disturbed forest surrounded by a large savannah; (**D**): Anjiabe, dominance of intact forest; (**E**): Ampombofofo, primary and relatively intact forest.

2.2. Sampling Techniques

We employed (i) pitfall trapping with drift fences, (ii) visual searching, and (iii) refuge examination [15] methods to sample animals.

2.2.1. Pitfall Traps

This method is intended for the capture of burrowing and terrestrial reptiles [11]. The pitfall traps were buckets (275 mm deep, 290 mm top internal diameter, 220 mm bottom internal diameter) with the handles removed and small holes (2 mm diameter) punched in the bottom to allow water drainage. Buckets were sunk into the ground below a drift fence made from plastic sheeting (0.5 m high) stapled in a vertical position to thin wooden stakes, with the fence bottom buried 50 mm deep into the ground using soil and leaf litter (Figure 3). The drift fence (100 m in length) was positioned to run across the middle of each pitfall trap. A pitfall trap was positioned at both ends of the drift fence, with the other nine traps at 10 m intervals. At each site, three lines of tarps were used, except in Anjiabe, where we deployed four to increase our effort proportionally with the size of the forest fragment. Lines were placed in each of the following forest types: ridge (along the crest of a ridge), slope (on a gradient, intermediate between ridge top and valley bottom) and valley (within 20 m of a stream in a valley bottom). At each site, sampling was conducted for five days. The traplines were checked every morning (06:00 a.m.) and late afternoon (06:00 p.m.).

Figure 3. Pitt fall trap line used to capture reptiles.

2.2.2. Visual Searching

Animals were sampled throughout transects of 150 m length which are composed of three parallel lines of 50 m, separated 20 m apart and side by side [16]. Two transects are separated at least by a distance of 200 m. Each transect is visited once during the day and once during the night for five consecutive days. Diurnal searches were conducted between 7:30 a.m. and 12:00 p.m., periods when reptiles are more active [16]. Night searches were carried out between 07:00 to 10:00 p.m. using headlamps (Petzl MYO RXP) to spot animals. The survey team was composed of three experienced herpetologists.

2.2.3. Refuge Examination

Several species retreat to refuge when inactive [17]. Refuge examination was performed along a transect line during the day. Microhabitats likely to be a refuge were examined: under and in fallen logs and rotten tree stumps, removal bark, rocks crevasses, in leaf litter, root mats and soil, among dead wood, in leaf axils of *Pandanus* screw palms and *Ravenala* traveler's palm.

For each encountered reptile, the following variables were recorded: time of sighting, GPS coordinates, substrate type and species names. At least two individuals of each species were pictured for documentation of the natural coloration of the animal in its habitat and to serve as later identification [18]. Species identification was based on morphology and on an expert-based assessment according to the key descriptions by Glaw and Vences [18] and the nomenclature based on current taxonomy. Two specimens for each species that are difficult to identify were collected and preserved in 70% ethanol [19].

2.3. Statistical Analysis

To have global insight into the populations within the community, raw counts for each taxon were given at each site. The species recorded from the area are classified using a system similar to that used by Wilson and McCranie [20] and can be summarized as follows: abundant (large numbers encountered on a regular basis); common (encountered on a regular basis); infrequent (unpredictable, few individuals seen); or rare (rarely seen). These classifications are based on data collected using active searching, refuge examination and pitfall traps.

3. Results

3.1. Species Richness and Composition

A total of 42 reptile species were recorded during a total effort of 25 days of active search (Table 2). Anjiabe and Ampombofofo had the highest number of encountered species with 31 and 29 species, respectively, followed by Beantely with 26 species. Ambanililabe and Antsisikala had the lowest number with 18 and 17 species, respectively (Figure 4).

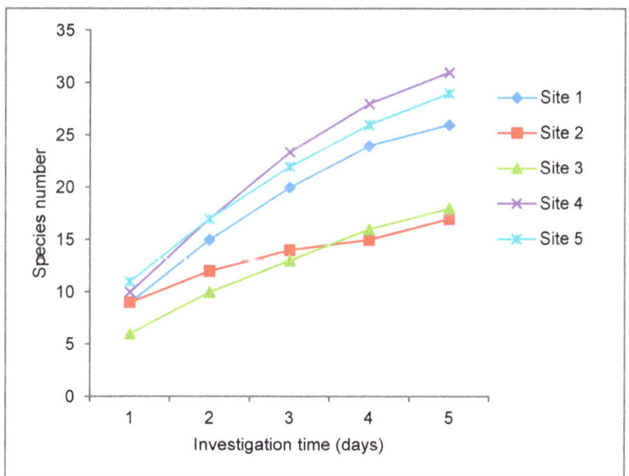

Figure 4. Accumulation curves of encountered reptile species in Bobaomby Complex.

None of the curves reached the plateau before the end of the investigation period, meaning that we may have not observed some species present at the study sites (Figure 4). Therefore, the Bobaomby Complex could contain more species of reptiles than we encountered during this study. The effort undertaken was apparently not sufficient to identify all the species in the study sites.

The reptile community in this area was dominated by lizards with 21 species (50%), 16 species of snakes (38.09%), and 5 species of chameleon species representing 11.9% of our total species (Table 2).

Most species are inventoried with multiple survey methods. However, some species were only recorded with the refuge examination. These species include *Ithycyphus miniatus*, *Ebenavia inunguis*, *Paracontias minimus*, *Paracontias* sp. aff. *rothschildi*, and *Paracontias* sp. Four species were collected only by pitfall trap: *Flexiseps ardouini*, *Heteroliodon fohy*, *Madascincus arenicola*, and *Liophidium therezieni*. The combination of these three techniques was useful for investigating the reptile community across different habitat types and species behaviors.

Table 2. List and raw counts of reptiles in the Bobaomby Complex. Captured by: VS = visual survey, PT = pitfall trapping, RE = refuge examination. IUCN Status: DD = data deficient, LC = least concern, NE = not evaluated, NT = Near Threatened, VU = Vulnerable, EN = Endangered, CR = Critically Endangered, Endemicity: E = endemic to Madagascar, Er = endemic regional, N = not endemic. Ecological distribution: TE = terrestrial, AB = arboreal, BR = burrowing.

Taxa	Captured by	IUCN Status	Endemicity	Ecological Distribution	Species Raw Counts					Distribution	New Distribution	Voucher Code
					Site 1	Site 2	Site 3	Site 4	Site 5			
SANZINIIDAE												
Acrantophis madagascariensis	VS	LC	E	TE	1	1		1	1	North		
Sanzinia volontany	VS	NE	E	AB, TE					1	West, north, south		
CHAMAELEONIDAE												
Brookesia ebenaui	VS	VU	Er	AB	3					North and northwest	Bobaomby Complex	UADBA-R-71739
Brookesia stumpffi	VS	LC	E	AB	20			8	21	North and northwest		UADBA-R-71736
Furcifer oustaleti	VS	LC	E	AB	2	1	1	1	2	All of Island, except central south		
Furcifer pardalis	VS	LC	E	AB	4	6	1	1	5	Northwest, north, east		
Furcifer petteri	VS	VU	Er	AB	2			18	4	North and northwest		UADBA-R-71740
LAMPROPHIIDAE												
Dromicodryas quadrilineatus	VS	LC	E	TE	1		1	1	1	North and east		
Heteroliodon fohy	PT	EN	E	TE	2			3	1	North		UADBA-R-71747
Ithycyphus miniatus	RE	LC	E	AB					1	West and north		
Leioheterodon madagascariensis	VS	LC	E	TE	1	1	1	1	1	Much of Island		
Leioheterodon modestus	PT, VS	LC	E	TE			1	1		North, west, south		
Liophidium therezieni	VS	VU	E	TE				1		North	Bobaomby Complex	UADBA-R-71745
Liophidium torquatum	VS	LC	E	TE		1	2	2	2	Much of Island, except south		UADBA-R-71748
Lycodryas granuliceps	VS	LC	E	AB	1		2	1	1	North		
Lycodryas inopinae	VS	EN	ER	AB				1	1	North		UADBA-R-71744
Madagascarophis colubrinus	VS	LC	E	TE	1	4	1		1	Much of Island		
Phisalixella arctifasciata	VS	LC	E	AB					R	East		
Phisalixella variabilis	VS	EN	E	AB	1				1	North and northwest		UADBA-R-71741
Pseudoxyrhopus ambreensis	VS	NT	Er	TE			2		1	North	Bobaomby Complex	UADBA-R-71749

Table 2. *Cont.*

Taxa	Captured by	IUCN Status	Endemicity	Ecological Distribution	Species Raw Counts					Distribution	New Distribution	Voucher Code
					Site 1	Site 2	Site 3	Site 4	Site 5			
PSAMMOPHIIDAE												
Mimophis occultus	VS	-	E	TE	1	1	1			North, northwest, west		
GEKKONIDAE												
Blaesodactylus boivini	VS	VU	Er	AB	6	32	3	4	2	North		UADBA-R-71731
Ebenavia inunguis	RE, VS	LC	N	AB				1		Northwest		
Geckolepis maculata	RE, VS	LC	E	AB	13	17	28	11	6	Northern and south		UADBA-R-71728
Hemidactylus frenatus	VS	LC	N	AB	2		1		1	North, west, south		UADBA-R-71727
Lygodactylus heterurus	VS	LC	Er	AB	1	2				North and northwest		UADBA-R-71724
Paroedura lohatsara	PT, VS	CR	Er					2	1	North	Bobaomby Complex	UADBA-R-71725
Paroedura stumpffi	VS	LC	E	TE, AB	2	14	3	1	1	North and northwest		UADBA-R-71726
Phelsuma abbotti	VS	LC	N	AB	1	12		3	4	North, west, Seychelles		UADBA-R-71721
Phelsuma grandis	VS	LC	E	AB	3	4	3	5	5	North		UADBA-R-71722
Uroplatus ebenaui	VS	VU	E	AB	2					North and northwest		UADBA-R-71718
Uroplatus henkeli	VS	VU	E	AB	1		1	1	3	Northwest		UADBA-R-71720
SCINCIDAE												
Flexiseps ardouini	PT	VU	E	TE				1	3	North		UADBA-R-71732
Madascincus polleni	PT	LC	E	TE	1	1	1	1		West		UADBA-R-71705
Madascincus arenicola	PT	CR	Er	TE, BR				13	10	North		UADBA-R-71704
Paracontias minimus	RE	CR	Er	BR				8	4	North		UADBA-R-71756
Paracontias sp.	RE		Er	BR				1		North		UADBA-R-71753
Paracontias sp. aff. *rothschildi*	RE		Er	BR				1		North		UADBA-R-71751
Trachylepis elegans	VS	LC	E	TE		2	8		3	Much of Island		UADBA-R-71707
Trachylepis gravenhorstii	VS	LC	E	TE	2	12		3	2	Much of Island		
TYPHLOPIDAE												
Madatyphlops mucronatus	VS	DD	E	BR				1		Northwest		UADBA-R-71703
Indotyphlops braminus	VS	-	E	BR	1	2		2		Much of Island		UADBA-R-71719
Total reptiles: 42					26	17	18	31	29			

3.2. Distribution and Conservation Status

Among the encountered reptile species, 13 were classified as threatened according to the IUCN Red List (Table 2), including seven listed as Vulnerable: *Brookesia ebenaui, Furcifer petteri, Blaesodactylus boivini, Uroplatus ebenaui, Uroplatus henkeli, Flexiseps ardouini,* and *Liophidium therezieni*; three are Endangered: *Heteroliodon fohy, Lycodryas inopinae,* and *Phisalixella variabilis*; and three are Critically Endangered: *Paracontias minimus, Madascincus arenicola* and *Paroedura lohatsara*. One species recorded is listed as Near Threatened: *Pseudoxyrhopus ambreensis*. These findings indicate that the Bobaomby Complex represents a potential refuge that can provide protection for these threatened species. Seven species with microhabitats in northern Madagascar were inventoried in the Bobaomby Complex: *Heteroliodon fohy, Lycodryas inopinae, Liophidium therezieni, Madascincus arenicola, Paracontias minimus, Paroedura lohatsara* and *Pseudoxyrhopus ambreensis*.

3.3. Species Raw Counts

The raw counts of reptile species varied among sites (Table 2). Overall, reptiles in the Bobaomby Complex are rare species represented by less than five individuals. One species (*Geckolepis maculata*) was found to be common or abundant at all of the study sites. Some species were found to be rare for some sites and common for others (Table 2): *Blaesodactylus Boivini, Brookesia stumpffi* and *Paroedura stumpffi*.

4. Discussion

4.1. Species Richness and Composition

Each reptile group is well represented at Bobaomby Complex. With its 42 species of reptiles, the Bobaomby Complex hosts more species compared to other sites in the northern of Madagascar. Orangea forest, located 10 km east of the Bobaomby Complex, is home to 22 species of reptiles [11], of which 81.8% are found in the Bobaomby Complex. The forest fragments between the national parks of Montagne d'Ambre and Ankarana host only 34 species of reptiles [7], and the Analamerana Special Reserve contains 32 species [12]. However, the Bobaomby Complex contains lower reptile richness than Montagne des Français hosting 52 reptile species [21]. We are conscious that differences in sampling techniques and effort may influence the number of species recorded across these studies.

In our study, the shape of the cumulative curves of the species at each site clearly shows that additional species could be added and the plateau is still far from being reached. Ramanamanjato et al. [22] reported that the sampling effort for herpetofauna surveys was important for accurate species inventories and suggested increasing the sampling period to nine days to ensure that all species at a given site are surveyed. Other species such as *Hemidactylus mercatorius, Zonosaurus boettgeri, Trachylepis tavaratra, Langaha madagascariensis, Xenotyphlops grandidieri* and *Pelusios castanoides* have been recorded in the Ampombofofo site [11], but were not recorded during the present study. As per inventory, no new species of reptile was encountered during our investigation. In fact, some of the habitat was not visited during our investigation due to logistical challenges. In addition, the weather conditions during this study, which included the passage of two cyclones, likely influenced our sampling.

The results of this investigation constitute a first overview of the reptile community for three of our study sites (Beantely, Antsisikala and Ambanililabe) and serve as a database that will fill the knowledge gaps on the reptiles of the northern part of Madagascar.

4.2. Ecological Characteristics of the Reptile Community

The reptile community of the Bobaomby Complex is composed of chameleons, lizards, ophidians, skinks and blind snakes. This indicates that there is a diversification of the habitats existing in this ecosystem. The reptile community is represented by arboreal (52.4%), terrestrial (33.3%) and burrowing (9.5%) species. Differences in the occurrence of different taxa between sites suggest a close relationship between the diversity of the ecological environment and the sensitivity of some groups to habitat disturbance. The majority of burrowing species were encountered in Anjiabe and Ampombofofo. These sites are characterized by a particularity of the substrate formed by sandy and soft soil with thick litter which constitutes a unique and suitable habitat for burrowing species, especially skinks [23]. Some species were observed only in more specific habitats. This is the case of arboreal and nocturnal snakes *Phisalixella variabilis, P. arctifaciata, Lycodryas inopinae* and *L. granuliceps* which were only encountered in forests with rock crevices or karstic formations [18]. Other species have the ability to adapt to different environments and biotopes such as *Blaesodactylus boivini* and *Phelsuma grandis* which have been observed in the forest and also in some very disturbed habitats.

4.3. Endemicity and Species Conservation Status

The majority of reptile species recorded in the Bobaomby Complex (92.8%) are all endemic to Madagascar except *Hemidactylus frenatus, Ebenavia inunguis, Phelsuma abbotti* [17] and *Indotyphlops braminus*. Six inventoried species *Flexiseps ardouini, Lycodryas inopinae, Heteroliodon fohy, Uroplatus ebenaui, Phisalixella variabilis* and *Furcifer petteri* are regionally endemic to northern Madagascar. Also, some species with restricted range and only known from a few localities are recorded in the Bobaomby Complex: *Paracontias minimus*, which was previously collected from Orangea [11], *Paroedura lohatsara* and *Pseudoxyrhopus ambreensis* in the Montagne des Français [22], and *Madascincus arenicola* in Baie des Dunes, Ramena [24]. The tree snake *Lycodrias inopinae*, recorded in Anjiabe and Ampombofofo, was previously known in the Montagne des Français [21].

4.4. Extension of Distribution Area

The Bobaomby Complex represents an extension of the distribution area for *Liophidium therezieni, Brookesia ebenaui, Heteroliodon fohy, Paroedura lohatsara* and *Pseudoxyrhopus ambreensis*. These species have been reported among the four localities including the Montagne des Français, Anatelo, and Forêt d'Orangea [9,11,21] and the identification of *B. ebenaui* at Beantely forest represented an extension of the distribution area for this species [25]. *B. ebenaui* was previously known in the Amber Forest [15], in the Montagne des Français [21] and in Ankarana [7].

4.5. Conservation Value

Given that herpetofauna constitutes an element in the choices of process identifying priority sites in terms of conservation [26,27], Beantely, Anjiabe and Ampombofofo forests are the potential sites for the main conservation zone of the future protected area because they are home to numerous threatened and endemic reptiles. They also constitute a remaining bloc of forest in the northern part of Madagascar and represent a new recorded range extension for some species. In total, thirteen species of reptiles (30.9%) from the Bobaomby Complex are on the IUCN Red List, of which three species are classified as Critically Endangered, three as Endangered and seven as Vulnerable. In addition, the records of four species indicate a range extension more than 20 km north of their previously known distribution [15,21]. The Bobaomby Complex may play an essential role in maintaining the population of reptiles in dry deciduous forests [24]. Reptiles are also among the groups most vulnerable to extinction due to their environmental requirements and their dependence on forest ecosystems [28]. Therefore, the creation of a new protected area of this site is strongly recommended with the inclusion of Beantely, Anjiabe and Ampombofofo as a core area according to its criteria for conservation prioritization [28,29]; as mentioned by Andreone, due to its natural richness, northern Madagascar is already a key area of conservation in the region [30].

5. Conclusions

The investigation we carried out in the Bobaomby Complex provides new insights into the reptilian diversity of the area. This investigation was the first to catalog the reptile communities of Beantely, Antsisikala and Ambanililabe. Among the 42 species recorded in the Bobaomby Complex, 13 species are threatened with extinction and 11 are endemic to the area. The Bobaomby Complex is rich in reptiles and plays an important role in the conservation of biodiversity representative of the northern part of Madagascar. The reptilian community of the Bobaomby Complex is characterized by the abundance of arboreal species. The geographical location, the heterogeneity of natural habitats, and the presence of micro-endemic and threatened species highlights the ecological importance of this site. The integration of the Bobaomby Complex into the network of protected areas of Madagascar will contribute to the maintenance of natural ecosystems and the conservation of reptilian diversity. We recommend the inclusion of Beantely, Anjiabe and Ampombofofo as the core area of the new protected area. The results of this investigation will serve as a database to support the establishment of a future protected area and fill in a knowledge gap on reptiles in the region. However, molecular analysis is recommended to confirm the taxonomy of some species which might be a new species in this region.

Photos of Reptiles from Bobaomby Complex, Captured during Fieldwork

A photographic repertoire of some species recorded is provided (Figures 5–7) to illustrates the repitles species within Bobaomby Complex during this investigation.

Figure 5. Reptile species of Bobaomby Complex: (**A**) *Acrantophis madagascariensis*, (**B**) *Blaesodactylus boivini*, (**C**) *Brookesia ebenaui*, (**D**) *Brookesia stumpffi*, (**E**) *Dromicodryas quadrilineatus*, (**F**) *Flexiseps ardouini*, (**G**) *Furcifer pardalis*, (**H**) *Furcifer petteri*, (**I**) *Geckolepis maculata*, (**J**) *Heteroliodon fohy*, (**K**) *Indotyphlops braminus*, (**L**) *Ithycyphus miniatus*.

Figure 6. Reptile species of Bobaomby Complex: (**A**) *Leioheterodon madagascariensis*, (**B**) *Leioheterodon modestus*, (**C**) *Liophidium therezieni*, (**D**) *Liophidium torquatum*, (**E**) *Lycodryas granuliceps*, (**F**) *Lycodryas inopinae*, (**G**) *Lygodactylus heterurus*, (**H**) *Madagascarophis colubrinus*, (**I**) *Madascincus arenicola*, (**J**) *Madascincus intermedius*, (**K**) *Madatyphlops mucronatus*, (**L**) *Mimophis occultus*.

Figure 7. Reptile species of Bobaomby Complex: (**A**) *Paracontias minimus*, (**B**) *Paracontias* sp., (**C**) *Paroedura lohatsara*, (**D**) *Paroedura stumpffi*, (**E**) *Phelsuma abbotti*, (**F**) *Phelsuma grandis*, (**G**) *Phisalixella arctifasciata*, (**H**) *Phisalixella variabilis*, (**I**) *Pseudoxyrhopus ambreensis*, (**J**) *Trachylepis gravenhorstii*, (**K**) *Uroplatus ebenaui*, (**L**) *Uroplatus henkeli*.

Author Contributions: Conceptualization, R.R.A. and R.; methodology, R.R.A.; software, R.; validation, J.R. and A.P.R.; formal analysis, R.; investigation, R.R.A.; resources, R.R.A.; data curation, R., writing—original draft preparation, R.; writing—review and editing, R.R.A., J.A.R., R.T.N., A.J.V.R., J.R. and A.P.R.; visualization, J.R. and R.R.A.; supervision, A.P.R.; project administration, R.R.A.; funding acquisition, R.R.A. All authors have read and agreed to the published version of the manuscript.

Funding: This research was funded by the Rainforest Trust.

Institutional Review Board Statement: Not applicable.

Informed Consent Statement: Not applicable.

Data Availability Statement: https://bobaombyreptile.org/01.2018/001 (accessed on 9 May 2023).

Acknowledgments: We acknowledge the Ministry of Environment and Sustainable Development, with their representative Gestion des Ressources Naturelles Renouvelable et des Ecosystèmes, which issued research permit (n°310/17/MEEF/SG/DGF/DSAP/SCB.Re) for the study. The Anjiabe Complex study was carried out in collaboration with the Mention Zoologie et Biodiversité Animale of the University of Antananarivo, the University of Antsiranana and Madagasikara Voakajy Association. We extend our thanks to local cookers and local guides for their precious support in the field and all students from two university susmentionned, without whom this project could not have been possible.

Conflicts of Interest: The authors declare no conflict of interest. The funders had no role in the design of the study; in the collection, analyses, or interpretation of data; in the writing of the manuscript; or in the decision to publish the results.

References

1. Myers, N.; Mittermeier, C.G.; da Fonseca, G.A.B.; Kent, J. Biodiversity hotspots for conservation priorities. *Nature* **2000**, *403*, 853–858. [PubMed]
2. Glaw, F.; Vences, M.; Raxworthy, C.J. Diversity and exploration of the Malagasy reptilia fauna. In *The New Natural History of Madagascar*; Goodman, S.M., Ed.; Princeton University Press: Princeton, NJ, USA, 2022; Volume 2, pp. 1423–1552.
3. Uetz, P.; Freed, P.; Aguilar, R.; Reyes, F.; Hošek, J. (Eds.) The Reptile Database. 2023. Available online: http://www.reptile-database.org (accessed on 9 May 2023).
4. Jenkins, R.K.B.; Tognelli, M.F.; Bowles, P.; Cox, N.; Brown, J.L.; Chan, L.; Andreone, F.; Andriamazava, A.; Andriantsimanarilafy, R.R.; Anjeriniaina, M.; et al. Extinction risks and the conservation of Madagascar's reptiles. *PLoS ONE* **2014**, *9*, e100173.
5. IUCN. The IUCN Red List of Threatened Species. 2020, Version 2022-2. Available online: https://www.iucnredlist.org (accessed on 25 May 2018).
6. Defries, R.; Hansen, A.; Turner, L.; Reid, R.; Liu, J. Land use change around protected areas: Management to balance human needs and ecological function. *Ecol. Appl.* **2007**, *17*, 1031–1038. [CrossRef] [PubMed]
7. Durkin, L.; Mark, D.S.; Elise, M.S.B. Herpetological surveys of forest fragments between Montagne d'Ambre national park and Ankarana special reserve, northern Madagascar. *Herpetol. Conserv. Biol.* **2011**, *6*, 114–126.
8. Mitchell, P.; Eaton, S.; Marsh, C. *Biological and Socio-Economic Research of the Dry Deciduous Forests of Ampombofofo, Madagascar. Report 13. Frontier-Madagascar*; Society for Environmental Exploration: London, UK; Institut Halieutique et des Sciences Marines, University of Toliara: Toliara, Madagascar, 2007.
9. D'Cruze, N.; David, H.; Annette, O.; David, E. The Importance of Herpetological Survey Work in Conserving Malagasy Biodiversity: Are We Doing Enough? *Herpetol. Rev.* **2009**, *40*, 19–25.
10. Key Biodiversity Areas Partnership. The World Database of Key Biodiversity Areas. *Birdlife International.* 2022. Available online: https://www.keybiodiversityareas.org/kba-news/wdkba (accessed on 29 June 2023).
11. Megson, S.; Mitchell, P.; Köhler, J.; Marsh, C.; Franzen, M.; Glaw, F.; D'Cruze, N. A comprehensive survey of amphibians and reptiles in the extreme north of Madagascar. *Herpetology* **2009**, *2*, 31–44.
12. Rakotondravony, H.A. Reptiles et amphibiens de la Réserve Spéciale de l'Analamerana et de la forêt classée d'Andavakoera dans l'extrême Nord de Madagascar. In *Inventaires de la Faune et de La flore du Nord de Madagascar Dans la Région Loky-Manambato, Analamerana et Andavakoera*; Goodman, S.M., Wilmé, L., Eds.; Recherches pour le Développement; Série Sciences Biologiques; Centre d'Information et de Documentation Scientifique et Technique: Antananarivo, Madagascar, 2006; Volume 23, pp. 149–173.
13. Goodman, S.M. *A Flora and Faunal Inventory of the Reserve Naturelle Integrale d'Andohahela, Madagascar, with Reference to Elevational Variation*; Fieldiana: Zoology, New Series, No. 94; Field Museum of Natural History: Chicago, IL, USA, 1999.
14. Moat, J.; Smith, P. *Atlas of the Vegetation of Madagascar*; Royal Botanic Gardens, Kew: London, UK, 2007.
15. Raxworthy, C.J.; Nussbaum, R.A. A Rainforest Survey of Amphibians, Reptiles and Small Mammals at Montagne d'Ambre, Madagascar. *Biol. Conserv.* **1994**, *69*, 65–73. [CrossRef]
16. Jenkins, R.K.B.; Brady, L.D.; Bisoa, M.; Rabearivony, J.; Griffiths, R.A. Forest disturbance and river proximity influence chameleon abundance in Madagascar. *Biol. Conserv.* **2003**, *109*, 407–415. [CrossRef]

17. Raxworthy, C.J. Introduction aux reptiles. In *Paysages Naturels et Biodiversité de Madagascar*; Goodman, S.M., Wilmé, L., Eds.; Publications Scientifiques du Museum: Paris, France, 2008; pp. 341–366.

18. Glaw, F.; Vences, M. *A Fieldguide to the Amphibians and Reptiles of Madagascar*, 3rd ed.; Vences & Glaw Verlags GbR Hardtstr.2, 50939 köln (Cologne): Cologne, Germany, 2007.

19. Steve, W.G.; Jeremy, F.J.; George, R.Z. Preserving reptiles for research. In *Reptile Ecology and Conservation: A Hand Book of Techniques*; Dodd, C.K., Ed.; Oxford University Press: Oxford, UK, 2016; pp. 73–86.

20. Wilson, L.D.; McCranie, J.R. The herpetofauna of Parque Nacional El Cusuco, Honduras (Reptilia, Amphibia). *Herpetol. Bull.* **2004**, *87*, 13–24.

21. D'Cruze, N.; Sabel, J.; Green, K.; Dawson, J.; Gardener, J.; Robinson, C.; Starkie, G.; Vences, M.; Glaw, F. The first comprehensive survey of amphibians and reptiles at Montagne des Français, Madagascar. *Herpetol. Conserv. Biol.* **2007**, *2*, 87–99.

22. Ramanamanjato, J.B.; Peter, B.M.; Nussbaum, R.A. Reptile, amphibian, and lemur diversity of the Malahelo Forest: A biogeographical transition zone in southeastern Madagascar. *Biodiver. Conserv.* **2002**, *11*, 1791–1807. [CrossRef]

23. Köhler, J.; Vences, M.; Martina, E.; Glaw, F. Systematics of limbless scincid lizards from northern Madagascar: Morphology, phylogenetic relationships and implications for classification (Squamata: Scincidae). *Org. Divers. Evol.* **2010**, *10*, 2. [CrossRef]

24. Miralles, A.; Köhler, J.; Glaw, F.; Vences, M. A molecular phylogeny of the «Madascincus polleni», with description of new species scincid lizard from the coastal dune area of the northern Madagascar. *Zootaxa* **2011**, *2876*, 1–16. [CrossRef]

25. Andriantsimanarilafy, R.R.; Rabearivony, J. A new record, with range extension, for the dwarf chameleon Brookesia ebenaui Boettger, 1880 from Beantely forest, north of Madagascar. *Herpetol. Notes* **2021**, *14*, 551–553.

26. Andreone, F.; Cadle, J.E.; Cox, N.; Glaw, F.; Nussbaum, R.A.; Raxworthy, C.J.; Stuart, S.N.; Vallan, D.; Vences, M. Species review of amphibian extinction risks in Madagascar: Conclusions from the Global Amphibian Assessment. *Conserv. Biol.* **2005**, *19*, 1790–1802. [CrossRef]

27. Rakotondravony, H.A. Aspects de la Conservation des Reptiles et des Amphibiens dans la Région de Daraina. *Madag. Conserv. Dev.* **2006**, *1*, 15. [CrossRef]

28. Andreone, F. Crossroads of herpetological diversity: Survey work for an integrated conservation of amphibians and reptiles in northern Madagascar. *Ital. J. Zool.* **2004**, *71* (Suppl. S2), 229–235. [CrossRef]

29. Gardner, C.J.; Raxworthy, C.J.; Metcalfe, K.; Raselimanana, A.P.; Smith, R.J.; Davies, Z.G. Comparing Methods for Prioritizing Protected Areas for Investment: A Case Study Using Madagascar's Dry Forest Reptiles. *PLoS ONE* **2015**, *10*, e0132803. [CrossRef] [PubMed]

30. Chape, S.; Harrison, J.; Spalding, M.; Lysenko, I. Measuring the extent and effectiveness of protected areas as an indicator for meeting global biodiversity targets. *Philso. Trans. R. Soc. B* **2005**, *360*, 443–455. [CrossRef] [PubMed]

Article

In Search of Suitable Breeding Sites: Habitat Heterogeneity and Environmental Filters Determine Anuran Diversity of Western Madagascar

Nanäa Mausberg *, Kathrin H. Dausmann and Julian Glos

Institute of Cell and Systems Biology, Universität Hamburg, Martin-Luther-King Platz 3, 20146 Hamburg, Germany; kathrin.dausmann@uni-hamburg.de (K.H.D.); julian.glos@uni-hamburg.de (J.G.)
* Correspondence: nanea.mausberg@gmail.com

Simple Summary: Amphibian biodiversity is declining around the globe and habitat destruction and fragmentation impact two-thirds of all amphibians. The distinct characteristics of amphibians, such as low mobility and permeable skin, make them more susceptible to environmental influences than other taxa. We have shown that within the small area of the Kirindy Forest in Madagascar, environmental differences influenced the selection of breeding sites by frogs to such an extent that the composition of the anuran larvae assemblage differed vastly between individual ponds. The diversity of breeding sites increases species richness by serving the needs of multiple species. Characteristics such as pond size, vegetation, and time until desiccation of ponds should be considered when designing conservation action plans to protect a wide variety of species. If breeding sites become unavailable because of a shift in environmental gradients, not only amphibian diversity but the whole ecosystem could be harmed.

Abstract: Environmental filtering shapes animal communities by preventing the colonization and persistence of certain species in a given habitat. More heterogenous environments are presumed to support a greater number of species and, consequently, increased species diversity, as environmental filters are also likely more heterogenous. Amphibians are especially sensitive to environmental influences due to distinct characteristics like permeable skin and low mobility. By analyzing the species richness and assemblage composition of tadpoles in 132 breeding ponds, we examined how the interplay of environmental variables shapes anuran species assemblages in breeding habitats of the dry forest of Western Madagascar. We found that environmental filtering is prevalent and habitat heterogeneity not only increases larval species richness but also alters species composition between these assemblages. Our study highlights the need for conserving heterogenous habitats to maintain local diversity. Furthermore, we recommend including multivariate modelling approaches to conservation efforts to acknowledge differences between specific habitats and beta diversity.

Keywords: community ecology; amphibians; oviposition; pond ecology; species assemblages

Citation: Mausberg, N.; Dausmann, K.H.; Glos, J. In Search of Suitable Breeding Sites: Habitat Heterogeneity and Environmental Filters Determine Anuran Diversity of Western Madagascar. *Animals* **2023**, *13*, 3744. https://doi.org/10.3390/ani13233744

Academic Editor: José Martín

Received: 30 September 2023
Revised: 20 November 2023
Accepted: 30 November 2023
Published: 4 December 2023

1. Introduction

Conservation of pond-breeding amphibians requires monitoring and protection of their terrestrial as well as their aquatic breeding habitats [1–4]. The degradation and loss of the latter are the main factors leading to declines in amphibian diversity [5]. Even subtle differences in environmental factors can result in major declines in species, if they affect crucial individual needs [6,7]. Alarmingly, two out of three amphibian species are impaired by habitat alterations, such as destruction and fragmentation [5,6]. Nevertheless, limnic habitats are usually not sufficiently considered in conservation plans [4], although their importance for many and in particular rare species has been reported by numerous studies [8–10]. Therefore, the IUCN Amphibian Conservation Action Plan highlighted the need for identification, investigation, and protective measures of amphibian habitats [6].

Madagascar has an especially high level of local endemism, with few remnants of undisturbed forest. Due to ongoing deforestation and degradation, many of Madagascar's species are under threat, particularly in the dry forests of Western Madagascar [4]. Amphibians there fulfill an important functional role in this ecosystem, which makes this area a conservation priority.

Due to their biphasic life cycle as tadpoles and frogs, many amphibians have complex ecological requirements regarding their habitats and the occurrence of these two ecologically very different stages can depend on their surroundings [11,12]. Biotic (e.g., predators, food availability) and abiotic (e.g., pond size, hydroperiod) properties of ponds are among the most important forces affecting species assemblages because they constrain the distribution of anuran species. These properties of ponds can act as filters, preventing or allowing species to colonize and persist in a given habitat [13,14]. Thus, they influence species' occurrences and, thereby, the assemblage composition of a habitat [7]. The distinct characteristics of amphibians, such as low mobility and permeable skin, make them especially sensitive to environmental filters and vulnerable to changes [15–17]. A high variability in environmental filters is equivalent to high heterogeneity between habitats. More heterogenous environments are presumed to support a greater number of species and, consequently, increased species diversity [18–21]. Identifying environmental characteristics that structure amphibian assemblages and determining suitable habitats are a prerequisite for understanding the ecological underpinnings of amphibians' lifestyle, and for proposing pertinent conservation plans [1,22–27].

As greater habitat heterogeneity offers a greater variety of microhabitats and thus niche opportunities that serve the needs of multiple species, we hypothesized that the environmental heterogeneity of breeding ponds a) increases the number of species present within a pond, in the following called species richness, and b) affects which species occur within a pond, here called species composition. Thereby, environmental filters limit the occurrence of some species that cannot cope with the environmental conditions of a certain habitat. They lack the ability to persist in this habitat. Species turnover between ponds that differ in environmental descriptors would indicate for that. Therefore, we tested for differences in environmental variables and species composition between two naturally occurring habitat types within the same forest in Western Madagascar, namely frog breeding ponds within the dry forest and ponds within a riverbed that intersects the forest (riverbed ponds Figure A1 Appendix A). Riverbed ponds usually form after the first heavy rains in November or December and last for up to one and a half months before ongoing rainfall leads to the connection of riverbed ponds before the river starts to flow [28]. Forest ponds usually arise later than ponds in the riverbed, generally in late November or mid-December. Their hydroperiod is very variable (between three days and five months), depending on rainfall. Differences in tadpole assemblage composition between ponds were defined as ß-diversity. We then performed multivariate analyses to evaluate which environmental descriptors of breeding ponds influence species richness and assemblage composition within and between these habitat types.

2. Material and Methods

2.1. Study Area

Central Menabe, the area between Morondava and the Tsiribinha River in Western Madagascar, has been a "Site de Conservation" since 2006 and covers an area of 1250 km². Located in the center of Menabe, 50 km northeast of Morondava and 20 km inland (44°39′ E, 20°03′ S; 18–40 m above sea level; [29]), the dry, deciduous Kirindy Forest/CNFEREF (Centre National de Formation, d'Etudes et de Recherche en Environnement et Foresterie Morondava) comprises approximately 120 km² [30]. The climate in Kirindy Forest is highly seasonal with a rainy season of three to five months from November/December to February/March, followed by a dry season with virtually no precipitation of seven to nine months. The mean annual precipitation is approximately 800 mm (range 390–1511 mm; data from 1906 to 1993; [29]). The soil has a low capacity to retain water, as it mainly consists of sandy soils [4]. Consequently, most breeding sites for amphibians desiccate completely during the dry season, allowing reproduction only during the rainy season [29]. Ponds in the Kirindy Forest are highly dynamic and heterogenous, with regard to predators and pond size [31], e.g., pond size varies from <5 m^2 to $>10,000$ m^2.

2.2. Data Acquisition

2.2.1. The Presence of Anuran Species at Breeding Ponds

A total of 132 potential breeding ponds were sampled (n = 47 forest ponds, n = 85 riverbed ponds) over four consecutive rainy seasons, from 1998 to 2002. These ponds were within a range of 3 km^2. The use of ponds as breeding sites by anurans was determined by the presence of tadpoles. This method directly determines breeding success and does not rely on the assumption that frog presence or calling activity corresponds to actual breeding at a breeding site.

Tadpoles were sampled by standardized dip-netting [32]: In each pond and in each sampling event, 30 dip net strokes were performed, randomly distributed over the pond. The dip net was triangularly shaped with a base of 400 cm^2 ($30 \times 30 \times 30$ cm; mesh size 1 mm). Each dip net stroke was 1 m long and touched the ground substrate. All tadpoles were identified to the species level in the field camp, using a stereomicroscope and existing literature [33,34]. Most tadpoles were subsequently returned into their natal ponds. Only a few voucher specimens were retained for comparison with a reference collection.

All ponds were repeatedly sampled every year. Sampling intervals were adapted to detect all species at a certain site irrespective of their breeding mode (i.e., explosive vs. prolonged breeders) and length of larval development, leading to a variable total number of sampling events at each site in one year. To detect tadpoles of explosively breeding species that tend to breed directly after heavy rainfall [31], a sampling event was repeated at each pond approximately one week after each rainfall event that exceeded 30 mm, and after each refilling of ponds after they had dried out. To detect tadpoles of prolonged breeders at breeding ponds with longer hydroperiods, additional sampling was performed in intervals of approximately one month throughout the rainy season.

Environmental variables and tadpole species composition were consistent between seasons and therefore data from all four years were pooled. Thus, species richness was defined as the total number of species at each breeding site and assemblage composition was defined as the species identities within one pond, irrespective of year and sampling date.

2.2.2. Environmental Variables of Breeding Ponds

For all ponds in both habitat types (forest and riverbed ponds), a set of nine abiotic and biotic variables was measured at each breeding pond (Table 1), representing measures of important aspects in tadpole ecology such as pond hydroperiod, microhabitat niches and predation. These variables were measured at each pond each year and also repeatedly within one year. All measurements of one variable were pooled and divided by the number of measurements for this variable. These means were used for further calculations. To keep an observer bias at minimum, all variables were measured always by the same person (JG).

2.3. Statistical Analysis

Statistical analyses were carried out using the statistical computing software *R* (V4.1.2, R Core Team 2021, Vienna, Austria) and the packages Vegan [35], MASS [36], permute [37] and lattice [38] for bivariate correlations, Wilcoxon-test, Generalized linear model (GLM), Non-metric multidimensional scaling (NMDS) and Mantel tests. Differences in environmental variables between habitat types "forest ponds" and "riverbed ponds" were calculated using non-parametric Wilcoxon signed-rank (W) tests.

Table 1. List of independent habitat variables recorded at each pond within the Kirindy Forest, including unit, detailed nomenclature and method.

Variable Name	Variable Definition	Unit	Method
Desiccation risk	Desiccation intensity of ponds.	Five categories: 1 (not at risk of desiccation = long hydroperiod) to 5 (extreme risk of desiccation = short hydroperiod)	Estimated by the number of days before desiccation, after a complete filling of the pond and a successive period of no rainfall.
Pond size	Absolute surface area of pond when maximally water filled.	m^2	Measured using length, width and shape.
Maximum depth	Maximum pond depth when maximally water filled.	cm	Measured at deepest point of water body.
Shallow water	Relative pond area shallower than 10 cm.	%	Visual estimation.
Submerged vegetation	Relative volume of pond filled with submerged water plants.	Six categories: 0 (0%), 1 (1–20%), 2 (21–40%), 3 (41–60%), 4 (61–80%), 5 (81–100%)	Visual estimation.
Surrounding vegetation	Density and complexity of vegetation structure $\leq$ 2 m from pond edge.	Six categories: 0 (no vegetation) to 5 (high density)	Visual estimation.
Leaf litter	Relative area of pond bottom covered with dead leaves.	Six categories: 0 (0%), 1 (1–20%), 2 (21–40%), 3 (41–60%), 4 (61–80%), 5 (81–100%)	Visual estimation.
Turbidity	Turbidity of the pond water influenced by dissolved and suspensed matters.	Three categories: 1 (clear water), 2 (slightly dull), 3 (very turbid water)	Visual estimation.
Predator occurrence	Visual estimation of invertebrate predators, including: dytiscid beetles > 0.5 cm, larval dytiscid beetles > 1 cm, water bugs (Belostomatidae) > 0.5 cm, water scorpions (Nepidae) > 1 cm, larval dragonflies (Anisoptera) > 0.5 cm, larval damselflies (Zygoptera) > 2 cm.	Six categories: 0 (no predators) to 5 (high density; >5 predators per dip net stroke)	Measured using dip-netting and the box-method [32] and averaged; determined to higher taxonomic levels in the field and subsequently released.

2.3.1. Influence of Environmental Variables on Species Richness

Variable Selection—Spearman correlations were used to identify environmental variables that were highly correlated (r > 0.7, as suggested by [39]), and one variable from a variable pair with high collinearity was excluded from further analysis. This was the case for desiccation risk and maximum depth of water bodies (r = −0.87, $p < 0.001$) and for desiccation risk and the percentage of shallow water (<10 cm) (r = 0.72, $p < 0.001$). Additionally, the shading of ponds, as percentage of the pond shaded at noon, correlated with pond surrounding vegetation (r = 0.57, $p < 0.001$), submerged vegetation (r = −0.3, $p = 0.03$), and leaf litter (r = 0.57, $p < 0.001$). Desiccation risk and the different types of vegetation were considered as the biologically more relevant variables. Therefore, maximum pond depth, the percentage of shallow water (<10 cm), and the ponds' shading were not included in the statistical models.

Modelling—In order to extract the key habitat factors that predict the choice of breeding waters, a GLM was designed [40]. A model was generated that showed which environmental variables best described a change in species richness. The habitat type and seven independent variables were included in the initial model as predictor variables. A Poisson error distribution with a log link function was used with species richness of tadpoles as the dependent variable [41]. Additionally, interactions between environmental variables and habitat type were included in the initial GLM. Variables and interactions were removed using a stepwise Akaike Information Criterion (stepAIC) approach. Thereby, the initial model was simplified and the best fitted model with the lowest AIC was identified.

2.3.2. The Influence of Environmental Variables on Assemblage Composition

To analyze for differences in assemblage composition (i.e., species-turnover) between sites and between habitat types (i.e., ß-diversity), data were arranged in matrices including absence (coded as "0") or presence (coded as "1") of each species at a respective breeding site. A multivariate analysis of permutational variance (perMANOVA) was performed.

One-way permutation analysis was set to 9999 permutations and used to determine significant dissimilarities. Graphical representation was obtained from NMDS [42]. Here, the Bray–Curtis dissimilarity index [43,44] was used as a distance measure and three dimensions were used for illustration.

To analyze the correlation of environmental variables of breeding sites with their tadpole assemblage composition, data of environmental variables were also arranged in matrices including the respective values (i.e., means of multiple sampling events, see above) of each breeding site. Correlations between the matrices of assemblage composition and environmental variables were identified using a Mantel test [45], and using the Bray–Curtis index [43] as a distance measure for both data matrices. Pearson's product-moment was used to determine correlations. Significance of Mantel correlations was computed through 9999 permutations.

3. Results

3.1. Comparison of Forest and Riverbed Ponds

Forest ponds were on average larger and much more variable in size but nevertheless, had a higher desiccation risk and contained relatively more submerged vegetation. The vegetation directly at the edges of the ponds were structurally richer and denser (Table 2). The habitat types did not differ significantly in water turbidity, predator occurrence and leaf litter at the bottom of the ponds.

Table 2. Environmental differences between forest ponds and riverbed ponds, indicated by means ± standard deviation and results of Wilcoxon-tests. Variables marked with an asterisk are considered to differ significantly between the habitat types.

Variables	Forest Ponds	Riverbed Ponds	Wilcoxon-Test
Desiccation (1 to 5)	3.5 ± 1.2	2.9 ± 0.9	2653.5, $p = 0.001$ *
Pond size (m^2)	1185.3 ± 3617.6	29.8 ± 62.3	3038.5, $p < 0.001$ *
Submerged vegetation (0 to 5)	1.4 ± 1.2	0.1 ± 0.2	3594.0, $p < 0.001$ *
Surrounding vegetation (0 to 5)	4.2 ± 0.9	2.4 ± 0.8	3630.0, $p < 0.001$ *
Turbidity (1 to 3)	1.4 ± 0.5	1.5 ± 0.7	1997.5, $p = 1$
Predator occurrence (0 to 5)	2.5 ± 1.3	2.1 ± 1.4	1202.5, $p = 0.25$
Leaf litter (0 to 5)	3.2 ± 1.3	3.6 ± 1.0	1595.5, $p = 0.07$

3.2. Distribution of Tadpole Species

Fifteen species were recorded within 132 ponds, over the duration of four years. We identified tadpoles belonging to the families of Mantellidae (*Aglyptodactylus laticeps, A. securifer, Blommersia wittei, Boophis doulioti, Boophis xerophilus, Laliostoma labrosum, Mantella betsileo*), Microhylidae (*Dyscophus insularis, Scaphiophryne brevis, S. calcarata, S. menabensis*), Hyperoliidae (*Heterixalus carbonei, H. luteostriatus, H. tricolor*) and Ptychadenidae (*Ptychadena mascareniensis*) (Figure A2 Appendix A). Species richness in ponds ranged from zero to eleven species (3.27 ± 2.03; mean ± SD). Tadpoles of 14 species in total were recorded in forest ponds (3.96 ± 2.52); nine species in riverbed ponds (2.88 ± 1.58). Tadpoles of six species were only recorded in forest ponds (*Blommersia wittei, Heterixalus carbonei, H. luteostriatus, H. tricolor, Ptychadena mascareniensis, Scaphiophryne menabensis*) and one species (*A. securifer*) only in riverbed ponds. Eight species occurred in both habitat types. Tadpoles of *D. insularis* were the overall most frequent species, occurring in over 50% of the waterbodies. In forest ponds, tadpoles of *B. doulioti* (n = 33) were the most frequent whereas *A. securifer* was the most frequent species in riverbed ponds (n = 71).

3.3. Environmental Variability and Species Richness

When only comparing species richness between habitat types (i.e., not including other environmental variables of the ponds), forest ponds had significantly more species than riverbed ponds (W = 2416.5, p = 0.04 (Table 3)). In the model selection process starting with a model including all environmental variables and interactions, six interactions and two variables were removed stepwise from the GLM: Habitat:Surrounding Vegetation (step 1), Habitat:Pond size (step 2), Pond size (step 3), Habitat:Desiccation (step 4), Surrounding Vegetation (step 5), Habitat:Turbidity (step 6) and Habitat:Leaf litter (step 7).

Table 3. Initial and final GLM according to a stepAIC approach. Initial model (AIC = 362.3) contained the habitat type, seven environmental variables and interactions of habitat type and each independent variable, respectively. The final model (AIC = 349.9) included habitat type, six environmental variables and two interactions. Variables marked with asterisks are considered significant variables within the model, with * = p < 0.05 and ** = p < 0.01.

Coefficients	Estimate	Standard Error	z-Value	p-Value
Initial generalized linear model (AIC = 362.3)				
Habitat	−0.31	0.11	−0.28	0.78
Desiccation	−0.22	0.12	−1.90	0.06
Leaf litter	0.11	1	1.11	0.27
Predator occurrence	0.17	0.73	2.39	0.02 *
Pond size	−0.38	0.23	−0.16	0.87
Submerged vegetation	0.65	1	0.70	0.51
Surrounding vegetation	−0.33	0.15	−0.22	0.83
Turbidity	1	0.19	0.52	0.60
Habitat:Desiccation	0.6	0.15	0.39	0.7
Habitat:Leaf litter	0.95	0.15	0.65	0.52
Habitat:Predator occurrence	−0.3	0.98	−3.05	<0.001 **
Habitat:Pond size	0.15	0.42	0.35	0.73
Habitat:Subermerged vegetation	0.53	0.42	1.25	0.21
Habitat:Surrounding vegetation	−0.74	0.22	−0.34	0.73
Habitat turbidity	0.2	0.22	0.85	0.73
Final generalized linear model (AIC = 349.9)				
Habitat	0.47	0.3	1.58	0.12
Desiccation	−0.18	0.07	−2.71	<0.01 **
Leaf litter	0.13	0.05	2.44	0.01 *
Turbidity	0.2	0.1	2.09	0.04 *
Predators	0.18	0.07	2.61	0.01 **
Submerged vegetation	0.09	0.08	1.19	0.23
Habitat:Submerged vegetation	0.61	0.37	1.66	0.1
Habitat:Predator occurrence	−0.3	0.09	−3.28	<0.001 **

Based on the final GLM, four variables and one interaction were significantly associated with tadpole species richness (Table 4). Species richness declined with desiccation risk in both habitat types, although the effect was stronger in forest ponds (Figure 1a). Species richness increased with the amount of leaf litter as substrate, in particular in forest ponds (Figure 1b). Also, the interaction of predator occurrence and habitat type had a significant influence, increasing species richness with predator density in forest ponds but decreasing species richness in riverbed ponds (Figure 1c). Finally, species richness increased with turbidity in riverbed ponds, but decreased in forest ponds (Figure 1d).

Table 4. Mantel test correlations of environmental variables and tadpole beta diversity of assemblages. Variables marked with an asterisk are considered to significantly correlate with a change in species composition between habitat types.

Variables	Mantel Statistic r	Significance
Habitat	0.43	$p < 0.001$ *
Desiccation	0.13	$p < 0.001$ *
Pond size	0.19	$p < 0.001$ *
Submerged vegetation	0.27	$p < 0.001$ *
Surrounding vegetation	0.28	$p < 0.001$ *
Turbidity	−0.01	0.61
Predator occurrence	0.08	0.02 *
Leaf litter	0.12	$p < 0.001$ *

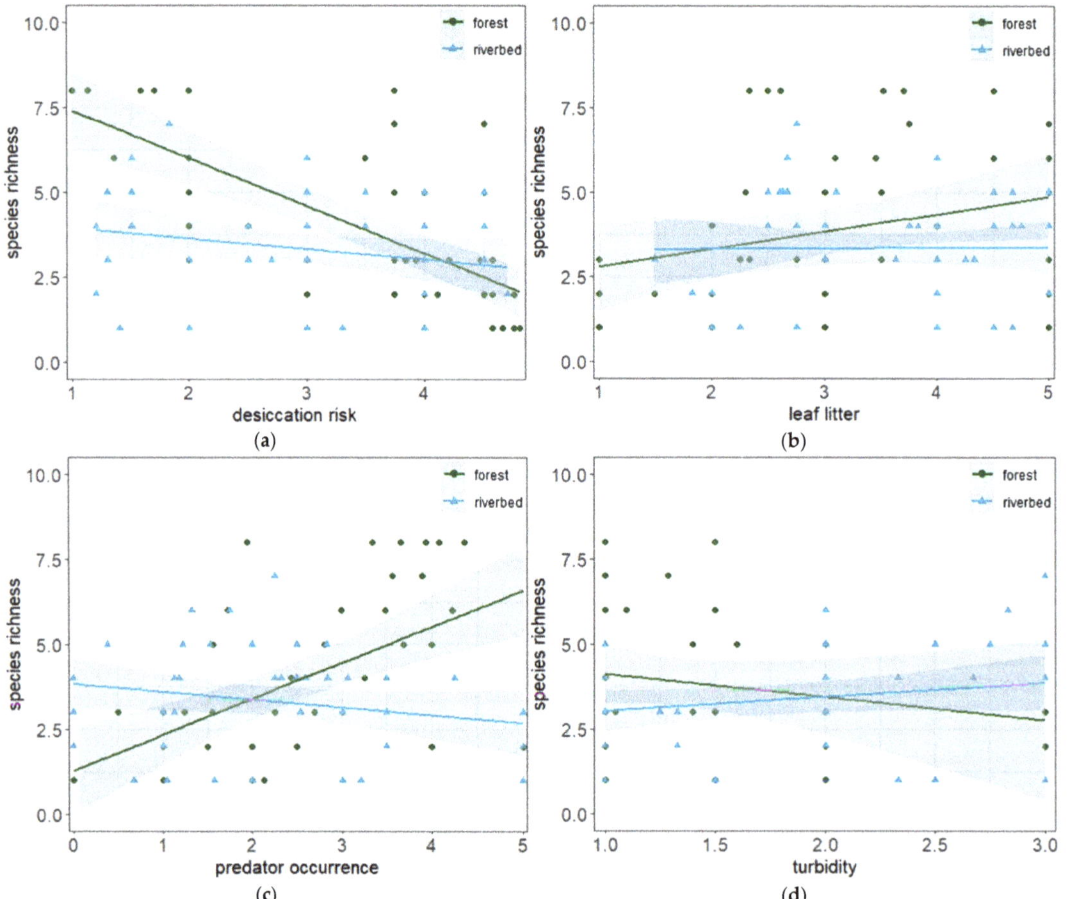

Figure 1. Variation in richness of tadpole species between two habitat types in relation to (**a**) desiccation risk, (**b**) leaf litter, (**c**) predator occurrence and (**d**) pond turbidity. Green dots indicate forest pond plots, blue triangles indicate riverbed ponds. Confidence intervals (95%) are indicated by grey shadows.

3.4. Species Composition in Different Habitat Types

Species composition differed significantly between forest ponds and riverbed ponds (perMANOVA: $R^2 = 0.55$, F = 155.81; $p < 0.001$; Figure 2).

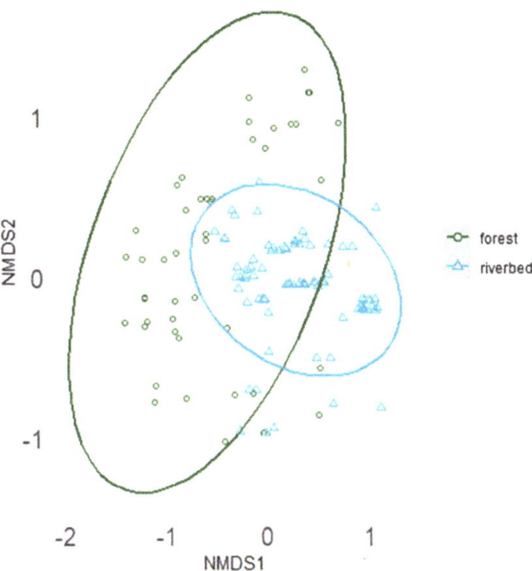

Figure 2. Non-metric multidimensional scaling of tadpole species compositions of forest ponds (green dots) and riverbed ponds (blue triangles); axes 1 and 2 of a three-dimensional analysis. Species communities incorporate absence/presence data. Ellipses indicate 95% confidence intervals fitted into the spatial ordination. $R^2 = 0.55$, $p < 0.001$, stress = 0.13.

Differences in species composition between breeding ponds were correlated with habitat type, desiccation risk, pond size, submerged vegetation, surrounding vegetation, predator occurrence and leaf litter (Table 4; Mantel test statistics). Only pond turbidity did not show a correlation with species composition.

4. Discussion

Frogs have to select breeding ponds to improve survival of their tadpoles. Amphibians of the Kirindy Forest show exceptional responses to their environment [31] and most of them are endemic to this habitat [4,46]. There are currently 15 known amphibian species that use forest and riverbed ponds of the Kirindy Forest as breeding sites [31]. Species such as *A. laticeps* and *S. menabensis* occur exclusively in relatively undisturbed forest patches [4]. Their presence is accompanied by low water permanency and diverse vegetation surrounding the pond, both of which indicate an undisturbed forest. Thus, they function as indicators for an intact environment [4]. They also represent umbrella species ("umbrella effect" [47]) for their habitats because other species including *B. doulioti*, *D. insularis*, *L. labrosum*, *M. betsileo* and *S. calcarata* [33] as well as aquatic species of other taxa, benefit from their protection [4]. However, the availability and nature of breeding ponds depends on environmental conditions and habitat types. Even within the relatively small area of Kirindy Forest, adjacent habitat types contain sufficient environmental heterogeneity to cause species turnover (change in species) between ponds. Thus, adult anurans that select either forest or riverbed ponds will encounter different environmental conditions at their breeding sites. The clear correlations of most environmental variables with assemblage composition indicate that environmental filters we measured were effective in both habitat types, limiting the occurrence of some species and making the site suitable for others. While most species breed in both habitats, a number of species are found almost exclusively either at forest ponds (*Aglyptodactylus laticeps*, *Heterixalus* spp.) or riverbed ponds (*Aglyptodactylus securifer*).

Pond characteristics that were shown to be important variables in determining amphibian assemblage composition included habitat heterogeneity, predation, pond size, surrounding vegetation, submerged vegetation, leaf litter and desiccation risk. Forest ponds with a long hydroperiod, clear water, an abundance of leaf litter and many invertebrates contained more species than others. In riverbed ponds, however, species richness was either un-correlated or only weakly correlated with hydroperiod, leaf litter and predators, but positively correlated with water turbidity.

Previous studies on anuran assemblages showed different effects of environmental heterogeneity at breeding ponds on assemblage composition. Some studies did not detect any environmental influence on assemblage composition [48]. Others not only linked greater habitat heterogeneity to increased community diversity of anurans in aquatic, but also in terrestrial habitats [49,50]. Many studies found that homogenous areas contained fewer species compared to heterogenous sites [51–53]. For conservation purposes, it is therefore necessary to develop individual approaches for each area, as no uniform pattern can be assumed.

Among the variables tested in this study, the size of the breeding pond and the length of the hydroperiod have been previously shown to influence which species select a certain breeding pond [11,20,53,54]. Smaller water bodies usually have shorter hydroperiods and so dry out faster [54]. Therefore, assemblage composition might be related to different lengths of developmental time of the tadpoles. For example, *B. xerophilus*, which has a longer developmental time (approximately 30 days larval duration) and is generally larger as tadpole, is restricted to breeding in larger, more permanent ponds, whilst *B. doulioti*, which shows high developmental plasticity, also uses smaller and more ephemeral ponds for oviposition [55]. Developmental time might depend on the type of reproductive pattern of the species. Most anuran species in the Kirindy Forest are explosive breeders, reproducing only after heavy rainfalls [56]. Their reproductive success depends on fine-tuning the time of oviposition to environmental conditions [57]. Accordingly, species with short larval development such as *A. laticeps* and *Scaphiophryne* spp. ($\leq$10 days; [31,33]) were found predominantly in temporary ponds, which have the advantage of fewer predators compared to permanent ponds. Some prolonged breeders reproduce over longer periods in the rainy season [56]. For example, *H. tricolor* and *H. carbonei*, both prolonged breeders, are specialized to breed in more permanent ponds [58]. Therefore, it is likely that explosive breeders mainly choose ponds with short hydroperiods and a lack of predators, whilst prolonged breeders prefer ponds with longer hydroperiods. This distinction between explosive and prolonged breeders is consistent with several other studies [59–62]. The dispersal of amphibians as well as the distribution of other species is related to the hydroperiod of available waterbodies [61,63]. Yet, the number of species that were present within a breeding pond was irrespective of the pond size within this study. This suggests that pond size may act as an environmental filter only for some species, namely the ones that require a specific pond size, but not for others that then take the spot of those that cannot cope with this certain pond size.

In addition to beta diversity, species richness can also be related to the risk of desiccation of breeding sites [14,19]. As desiccation risk decreases, more species can exploit a breeding site [60]. Thus, persistent ponds are assumed to have higher species richness than ponds with a shorter hydroperiod [64]. In Kirindy Forest, the choice of breeding sites was considerably more affected by desiccation risk in forest ponds than in riverbed ponds. Riverbed ponds are mainly rock pools with a better capacity to retain water, and thus a lower risk for desiccation. The chance of these drying before the tadpoles have metamorphosed is slim. Indeed, riverbed ponds were usually the first to be used for spawning [28].

However, the riverbed ponds present other challenges to tadpoles: strong rainfall raise the water level and eventually the separate ponds merge and the river starts to flow. Invasive cichlid fish (*Oreochromis* sp.) that have spent the dry season in the few permanent ponds of the riverbed are able to move freely throughout the riverbed and increase the risk of predation for tadpoles [28]. Predation can affect communities via (a) direct predation [65,66]; (b) elimination of competitors [67]; and (c) differential predation, depending on prey size [68]. Size-directed predation can favor survivors by reducing competition in an early larval stage [68]. Adult amphibians may then choose breeding sites that show lower competition over sites with low predation which in turn, might lead to a higher species richness in these habitats [68]. Therefore, predation in riverbed ponds reduces the occurrence of all tadpole species equally. Forest ponds, however, stay free from fish predation [28]. In forest ponds, high species richness was observed together with a high occurrence of invertebrate predators (mainly larvae of dragonflies, dytiscid beetles and belostomatid water bugs). This is not an unusual finding. Size-directed predation can favor survivors by limiting individuals in their early metamorphosis, thus reducing competition. Adult amphibians may then choose breeding sites that show lower competition over sites with low predation which in turn, might lead to a higher species richness in these habitats [68].

Vegetation in and around waterbodies and riparian areas can decrease desiccation by maintaining humidity and has been considered an important influence affecting assemblage composition [20,49,51,54]. Vegetation supports sites for vocalization, mating, and oviposition [20] and increases the structural complexity of an environment. Hence, the increased availability of reproductive sites and niches serves the multiple needs of different species of breeding frogs [18,19,69,70]. In forest ponds, vegetation was generally higher compared to riverbed ponds and beta diversity was strongly influenced by these differences. Vegetation structure might influence the choice of breeding sites of anurans in the terrestrial as well the aquatic habitat [71,72]. As a reaction to high predation risks, many species (as frogs) tend to prefer structurally complex pond surroundings for oviposition, but also (as tadpoles) structurally complex sites within ponds [65,73], because most predators hunt visually. In a structurally complex habitat, foraging efficiency of predators is reduced when vegetation functions as shelter for the prey individuals, and thus, the survival of tadpoles significantly increases with denser aquatic vegetation [74,75].

Like vegetation, pond turbidity can also provide visual cover for tadpoles. Water turbidity has a camouflaging effect that decreases the risk of predation, making the pond available for a greater variety of tadpole species [76]. Additionally, turbidity results from an increase in organic sediments, which may serve as a food resource for tadpoles [77]. Indeed, in riverbed ponds, turbidity was accompanied by a high species richness. However, turbidity negatively influenced species richness in forest ponds. This might be an effect of water turbidity also negatively affecting growth rates of tadpoles and survival to metamorphosis [77]. The conflicting effects observed here could demonstrate a trade-off of using the advantages of turbidity as a shelter and/or food resource and its negative effect on growth and/or survival of tadpoles.

Six out of seven environmental variables that were analyzed here were correlated with species composition, whilst four variables were correlated with species richness. As comparisons with studies in other anuran habitats have shown, relevant variables differ between different habitats and species assemblages. Therefore, different environmental influences must be considered in conservation management to protect unique anuran assemblage in each type of breeding pond.

5. Conservation Implications

The heterogeneity between breeding sites was able to support the individual requirements of a vast variety of anuran species. But this also shows that environmental filters were effective in both habitat types and thus limited the occurrence of some species but made the site suitable for those that were capable of coping with the prevailing conditions. Several species of this study, such as *A. laticeps* and *S. menabensis*, are endangered and have a very limited area of occurrence within Western Madagascar. Additionally, they occur exclusively in relatively undisturbed forest patches within their distribution range [4]. These species live on the edge, and their life-history is finely tuned to conditions typical for undisturbed forests. They breed in ponds with occasionally very short hydroperiods of only ten days or less. If such ponds are under the forest canopy and have micro-climatic conditions of an undisturbed forest, the larvae regularly metamorphose just before the pond dries out. In conservation efforts, these species may function as indicators of an intact habitat [4]. They may also represent umbrella species ("umbrella effect" [47]) as several other frog species (such as *B. doulioti*, *D. insularis*, *L. labrosum*, *M. betsileo*, and *S. calcarata*), as well as terrestrial and aquatic species of other taxa, may benefit from their conservation [4].

Due to ongoing deforestation and degradation, many of the species in the Menabe Region are under threat [4]. The development of priority conservation plans is thus a critical step to guide pond conservation strategies, either through regular review of the assignment of important sites, management, and acquisition policy, or implementation of conservation restrictions and other land-use agreements that preserve the variety of breeding sites. This study highlights the need to preserve a variety of ponds with differences in hydroperiod, pond size and vegetation (including surrounding vegetation, submerged vegetation and leaf litter) which provide diverse microhabitats and refugia. These habitat characteristics should be considered when attempting to identify breeding sites that could provide core sites in conservation reserves designed for the conservation of water-breeding amphibians. If breeding sites become unavailable because of a shift in these environmental gradients, not only amphibian diversity but perhaps the whole ecosystem might be harmed.

Author Contributions: Conceptualization, K.H.D. and J.G.; methodology, J.G.; validation, J.G.; formal analysis, N.M. and J.G.; investigation, K.H.D. and J.G.; data curation, N.M. and J.G.; writing—original draft, N.M.; writing—review and editing, K.H.D. and J.G.; visualization, N.M. All authors have read and agreed to the published version of the manuscript.

Funding: This research was funded by a scholarship of the German Academic Exchange Service (DAAD) for Julian Glos.

Institutional Review Board Statement: Permits for this study was approved and issued by the Ministère pour la Production Animale, and the Département des Eaux et Forêts Madagascar (Research permit # 83/MEF/SG/DGEF/DGDRF/SCB; collection and export permit # 0105N/EA02/MG02).

Informed Consent Statement: Not applicable.

Data Availability Statement: Data will be deposited in Dryad at acceptance of the manuscript.

Acknowledgments: We would like to thank the following persons and institutions for their help and collaboration, for providing permits and for funding: J. Ganzhorn (Universität Hamburg, Germany), William J. Foley (Australian National University, Canberra), Département de Biologie Animale (Université d'Antananarivo), Ministère des Eaux et Forêts (Madagascar), German Primate Center (Göttingen, Germany), German Academic Exchange Service (DAAD), and CNFEREF (Centre National de Formation, d'Etudes et de Recherche en Environnement et Foresterie, Morondava).

Conflicts of Interest: The authors declare no conflict of interest.

Appendix A

Figure A1. Forest pond (**left**) and riverbed pond (**right**) in Kirindy Forest. Species richness and composition of tadpole assemblages differ between these habitat types.

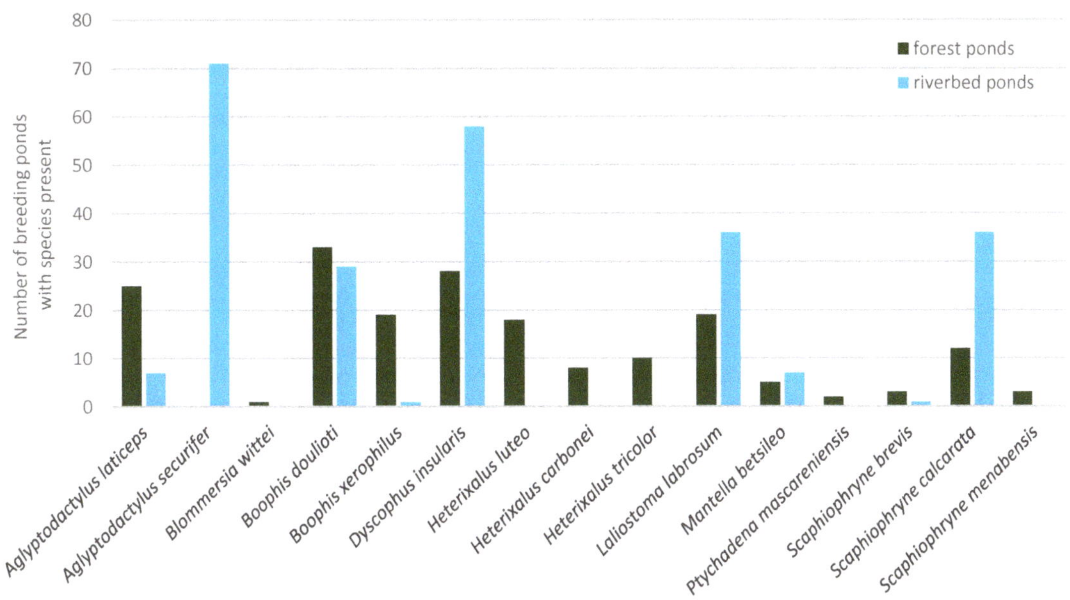

Figure A2. All tadpole species found at 132 ponds within the Kirindy Forest, according to the habitat types: forest ponds (green) and riverbed ponds (blue).

References

1. Araújo, M.B.; Williams, P.H.; Fuller, R.J. Dynamics of extinction and the selection of nature reserves. *Proc. R. Soc. Biol. Sci.* **2002**, *269*, 1971–1980. [CrossRef] [PubMed]
2. Stuart, S.N.; Chanson, J.S.; Cox, N.A.; Young, B.E.; Rodrigues, A.S.L.; Fischman, D.L.; Waller, R.W. Status and trends of amphibian declines and extinctions worldwide. *Science* **2004**, *306*, 1783–1786. [CrossRef] [PubMed]
3. Andreone, F.; Cadle, J.E.; Cox, N.; Glaw, F.; Nussbaum, R.A.; Raxworthy, C.J.; Stuart, S.N.; Vallan, D.; Vences, M. Species review of amphibian extinction risks in Madagascar: Conclusions from the Global Amphibian Assessment. *Conserv. Biol.* **2005**, *19*, 1790–1802. [CrossRef]
4. Glos, J.; Volahy, A.T.; Bourou, R.; Straka, J.; Young, R.; Durbin, J. Amphibian conservation in central Menabe. *A Conserv. Strategy Amphib. Madag.* **2008**, *48*, 107–124.
5. Sodhi, N.S.; Bickford, D.; Diesmos, A.C.; Lee, T.M.; Koh, L.P.; Brook, B.W.; Sekercioglu, C.H.; Bradshaw, C.J.A. Measuring the meltdown: Drivers of global amphibian extinction and decline. *PLoS ONE* **2008**, *3*, e1636. [CrossRef] [PubMed]
6. Bishop, P.J.; Angulo, A.; Lewis, J.P.; Moore, R.D.; Rabb, G.B.; Garcia Moreno, J. The Amphibian Extinction Crisis—What will it take to put the action into the Amphibian Conservation Action Plan? *Surv. Perspect. Integr. Environ. Soc.* **2012**, *5*, 97–111.
7. Jorge, R.F.; Simões, P.I.; Magnusson, W.E.; Lima, A.P. Fine-scale habitat heterogeneity explains the local distribution of two Amazonian frog species of concern for conservation. *Biotropica* **2016**, *48*, 694–703. [CrossRef]
8. Della Bella, V.; Bazzanti, M.; Chiarotti, F. Macroinvertebrate diversity and conservation status of Mediterranean ponds in Italy: Water permanence and mesohabitat influence. *Aquat. Conserv. Mar. Freshw. Ecosyst.* **2005**, *15*, 583–600. [CrossRef]
9. Davies, B.; Biggs, J.; Williams, P.; Whitfield, M.; Nicolet, P.; Sear, D.; Bray, S.; Maund, S. Comparative biodiversity of aquatic habitats in the European agricultural landscape. *Agric. Ecosyst. Environ.* **2008**, *125*, 1–8. [CrossRef]
10. Silver, C.A.; Vamosi, S.M.; Bayley, S.E. Temporary and permanent wetland macroinvertebrate communities: Phylogenetic structure through time. *Acta Oecologica* **2012**, *39*, 1–10. [CrossRef]
11. Keller, A.; Rödel, M.-O.; Linsenmair, K.E.; Grafe, T.U. The importance of environmental heterogeneity for species diversity and assemblage structure in Bornean stream frogs. *J. Anim. Ecol.* **2009**, *78*, 305–314. [CrossRef] [PubMed]
12. Von May, R.; Jacobs, J.M.; Santa-Cruz, R.; Valdivia, J.; Huamán, J.M.; Donnelly, M.A. Amphibian community structure as a function of forest type in Amazonian Peru. *J. Trop. Ecol.* **2010**, *26*, 509–519. [CrossRef]
13. Badgley, C.; Fox, D.L. Ecological biogeography of North American mammals: Species density and ecological structure in relation to environmental gradients. *J. Biogeogr.* **2000**, *27*, 1437–1467. [CrossRef]
14. Werner, E.E.; Skelly, D.K.; Relyea, R.A.; Yurewicz, K.L. Amphibian species richness across environmental gradients. *Oikos* **2007**, *116*, 1697–1712. [CrossRef]
15. Welsh, H.H.; Ollivier, L.M. Stream amphibians as indicators of ecosystem stress: A case study from California's Redwoods. *Ecol. Appl.* **1998**, *8*, 1118. [CrossRef]

16. Vormisto, J.; Phillips, O.L.; Ruokolainen, K.; Tuomisto, H.; Vásquez, R. A comparison of fine-scale distribution patterns of four plant groups in an Amazonian rainforest. *Ecography* **2000**, *23*, 349–359. [CrossRef]

17. Browne, C.L.; Paszkowski, C.A.; Foote, A.L.; Moenting, A.; Boss, S.M. The relationship of amphibian abundance to habitat features across spatial scales in the Boreal Plains. *Écoscience* **2009**, *16*, 209–223. [CrossRef]

18. Tews, J.; Brose, U.; Grimm, V.; Tielbörger, K.; Wichmann, M.C.; Schwager, M.; Jeltsch, F. Animal species diversity driven by habitat heterogeneity/diversity: The importance of keystone structures. *J. Biogeogr.* **2004**, *31*, 79–92. [CrossRef]

19. Burne, M.R.; Griffin, C.R. Habitat associations of pool-breeding amphibians in eastern Massachusetts, USA. *Wetl. Ecol. Manag.* **2005**, *13*, 247–259. [CrossRef]

20. Afonso, L.G.; Eterovick, P.C. Microhabitat choice and differential use by anurans in forest streams in southeastern Brazil. *J. Nat. Hist.* **2007**, *41*, 937–948. [CrossRef]

21. Maseko, M.S.; Zungu, M.M.; Ehlers Smith, D.A.; Ehlers Smith, Y.C.; Downs, C.T. High microhabitat heterogeneity drives high functional traits in forest birds in five protected forest areas in the urban mosaic of Durban, South Africa. *Glob. Ecol. Conserv.* **2019**, *18*, e00645. [CrossRef]

22. Pawar, S.; Koo, M.S.; Kelley, C.; Ahmed, M.F.; Chaudhuri, S.; Sarkar, S. Conservation assessment and prioritization of areas in Northeast India: Priorities for amphibians and reptiles. *Biol. Conserv.* **2007**, *136*, 346–361. [CrossRef]

23. Glos, J.; Dausmann, K.H.; Linsenmair, E.K. Modeling the habitat use of *Aglyptodactylus laticeps*, an endangered dry-forest frog from Western Madagascar. *A Conserv. Strategy Amphib. Madag.* **2008**, *45*, 125–142.

24. Baldwin, R.F.; de Maynadier, P.G. Assessing threats to pool-breeding amphibian habitat in an urbanizing landscape. *Biol. Conserv.* **2009**, *142*, 1628–1638. [CrossRef]

25. Ochoa-Ochoa, L.; Urbina-Cardona, J.N.; Vázquez, L.-B.; Flores-Villela, O.; Bezaury-Creel, J. The effects of governmental protected areas and social initiatives for land protection on the conservation of Mexican amphibians. *PLoS ONE* **2009**, *4*, e6878. [CrossRef] [PubMed]

26. D'Amen, M.; Bombi, P.; Pearman, P.B.; Schmatz, D.R.; Zimmermann, N.E.; Bologna, M.A. Will climate change reduce the efficacy of protected areas for amphibian conservation in Italy? *Biol. Conserv.* **2011**, *144*, 989–997. [CrossRef]

27. de Pous, P.; Beukema, W.; Weterings, M.; Dümmer, I.; Geniez, P. Area prioritization and performance evaluation of the conservation area network for the Moroccan herpetofauna: A preliminary assessment. *Biodivers. Conserv.* **2011**, *20*, 89–118. [CrossRef]

28. Glos, J.; Erdmann, G.; Dausmann, K.H.; Linsenmair, K.E. A comparative study of predator-induced social aggregation of tadpoles in two anuran species from western Madagascar. *Herpetol. J.* **2007**, *17*, 261–268.

29. Ganzhorn, J.U.; Sorg, J.-P. Ecology and economy of a tropical dry forest in Madagascar. *Primate Rep.* **1996**, *46*.

30. Nelson, R.; Horning, N. AVHRR-LAC estimates of forest area in Madagascar, 1990. *Int. J. Remote Sens.* **1993**, *14*, 1463–1475. [CrossRef]

31. Glos, J. The amphibian fauna of the Kirindy dry forest in western Madagascar. *Salamandra Rheinb.* **2003**, *39*, 75–90.

32. Heyer, W.R.; Donnelly, M.A.; McDiarmid, R.W.; Hayek, L.-A.; Foster, M.S. *Measuring and Monitoring Biological diversity: Standard Methods for Amphibians*; Smithsonian Institution Press: Washington, DC, USA, 1994; pp. 57–60.

33. Glos, J.; Linsenmair, K.E. Descriptions of the tadpoles of *Aglyptodactylus laticeps* and *Aglyptodactylus securifer* from Western Madagascar, with notes on life history and ecology. *J. Herpetol.* **2004**, *38*, 131–136. [CrossRef] [PubMed]

34. Glaw, F.; Vences, M. *A Fieldguide to the Amphibians and Reptiles of Madagascar*; Vences, M., Glaw, F., Eds.; Verlags GbR.: München, Germany, 2007.

35. Oksanen, J.; Simpson, G.L.; Blanchet, F.G.; Kindt, R.; Legendre, P.; Minchin, P.R.; O'Hara, R.B.; Solymos, P.; Stevens, M.H.H.; Szoecs, E.; et al. Vegan: Community Ecology Package. 2022. Available online: https://CRAN.R-project.org/package=vegan (accessed on 3 March 2023).

36. Venables, W.N.; Ripley, B.D. *Modern Applied Statistics with S-PLUS*; Springer: Berlin/Heidelberg, Germany, 2002; Volume 4.

37. Simpson, G.L. Permute: Functions for Generating Restricted Permutations of Data. 2022. Available online: https://CRAN.R-project.org/package=permute (accessed on 26 February 2023).

38. Sarkar, D. *Lattice: Multivariate Data Visualization with R*; Springer: New York, NY, USA, 2008. Available online: http://lmdvr.r-forge.r-project.org (accessed on 8 March 2023).

39. Fielding, A.H.; Bell, J.F. A review of methods for the assessment of prediction errors in conservation presence/absence models. *Environ. Conversat.* **1997**, *24*, 38–49. [CrossRef]

40. Crawley, M.J. Generalized Linear Models. In *The Oxford Handbook of Quantitative Methods*; Oxford University Press: New York, NY, USA, 2013; pp. 557–578.

41. Quinn, G.P.; Keough, M.J. *Experimental Design and Data Analysis for Biologists*; Cambridge University Press: Cambridge, UK, 2002.

42. Clarke, K.R. Nonmetric multivariate analysis in community-level ecotoxicology. *Environ. Toxicol. Chem.* **1999**, *18*, 118.

43. Bray, J.R.; Curtis, J.T. An ordination of the upland forest communities of Southern Wisconsin. *Ecol. Monogr.* **1957**, *4*, 325–339. [CrossRef]

44. Legendre, P.; Legendre, L.F.J. Numerical Ecology: Developments in environmental modelling. *Elsevier Sci. Publ. Co.* **1998**, *24*, 201–205.

45. Mantel, N. The detection of disease clustering and a generalized regression approach. *Cancer Res.* **1967**, *27*, 209–220. [PubMed]

46. Vences, M.; Wollenberg, K.C.; Vieites, D.R.; Lees, D.C. Madagascar as a model region of species diversification. *Trends Ecol. Evol.* **2009**, *24*, 456–465. [CrossRef] [PubMed]
47. Simberloff, D. Flagships, umbrellas, and keystones—Is single-species management passe in the landscape era? *Cireat Br.* **1998**, *83*, 247–257. [CrossRef]
48. Vasconcelos, T.S.; Rossa-Feres, D.C. Diversidade, distribuição espacial e temporal de anfíbios anuros (Amphibia, Anura) na região noroeste do estado de São Paulo, Brasil. *Biota Neotrop.* **2005**, *5*, 137–150. [CrossRef]
49. Bastazini, C.V.; Munduruca, J.F.V.; Rocha Pedro Luís, B.; Napoli Marcelo, F. Which environmental variables better explain changes in anuran community composition? A case study in the Restinga of Mata de São João, Bahia, Brazil. *Herpetologica* **2007**, *63*, 459–471. [CrossRef]
50. Piatti, L.; Souza, F.L.; Filho, P.L. Anuran assemblage in a rice field agroecosystem in the Pantanal of central Brazil. *J. Nat. Hist.* **2010**, *44*, 1215–1224. [CrossRef]
51. Parris, K.M.; McCarthy, M.A. What influences the structure of frog assemblages at forest streams? *Austral Ecol.* **1999**, *24*, 495–502. [CrossRef]
52. Eterovick, P.C. Distribution of anuran species among montane streams in south-eastern Brazil. *J. Trop. Ecol.* **2003**, *19*, 219–228. [CrossRef]
53. Vasconcelos, T.S.; Santos, T.G.; Rossa-Feres, D.C.; Haddad, C. Influence of the environmental heterogeneity of breeding ponds on anuran assemblages from southeastern Brazil. *Can. J. Zool.* **2009**, *87*, 699–707. [CrossRef]
54. Parris, K.M. Environmental and spatial variables influence the composition of frog assemblages in sub-tropical eastern Australia. *Ecography* **2004**, *27*, 392–400. [CrossRef]
55. Julian Glos, K. Eduard Linsenmair. Description of the tadpoles of *Boophis doulioti* and *B. xerophilus* from Western Madagascar with notes on larval life history and breeding ecology. *Amphib. Reptil.* **2005**, *26*, 459–466. [CrossRef]
56. Glos, J.; Wegner, F.; Dausmann, K.H.; Linsenmair, K.E. Oviposition-site selection in an endangered Madagascan frog: Experimental evaluation of a habitat model and its implications for conservation. *Biotropica* **2008**, *40*, 646–652. [CrossRef]
57. Hartel, T.; Bancila, R.; Cogalniceanu, D.A. Spatial and temporal variability of aquatic habitat use by amphibians in a hydrologically modified landscape. *Freshw. Biol.* **2011**, *56*, 2288–2298. [CrossRef]
58. Glos, J.; Metzner, A.; Schmidtke, C.; Rödel, M.-O. The mating system of the reed frog *Heterixalus tricolor* (Anura: Hyperoliidae) from western Madagascar's dry forests. *Salamandra* **2019**, *55*, 151–159.
59. Snodgrass, J.W.; Komoroski, M.J.; Bryan, A.L.; Burger, J. Relationships among isolated wetland size, hydroperiod, and amphibian species richness: Implications for wetland regulations. *Conserv. Biol.* **2000**, *14*, 414–419. [CrossRef]
60. Paton, P.W.C.; Crouch, W.B., III. Using the phenology of pond-breeding amphibians to develop conservation strategies. *Conserv. Biol.* **2002**, *16*, 194–204. [CrossRef] [PubMed]
61. Babbitt, K.J.; Baber, M.J.; Tarr, T.L. Patterns of larval amphibian distribution along a wetland hydroperiod gradient. *Can. J. Zool.* **2003**, *81*, 1539–1552. [CrossRef]
62. Egan, R.S.; Paton, P.W.C. Within-pond parameters affecting oviposition by wood frogs and spotted salamanders. *Wetlands* **2004**, *24*, 1–13. [CrossRef]
63. Wellborn, G.A.; Skelly, D.K.; Werner, E.E. Mechanisms creating community structure across a freshwater habitat gradient. *Annu. Rev. Ecol. Syst.* **1996**, *27*, 337–363. [CrossRef]
64. Weyrauch, S.L.; Grubb, T.C., Jr. Patch and landscape characteristics associated with the distribution of woodland amphibians in an agricultural fragmented landscape: An information-theorie approach. *Biol. Conserv.* **2004**, *115*, 443–450. [CrossRef]
65. Kats, L.B.; Petranka, J.W.; Sih, A. Antipredator defenses and the persistence of amphibian larvae with fishes. *Ecology* **1988**, *69*, 1865–1870. [CrossRef]
66. Hero, J.-M.; Magnusson, W.E.; Rocha, C.F.D.; Catterall, C.P. Antipredator defenses influence the distribution of amphibian prey species in the Central Amazon Rain Forest. *Biotropica* **2001**, *33*, 131. [CrossRef]
67. Morin, P.J. Predation, competition, and the composition of larval anuran guilds. *Ecol. Monogr.* **1983**, *53*, 119–138. [CrossRef]
68. Wilbur, H.M. Experimental Ecology of food webs: Complex systems in temporary ponds. *Ecology* **1997**, *78*, 2279–2302. [CrossRef]
69. Lawton, J.H. Plant Architecture and the Diversity of Phytophagous Insects. *Annu. Rev. Entomol.* **1983**, *28*, 23–39. [CrossRef]
70. Haddad, C.F.B.; Prado, C.P.A. Reproductive modes in frogs and their unexpected diversity in the Atlantic Forest of Brazil. *BioScience* **2005**, *55*, 207. [CrossRef]
71. Werner, E.E.; Glennemeier, K.S. Influence of forest canopy cover on the breeding pond distributions of several amphibian species. *Copeia* **1999**, 1–12. [CrossRef]
72. Afonso, L.G.; Eterovick, P.C. Spatial and temporal distribution of breeding anurans in streams in southeastern Brazil. *J. Nat. Hist.* **2007**, *41*, 949–963. [CrossRef]
73. Magurran, A.E.; Higham, A. Information transfer across fish shoals under predator threat. *Ethology* **1988**, *78*, 153–158. [CrossRef]
74. Babbitt, K.J.; Tanner, G.W. Effects of cover and predator identity on predation of *Hyla squirella* tadpoles. *J. Herpetol.* **1997**, *31*, 128–130. [CrossRef]
75. Fouilloux, C.A.; Yovanovich, C.A.M.; Rojas, B. Tadpole responses to environments with limited visibility: What we (don't) know and perspectives for a sharper future. *Front. Ecol. Evol.* **2022**, *9*, 1–8. [CrossRef]

76. Gregory, R.S. Effect of Turbidity on the Predator Avoidance Behaviour of Juvenile Chinook Salmon (*Oncorhynchus tshawytscha*). *Can. J. Fish. Aquat. Sci.* **1993**, *50*, 241–246. [CrossRef]
77. Wood, S.L.R.; Richardson, J.S. Impact of sediment and nutrient inputs on growth and survival of tadpoles of the Western Toad. *Freshw. Biol.* **2009**, *54*, 1120–1134. [CrossRef]

Article

Revision of the Subgenus *Ochthomantis* Frogs from Madagascar (Amphibia: Mantellidae) with the Description of Four Species and Resurrection of *Mantidactylus catalai* and *M. poissoni* [†]

Nirhy H. C. Rabibisoa [1,*], Rachel S. Welt [2] and Christopher J. Raxworthy [2,*]

[1] Sciences de la Vie et de l'Environnement, Faculté des Sciences, de Technologies et de l'Environnement, Université de Mahajanga, Campus Universitaire d'Ambondrona, BP 652, Mahajanga 401, Madagascar

[2] Department of Herpetology, American Museum of Natural History, Central Park West at 79th Street, New York, NY 10024-5192, USA; rwelt@amnh.com

[*] Correspondence: nhcrabibisoa@gmail.com (N.H.C.R.); rax@amnh.org (C.J.R.); Tel.: +261-348-942-204 (N.H.C.R.)

[†] Zoobank link: urn:lsid:zoobank.org:pub:A2D6E0A9-F0D9-444C-9C90-82FF5294D0FB.

Simple Summary: The genus *Mantidactylus* spp. is one of the exceptionally diverse amphibian clades from Madagascar. Currently, 57 species in 6 subgenera are recognized. One subgenus, *Ochthomantis*, is the focus of the present study. Here, we revise this taxonomic group to recognize the presence of cryptic species through an assessment of morphological and, where available, molecular (16S mitochondrial gene) variation of 637 sexed adult specimens. Our results show that *Ochthomantis* contains eleven species, including the resurrected *Mantidactylus catalai* and *M. poissoni*, and four newly described species. We cannot confirm that *M. majori* should be considered as part of the subgenus *Ochthomantis*, so we do not include it in our descriptions. Following our taxonomic revision, we also present a practical simple key to identify all of the species belonging to this subgenus.

Citation: Rabibisoa, N.H.C.; Welt, R.S.; Raxworthy, C.J. Revision of the Subgenus *Ochthomantis* Frogs from Madagascar (Amphibia: Mantellidae) with the Description of Four Species and Resurrection of *Mantidactylus catalai* and *M. poissoni*. *Animals* **2023**, *13*, 2800. https://doi.org/10.3390/ani13172800

Academic Editors: Franco Andreone, Angelica Crottini, Andolalao Rakotoarison and Fandresena Rakotoarimalala

Received: 30 June 2023
Revised: 23 August 2023
Accepted: 31 August 2023
Published: 3 September 2023

Abstract: The subgenus *Ochthomantis* is an obligate forest and stream-dwelling group of mantellid frogs, endemic to Madagascar, with six species currently recognized. However, this group suffers from ongoing taxonomic confusion due to low numbers of examined specimens, and failure to consider morphological variation from development and sexual dimorphism. Here, we examined the morphology of 637 sexed adult specimens collected by us in the field and from other museum collections. We also sequenced a DNA fragment of the 16S mtDNA gene for each lineage to determine congruence between morphological and molecular data sets and to help delimit species. Our results demonstrate that the subgenus *Ochthomantis* includes eleven valid species: five already recognized, *M. catalai* and *M. poissoni* that we resurrect from synonymy, and four new species which we describe for the first time here. In some analyses, *Mantidactylus majori* groups with other *Mantidactylus* subgenera, so we do not consider it a member of the subgenus *Ochthomantis* in this study. All species have restricted distributions and elevational ranges in the humid forests of Madagascar. This study demonstrates the utility of assessing cryptic species using both diagnostic morphological characters and molecular data. The discovery of this new cryptic biodiversity, and the taxonomic revision herein, will likely require conservation activities for those species with the most restricted distributions.

Keywords: *Ochthomantis*; *Mantidactylus*; cryptic species; morphology; 16S; Madagascar

1. Introduction

Within the diverse radiation of *Mantidactylus* frogs (Anura: Mantellidae) in Madagascar, 57 species have been described [1] across 6 subgenera [2–4]. One of these is the subgenus *Ochthomantis* which currently contains six valid species: *Mantidactylus femoralis*, *M. mocquardi*, *M. ambreensis*, *M. ambony*, *M. majori*, and *M. zolitschka* [5,6]. *Ochthomantis* is characterized by the other *Mantidactylus* subgenera by the combination of the following unique set of characteristics: no webbing between fingers, webbing between toes, large tympanum (more than half the eye diameter), sexual dimorphism in size of tympanum (smaller in females) and body size (larger in females), presence of an inguinal pale streak marking, and living close to rivers [2,6,7]. The *Ochthomantis* group has been recorded from a broad range of elevations from sea level to 2600 m and occurs in most regions of Madagascar except the dry south and west.

The first species description for this group was made by Boulenger [8] for *Mantidactylus femoralis* and this author later also described the genus *Mantidactylus* [9]. Thereafter, additional species were described and some of them later synonymized [4,6]. The subgenus *Ochthomantis* was described by [7,10], with the most recently described species being *M. zolitschka* [10] and *M. ambony* [5], and a review was conducted by [2,3] who proposed a new taxonomy of Malagasy mantellines based on molecular results. The most recent molecular study tentatively suggests that *Mantidactylus argenteus* might also belong to the *Ochthomantis* clade [11].

Many specimens of the *M. femoralis* complex have been collected since 1989 by different researchers such as Ronald Nussbaum's team (University of Michigan, Christopher Raxworthy's team (American Museum of Natural History) and Nirhy Rabibisoa and others at the University of Antananarivo. In these collections, it has become clear that *M. femoralis* and other *Ochthomantis* species include distinct morphospecies that are strong candidates for unrecognized additional species. Similarly, other authors have indicated that the group likely includes unrecognized species based on molecular divergence for the 16S rRNA locus and morphological differences [10,11]. A past review of mantellid species using morphological characters has demonstrated that undescribed species likely exist [12]. Therefore, a study utilizing both morphological characters and molecular data is important to resolve the classification of the subgenus *Ochthomantis*. Thus, this study aims to address this issue of cryptic species and to validate candidate species by undertaking a morphological analysis of 637 adult *Ochthomantis* specimens combined with an analysis of genetic data and a detailed historical taxonomic review of all species in the group.

Based on the findings from our study, we here resurrect two species that were previously considered junior synonyms: *M. catalai* and *M. poissoni*; and describe four new species. According to these results, we conclude that the subgenus *Ochthomantis* contains at least 11 species.

2. Materials and Methods

Field surveys were carried out during the rainy season (January–May) at numerous sites in Madagascar, except in the Moramanga Region where fieldwork took place during the cooler dry season (July–September) and continued during the rainy season (November–December) to provide additional data on ecology, behavior, and reproduction. All specimens of *Ochthomantis* were collected between 1989 and 2010 within rainforest and deciduous forests in Northwestern and Northern Madagascar. Frogs and tadpoles were collected during both the day and the night using headlamps by searching on the ground and vegetation up to 2 m in height, along riverbanks, in rivers streams, ponds, and lakes. The following information was recorded at the time of capture for each individual: date, time, temperature, elevation (using GPS and altimeter), and microhabitat characteristics. Photographs of representative specimens were taken to record coloration in life. Vouchers specimens collected by the authors were euthanized using chlorotone, fixed using 10% formalin, and later stored in 70% ethanol. Liver and/or thigh muscle was removed from representative specimens before fixation and frozen into liquid nitrogen or preserved in alcohol for molecular analysis. The field tag series abbreviations used in this study are: APR, Achille P. Raselimanana; FAZC and FN, Franco Andreone; J.B. and MRJ, J. B. Ramanamanjato; JR, Jeannot Razafimanantsoa; LLS and LV, Olivier S. Ramilison; MA, Mirana Anjeriniaina; NR, Nirhy H. Rabibisoa; RD and SMG, Domoina Rakotomalala; RJS, Jasmin Randianirina; RAN, Ronald A. Nussbaum; RAX, Christopher J. Raxworthy. A total of 637 specimens were examined in detail, collected by us, and housed at the following collections: AMNH (American Museum of Natural History, Department of Herpetology), BM (The British Museum, London), MNHN (Muséum National d'Histoire Naturelle de Paris); UADBA (University of Antananarivo Department of Animal Biology), PBZT (Parc Botanique et Zoologique de Tsimbazaza Antananarivo), and UMMZ (University of Michigan, Museum of Zoology). Museum collections are abbreviated using standard institutional abbreviations as listed in [13], with the addition of UADBA and PBZT.

Sex and maturity were determined based on examination of the gonads and development of the femoral glands. Juveniles (J) were not included in the morphological studies reported here and males (M) and females (F) were analyzed separately due to the sexual dimorphism in this group. Morphological measurements were made by N.R. using calipers to 0.05 mm precision and a binocular microscope. Each measurement was made as described by [14]. The definition of each character measured followed [15]. Abbreviations used for morphological measurements are given in Appendix A, Table A1, and Figure A1. Webbing character has been diagnosed in counting of free phalanges (without webbing), then we have scored each free phalanges following the method described by [15] using the formula from [16] to facilitate comparisons with other species of *Mantidactylus* and most subsequent authors who published accounts on Madagascan anurans (0 = webbing reaching end of terminal phalange, 0.25 = 0.25 of terminal phalange free of webbing, 1 = terminal phalange free of webbing etc.). Extending the total sum of free phalanges for each limb is called webbing score (WS).

All genetic samples are listed in Supplemental Data with their Genbank accession numbers (see Appendix C; Table S1), including our newly sequenced specimens and additional *Ochthomantis* samples from Genbank (see [11]). Trees were rooted using *Mantidactylus peraccae*, *M. cowanii*, *M. lugubris*, and *M. guttulatus*; species related to *Ochthomantis* [2,11]. DNA from either frozen or ethanol-preserved (70%) tissue samples was isolated using the QIAGEN DNeasy spin columns. DNA from formalin-fixed museum specimens was extracted using a modified method from [17] and precautionary steps were taken to prevent contamination [18]. To allow for the inclusion of other *Ochthomantis* sequence data from previous studies (see [2,11]), our sequencing efforts focused on the mitochondrial 16S rRNA gene. PCR amplification was performed under locus-specific parameters. All sequences were aligned using MUSCLE. BLAST searches (NCBI) were performed for each contiguous sequence to identify any potential contamination. The data set was partitioned by stems and loops (for the 16S rRNA locus) for Bayesian analysis conducted in BEAST v.2.7.5. Markov chain was run for 3×10^8 generations and trees were sampled every 3000 generations. To determine that stationarity had been reached, we ensured that effective sample size of all parameters was greater than 200, using TRACER v1.6 [19]. The first 20% of trees were discarded as burn-in. The trees retained were combined to produce a maximum clade credibility tree.

The 16S rRNA clades identified from the molecular analyses were used as guide to establish congruent morphological groups that could be diagnosed using fixed morphological character states. The resulting recognized groups were thus supported by both molecular and morphological diagnostic criteria and we consider them here as species, based on these congruent dual species recognition criteria (see [20]). All recognized species also were more than 3% divergent to all other species based on the 16S rRNA gene (uncorrected distance), thus also meeting another species recognition criterion that has been widely applied in Malagasy amphibian taxonomy, e.g., [21]. All currently recognized *Ochthomantis* species were assigned to their appropriate congruent molecular and morphological group based on the character states given in their species descriptions, and directly observed by us through examining type specimens. For the unnamed congruent molecular and morphological groups, we compared their morphological diagnoses with character states in all other currently considered junior synonym *Ochthomantis* species using the same methods as described above for the senior synonyms. In cases of morphological correspondence, this was used as justification for resurrecting junior synonyms. For the remaining unnamed groups, we considered these as undescribed species. These species are formally described here. We also provide re-descriptions of species that we resurrect from junior synonymy. At the end, we also provide a key to identify the 11 species of the *Ochthomantis* subgenus.

3. Results

3.1. Molecular Analysis, and Diversity

Figure 1 shows the recovered 16S rRNA tree topology. Based on these clades and their corresponding morphological congruence, we consider the subgenus *Ochthomantis* to include 11 species. While *Mantidactylus majori* has been considered to be within *Ochthomantis* and is recovered as such in this tree, we posit that this species belongs to another subgenus according to morphological results from this study and [2]. We discuss this in more detail within the taxonomy chapter (Discussion). All eleven species of the subgenus *Ochthomantis* have >3% uncorrected *p*-distance for 16S, i.e., more than 3% divergent to each other based on the 16S rRNA gene (uncorrected distance), and are readily diagnosable based on their morphology (see Figure 1). Five species are currently recognized: *M. ambreensis*, *M. ambony*, *M. femoralis*, *M. mocquardi*, and *M. zolitschkia*; two are currently junior synonyms that require recognition as good species: *M. catalai* and *M. poissoni*; and four species are undescribed. The specimens labeled with asterisks (see Specimens Examined and Table S1) have been sequenced for 16S and were used to estimate the gene tree through their DNA sequences and alignment (see Appendix C).

3.2. Redescription of Currently Recognized Species

The following five species are redescribed here, based on morphological data for newly available specimens and previously reported voucher specimens, which considerably increase the samples sizes: *M. femoralis* (124 specimens), *M. ambreensis* (46 specimens), *M. ambony* (14 specimens), *M. mocquardi* (148 specimens), and *M. zolitschka* (10 specimens) (Figure 2). We provide these redescriptions to facilitate comparison with the descriptions provided for the new and resurrected species.

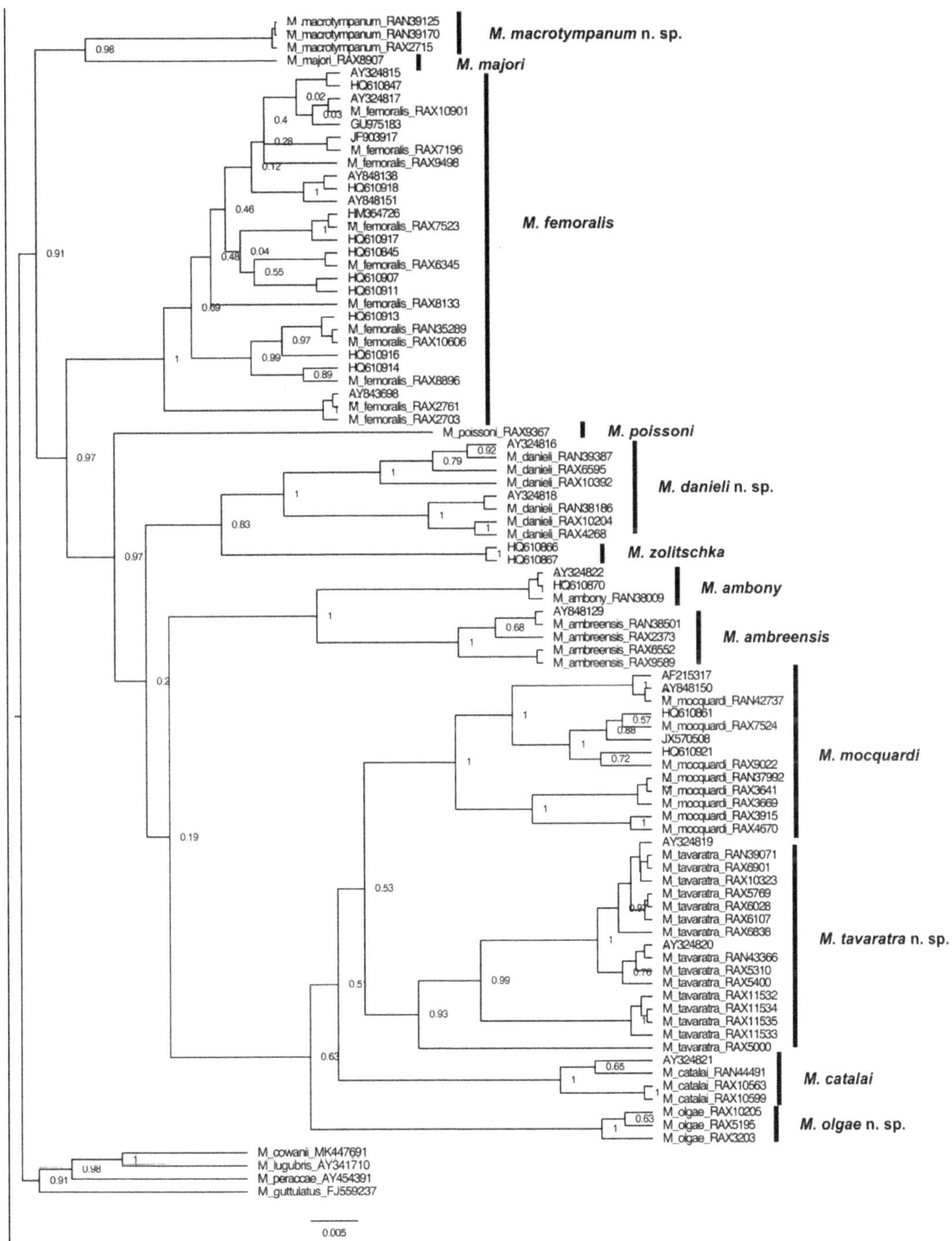

Figure 1. 16S rRNA gene tree of samples of the subgenus *Ochthomantis* from a BEAST analysis. Posterior probabilities support molecular species groupings. *Mantidactylus cowani*, *M. lugubris*, *M. peraccae*, and *M. guttulatus* were used as outgroup.

Figure 2. Photos of valid nominal species of the subgenus *Ochthomantis*: (**a**). *Mantidactylus femoralis*, Zahamena (CJR); (**b**). *Mantidactylus ambreensis*, Montagne d'Ambre (CJR), (**c**): *Mantidactylus ambony*, Montagne d'Ambre (CJR); (**d**). *Mantidactylus mocquardi*, Betampona (NR), (**e**). *Mantidactylus zolitschka*, An'ala (NR).

3.2.1. *Mantidactylus femoralis* [8]

Rana femoralis [8] (p. 462) (Syntypes: BMNH 1947.2.22.65–68, according to [22] (p. 26); BMNH 1947.2.22.65 designated lectotype by [10] (p. 85); *Rana flavicrus* [23] (p. 245); *Mantidactylus flavicrus* [9] (p. 450); *Mantidactylus femoralis*: [24] (p. 393); *Mantidactylus (Mantidactylus) flavicrus* [25] (p. 25); *Mantidactylus femoralis*: [26] (p. 235); *Mantidactylus (Hylobatrachus) femoralis*: [27] (p. 312); *Mantidactylus (Ochthomantis) femoralis*: [7] (p. 400); [3] (p. 3).

Lectotype: BMNH 1947.22.65, type locality: "East Betsileo", Madagascar;

Paratype: BMNH 1947.22.66–68. same location as the lectotype;

Paralectotypes: BMNH 1947.22.66–68. same location as the lectotype.

Specimens Examined: BMNH 1947.22.65–68: East Betsileo. AMNH A23781 and A50366: Moramanga, District Moramanga, Alaotra Mangoro Region, Madagascar. AMNH A157112 adult male: Ampanasana Ankolony (14°26,2′ S 49°46,5′ E), Marojejy National Park, Andapa District, Sava Region, Madagascar, October 1998, A. Raselimanana. AMNH A167521* (RAX 6345) subadult: Analapakila Trois Lacs, District Bealanana (14°26.233′ S 48°36.696′ E, 1400 m), 12 March 2003, N. Rabibisoa and S. Mahaviasy. AMNH A167580* (RAX 2703) juvenile: Antsahatelo (13°51.588′ S 48°51.979′ E, 800 m), 6 April 2001, S. Mahaviasy, N. Rabibisoa, C. J. Raxworthy, A Razafimanantsoa, and A. Razafimanantsoa. AMNH A167581* (RAX 2761) adult male: Ramena River, Tsaratanana Reserve, Ambanja District (13°55.071′ S 48°53.179′ E; 750 m), 8 April 2001, same collectors as previous. AMNH A174623* (RAX 7196) female adult: Ankafina Tsarafidy, Ambohimasoa District, Haute Matsiatra Region, Madagascar (21°12.598′ S 47°12.874′ E, 1420 m), 16 February 2004, N. Rabibisoa, M. Randriambahiniarime, F. Ranjanaharisoa, and C. J. Raxworthy. AMNH A174627* (RAX 7523) adult male: Betampona Reserve, Toamasina District, Atsinanana Region, Madagascar (17°55.866′ S 49°12.190′ E, 350 m); 2 February 2004, N. Rabibisoa, M. Randriambahiniarime, F. Ranjanaharisoa, and C. J. Raxworthy. AMNH A174646* (RAX

8133) adult female: Manasamena, Lakato, Moramanga District, Alaotra–Mangoro Region, Madagascar (19°02.637′ S 48°20.910′, 950 m), 27 March 2004, N. Rabibisoa and N. Rakotondrazafy. AMNH A174651* (RAX 8896) juvenile: Kianjavato–Vatovavy, Ranomafana District, Vatovavy–Fitovinany Region, Madagascar (21°22.791′ S 47°52.052′ E, 150 m), 18 February 2006, N. Rabibisoa and C.J. Raxworthy. AMNH A174654* (RAX 9498) juvenile: Ambohibehivavy–Vasiana, Betafo District, Vakinankaratra Region, Madagascar (19°41.387′ S 46°06.953′ E, 850 m), 28 March 2006, N. Rabibisoa, N. Rakotondrazafy, and J. Rafanomezantsoa. AMNH A181735* (RAX 10606): Beampingaratsy Pass Anosy Mts, District Tolagnaro, Anosy Region, Madagascar (24°28.244′ S 46°53.521′ E, 520 m), Feb. 13, 2009, C.J. Raxworthy. AMNH A187128* (RAX 10901): Ambatomenaloha/Itremo, Ambatofinandrahana District, Amorin'I Mania Region, Madagascar (20°37.130′ S 46°33.347′ E, 1650 m), 19 December 2009, C.J. Raxworthy. UADBA 4517–18, 4520 (RAN 52471, 52470, 52109): Emiminiy, Andohahela National Park, Tolagnaro District, Anosy Region, Madagascar (24°37.55′ S 46°45.92′ E, 500 m), 21 October 1995, A. Raselimanana and J. B. Ramanamanjato. UADBA 20478–20479: Andriankely, Anjozorobe–Angavo National Parc, Anjozorobe District, Analamanga Region, Madagascar (18°25.225′ S 47°56,245′ E, 1250 m), 2 February 2003, M. Anjeriniana. UADBA 26118, 26268–26270 (NR 1866, 1819, 1858, 1859): Ampanatovana Lakato, Moramanga District, Alaotra–Mangoro Region, Madagascar (19°02.637′ S 48°20.912′ E, 1025 m), 29 November 2003 and 6 December 2003, N. Rabibisoa, N.A. Rakotondrazafy. UADBA 26249, 26262, 26389 (RAX7198, 7197, 7212): Ankafina Tsarafidy, Ambohimasoa District, Amoron'I Mania Region, Madagascar (21°12.598′ S 47°12.874′ E, 1150 m), 16 February 2004, N. Rabibisoa, M. Randriambahiniarime, F. Ranjanaharisoa, and C.J. Raxworthy. UADBA 26250–26251, 26263–26265, 26402 (RAX 7961, 7553, 7945, 7555, 8002, 7618): Betampona Strict Natural Reserve, Toamasina II District, Atsinanana Region, Madagascar (17°55.866′ S 49°12.190′ E, 250–450 m), 28 February 2004–17 March 2004, N. Rabibisoa, M. Randriambahiniarime, F. Ranjanaharisoa, and C. J. Raxworthy. UADBA 26252 (RAN 44674) adult male: Sahavatoy River, Andringitra National Park, Ambalavao District, Ihorombe Region, Madagascar (22°13.667′ S 47°0.217′ E, 810 m), 24 November 1993, N. Rabibisoa, A. Razafimanantsoa, and C. J. Raxworthy. UADBA 26253–26254, 26374, 26378 (RAN 45537, 45684, 45538, 44833): Rangovalo Ridge, Zahamena National Park, Ambatondrazaka District, Alaotra Mangoro Region, Madagascar (17°40,5′ S 48°45.5′ E, 1150 m), 4–8 March 1994, J.B. Ramanamanjato, A. Raselimanana, C.J. Raxworthy, A. Razafimanantsoa, and A. Razafimanantsoa. UADBA 26255, 26380 (RAN47059, RAN47047): Andranomangoboka Ambohijanahary, Morafenobe District, Melaky Region, Madagascar (18°14.787′ S 45°21.419′ E, 730–950 m), 16 January 1995, J. B. Ramanamanjato, A. Raselimanana, C. J. Raxworthy, A. Razafimanantsoa, and A. Razafimanantsoa. UADBA 26257 (RAN 47663): Ambohimanana Tolongoina, Ikongo District, Vatovavy–Fitovavy Region, Madagascar (21°28.557′ S 47°33.759′ E, 600 m), 9 February 1995, J. B. Ramanamanjato, A. Raselimanana, C. J. Raxworthy, A. Razafimanantsoa, and A Razafimanantsoa. UADBA 26258 (RAN 47953): Sahamalio, Isalo National Park, Ranohira District, Ihorombe Region, Madagascar (22°26.315′ S 45°15.648′ E, 700 m), 18 February 1995, C. J. Raxworthy, A. Raselimanana, J. B. Ramanamanjato, A Razafimanantsoa, and A Razafimanantsoa. UADBA 26259, 26284, 26260–26261 (RAN 47995, 47996, 48219, 48221): Canyon Singe, Isalo National Park, Ranohira District, Ihorombe Region, Madagascar (22°29.138′ S 45°23.086′ E, 600 m), 20 February 1995, same collectors as previous. UADBA 26266, 26385 (RAX 8157, 8158): Manasamena Lakato, Moramanga District, Alaotra Mangoro Region, Madagascar (19°02.637′ S 48°20.910′ E, 950 m), 27 March 2004, N. Rabibisoa and N. Rakotondrazafy. UADBA 26267 (RAN 45891): Namarafana, Zahamena National Park, Ambatondrazaka District, Alaotra Mangoro Region, Madagascar (17°44′ S 48°58.5′ E, 420 m), 16 March 1994, J. B. Ramanamanjato, A. Raselimanana, C. J. Raxworthy, A. Razafimanantsoa, and A. Razafimanantsoa. UADBA 26281, 26376 (RAN 47280, 47248): Doany Ambohijanahary, Morafenobe District, Melaky Region, Madagascar (18°17.297′ S 45°33.289′ E, 1220 m), 24 January 1995, same collectors as previous. UADBA 26282 (RAN 47580) adult female: Ambohitantely Special Reserve, Ankazobe District, Analamanga Region, Madagascar

(18°11.158′ S 47°16.757′ E, 1580 m), 2 February 1995, same collectors as previous. UADBA 26363 (JB 127) adult female: Andranomay, Anjozorobe District, Analamanga Region, Madagascar (18°28.8′ S 47°57′ E, 1300 m), 12 December 1996, A. Raselimanana. UADBA 26377 (RAN 47631) same condition as UADBA 26257 except date of collection Feb. 8, 1995. UADBA 26375, 26379 (RAN 45395, 45364): Volontsagana River, Zahamena National Park, Ambatondrazaka District, Alaotra–Mangoro Ragion (17°42′ S 48°46′ E, 850 m), 28 February 1994, J. B. Ramanamanjato, A. Raselimanana, C. J. Raxworthy, A. Razafimanantsoa, and A. Razafimanantsoa. UADBA RAX10955*: Mandraka, Manjakandriana District, Analamanga Region, Madagascar (18°57.733′ S 47°55.050′ E, 1140 m), 18 March 2006, N. Rabibisoa, and J. Rafanomezantsoa. UMMZ 212835 (RAN 39921) adult male: Manantenina River, Marojezy National Park, Andapa District, Sava Region, Madagascar (14°26′ S 49°46′ E, 700 m), 25 November 1992, R. A. Nussbaum, A. Razafimanantsoa, A Razafimanantsoa, and C J Raxworthy. UMMZ 197651* (RAN 35289): Manantantely Forest, District Tolagnaro, Anosy Region (24°39′ S 46°55.083′ E, 125 m), 6 November 1990, J.B. Ramananjato, A Raselimanana, RA Naussbaum, and C J Raxworthy.

Additional specimens examined: see Appendix B.

Diagnosis: A medium- to large-sized *Ochthomantis* (adult male SVL 31–43.4 mm; adult female SVL 43.0–62.4 mm); tibiotarsal articulation beyond nostrils (rarely between eye and nostril); 1.5–2 free phalanges on the internal edge of toe 4; width of digit terminal disc $\geq$ 1.70 disc base; white stripe along the superior lip and prominent yellow patch in the inguinal region. Distinguished from other subgenus species by the following combination of characters: from *M. ambreensis* and *M. ambony* by yellow line or patch in the inguinal region; *M. poissoni* by absence under the eye of large white spot or partly fused white spots on upper lip; *M. mocquardi, M. catalai, M. olgae* n. sp., and *M. tavaratra* n. sp. by the presence of 1.5–2 free phalanges on the internal edge of toe 4, yellow patch in the inguinal region, and tibiotarsal articulation beyond eye; *M. zolitschka* by its large size (SVL $\geq$ 33 mm) and yellow patch in inguinal region; *M. danieli* n. sp. by the yellow patch in inguinal region and width of the digit terminal disc $\geq$ 1.70 base of disc; *M. macrotympanum* n. sp. by smaller adult male SVL (<60 mm). Character diagnostics are summarized in Tables 1 and 2.

Table 1. Comparison of morphological measurements of the currently valid nominal species of the subgenus *Ochthomantis*. All measurements in mm, F: adult female, M: adult male. For morphological abbreviations see Appendix A.

Size Range	*M. femoralis*		*M. ambreensis*		*M. ambony*		*M. mocquardi*		*M. zolitschka*	
	M	F	M	F	M	F	M	F	M	F
SVL	39.1–62.4	33.0–43.4	38.2–42.2	33.3–39	34.0–37.9	30.0–31.8	46.0–64.3	51.9–63.3	51.9–63.3	26.5–30.6
TD	2.8–6.2	3.0–5.1	2.7–3.7	3.3–5.9	2.2–3.3	4.4–5.0	2.6–5.2	3.5–4.7	3.5–4.7	2.8–3.2
ED	4.4–9.2	3.8–7.5	4.5–5.7	4.2–5.3	4.0–4.9	3.5–4.3	5.7–8.8	6.0–8.2	6.0–8.2	3.3–3.7
EN	2.9–7.0	2.5–4.1	2.8–3.8	2.5–4.0	2.1–3.6	2.3–2.8	3.1–5.6	3.6–5.4	3.6–5.4	1.9–2.7
EST	0.9–3.1	0.4–2.5	1.4–1.9	1.0–2.1	1.3–1.9	1.2–1.7	1.3–3.3	1.6–3.5	1.6–3.5	0.6–1.2
NS	1.0–3.3	1.0–2.8	1.5–2.7	1.1–2.6	1.1–2.0	1.4–2.0	1.2–4.3	1.3–3.8	1.3–3.8	1.3–2.0
NN	3.0–5.5	1.9–4.4	3.2–4.0	2.5–4.5	2.9–3.6	2.6–3.3	3.0–6.6	4.7–5.8	4.7–5.8	1.6–3.3
HW	10.9–20.3	10.7–15.2	11.6–13.8	9.9–12.4	10.6–11.8	9.0–11.0	13.8–23.4	17.1–21.3	17.1–21.3	8.0–10.3
HL	16.2–25.7	14.0–19.3	14.0–16.7	13.8–17.4	13.0–15.5	12.8–14.8	19.7–28.4	22.2–27.5	22.2–27.5	11.2–13.1
HDL	11.5–17.2	10.0–13.4	10.4–13.0	9.0–12.0	9.7–11.7	8.6–9.9	11.4–19.6	13.0–18.1	13.0–18.1	8.0–10.2
RC	8.4–10	5.6–8.7	7.0–8.6	5.5–7.3	6.3–7.8	5.6–6.2	7.7–12.0	9.2–12.1	9.2–12.1	5.2–5.3
FE	20.6–34	15.8–21.6	18.6–20.6	15.4–19.8	18.0–20.2	14.1–16.0	24.2–33.1	27.0–31.9	27.0–31.9	14.3–14.8
TI	21.6–32.7	17.0–23.0	19.8–23.5	17.1–21.3	19.3–20.5	16.0–17.2	24.0–35.4	28.1–32.1	28.1–32.1	14.5–14.7
FT	22–32.1	17.6–23.8	18.4–21.6	16.1–20.6	18.5–20.3	15.3–16.8	23.1–33.7	26.8–32.0	26.8–32.0	19.5–20.9
TA	9.8–14	8.1–11.5	9.7–11.5	8.0–10.6	8.7–10.1	7.6–8.0	9.5–14.6	11.3–14.2	11.3–14.2	6.2–6.8
EO	6.1–10.3	4.6–7.5	6.1–7.5	5.1–7.0	5.6–6.4	4.6–5.7	7.5–11.9	8.8–11.3	8.8–11.3	4.4–4.6
EM	2.2–3.6	1.6–3.2	1.7–2.6	1.7–2.4	1.9–2.4	1.6–1.9	2.2–4.3	2.8–4.1	2.8–4.1	1.9–2.0
EHEAD	5.3–9.5	4.3–8.4	5.4–7.2	4.5–6.4	5.0–5.6	4.4–5.0	6.8–10.8	6.7–9.3	6.7–9.3	4.9–5.4
T1	1.1–2.6	1.0–2.2	1.2–1.8	1.0–1.8	1.1–1.3	0.8–1.4	1.6–2.7	1.8–3.0	1.8–3.0	0.8–0.9
TO3	5.4–9.2	4.5–6.6	4.7–6.0	4.0–5.7	4.8–5.5	3.7–4.6	5.5–8.8	7.1–8.8	7.1–8.8	4.8
TO5	6.0–10	5.3–7.0	5.4–6.6	4.8–6.3	5.2–6.4	4.3–5.4	6.5–10.4	8.0–9.7	8.0–9.7	5.3

Table 2. Comparison of qualitative morphological characters and biogeography of the currently valid nominal species of the subgenus *Ochthomantis*.

Morphological Characters	*M. femoralis*	*M. ambreensis*	*M. ambony*	*M. mocquardi*	*M. zolitschka*
Snout tip very pointed	No	No	No	Yes	No
Tibiotarsal articulation position	Beyond nostrils	Between eye–nostril	Between eye–nostril	Between eye–nostril	Nostrils
Large tympanum	No	Yes	Yes	No	No
Large digit terminal disc	Yes	Yes	Yes	Yes	Yes
Body with striking granules	No	No	No	Yes	No
Body coloration	Dark brownish	Dark green, brown grayish, brown	Dark green, brown	Black, blackish	Brown
Throat with two parallel marks	Stripe	Bar	Bar	X-like pattern and +-like pattern	Y-like pattern
Dorsum with crossbar	No	No	No	V-like pattern or Y-like pattern	No
Mouth with whitish band	No	Yes	Yes	If present (interrupted)	Yes
Yellowish or whitish shape in groin area	Yes (band oblique)	Yes (band running along flank)	Yes (band running along flanks)	If present (spot)	Yes (narrower oblique band)
Ventral surface coloration	Yellowish	With white spot pigment	With white spot pigment	Blackish, silver pigment	Yellowish
Dorsal surface with black spot	No	No	No	Yes	No
Toe fully webbed	No	No	No	No	No
Elevation (m)	90–1600	200–1150	350–1150	350–1000	840
Distribution	North, Sambirano, east, central highland, southeast	North, Sambirano, northwest	Montagne d'Ambre, Analabe, Maevatanana	Northeast, central east, Betampona, Corridor Ankeniheny/Zahamena, Ambatovy, Mantadia, Anala	An'ala

Description of reference specimen UADBA 19662 (NR 1724): adult male (SVL = 37.45 mm) in excellent state of preservation. Measurements are presented in Table S2. Snout tip pointed in dorsal and lateral views and triangular in ventral view. Snout tip with 1.75 mm ventral extension beyond the mouth. Head 1.44 times longer than width. Head length 0.48 times SVL. Canthus rostralis weakly evident. Tympanum diameter 0.88 times eye. Ovoid tympanum distinct to supratympanic fold and which runs towards above forearm and shoulder girdle articulation. Internarial distance 0.28 times head width. Tongue ovoid anteriorly and bifid posteriorly. Round nostrils with lateral aperture. Eye to nostril distance 1.45 times nostril to snout distance. Forearm length 0.50 times SVL. Hand length (including discs) 0.30 times SVL. Fingers without webbing. Inner and outer tubercle metacarpals are very developed in granule-like patterns. Fingers without webbing. Relative finger length 1 < 2 < 4 < 3. Digits with large terminal discs (widest part twice of basal disc width). Tibiotarsal articulation beyond snout tip. Lateral metatarsal separated. Hindlimb 1.90 times SVL. Thigh length 0.95 times tibia length. Foot including tarsus 0.82 times SVL. Inner metatarsal tubercle shield-like pattern at the base of toe 1 (length 1.5 mm). Outer metatarsal tubercle with small granule. Toes with extending webbing and webbing formula: 1 (1), 2i (1), 2e (0.5), 3i (1), 3e (0.5), 4i (2), 4e (1.5), 5 (0). Total sum of free phalanges 7.50: Relative toe length 1 < 2 < 3 < 5 < 4. The lateral surface of the body with small granules and the dorsal surface of the body is almost without granules. Oblong femoral glands slightly developed and its medio-proximal area with pores surrounded by many granules, giving them a crater-like pattern. Internally, femoral glands type 3 [2,28]. In the preservative, the dorsal surface of the head, body, and limbs with dark brown color and a thick pale brown vertebral line running from the snout tip to the anal pore. Upper lip with white band on its lower edge. Lower lip with white spot. The ventral surface of the body with homogeneous pale brown color and darker brown spot pigments, but spots become lighter or less dense on the belly. Throat with dark brown parallel stripes. The inguinal region with prominent pale yellow oblique line. Outer sides of hindlimb and forelimb with darker brown transversal bands.

Variation: Morphometric variation is summarized in Table S2. Sexual dimorphism is evident: males have smaller SVL than females (33–43 mm vs. 43–62 mm), shorter tibia, larger toe 3, and the head relatively smaller. Ratio td/ed larger in males than in females (0.69–1.00 vs. 0.52–0.76). The femoral glands are more swollen in males and smaller and more circular in females. The cloacal area in females has black spots. The free phalanges on toes vary: toe 1 (0.25–1), internal edge of the toe 2 (1–1.50), external edge of the toe 2 (0–1) internal edge of the toe 3 (1–1.75), external edge of the toe 3 (0–1), internal edge of the toe 4 (1.50–2), external edge of the toe 4 (1–2), and in toe 5 (0–0.75).

Coloration in life: The iris has a golden ring on its outer area. A dark brown transversal bar is present between the eyes. A dorsal surface of the body may have or not have a pale yellow vertebral line. The inguinal region has a distinct oblique and large yellow patch, sometimes, it extends onto the ventral side. The ventral surface of the body is often pale brown on the anterior parts and brown yellowish color on the belly. The throat sometimes has a pair of dark brown parallel stripes. The inner surface of the thigh has a brown color and its outer part has yellow spots. Hindlimbs have alternate transversal bands, brown and black.

Habits: semi-aquatic rainforest species living close to rivers or small streams with slow-flowing water and rarely, either observed outside the rainforest or along the streams close to relict forests (e.g., Tsarafidy). This species was found on rocks, leaves, and branches or on the ground along river banks between 9.00 am and 23.30 pm. However, it is more active at night. During the day, it can be found hidden in holes and rock crevices along rivers. At night, the females are generally found sitting on shrubs living along streams and more rarely observed on the rocks far from the riverbank. Males can be found in all areas along rivers, except on branches and plant stems. The vertical distribution of individuals on shrubs might be different between the sexes: we observed males up to 1 m above the ground and females up to 2 m. Individuals prefer streams with stream beds between 1–10 m in width and water depth between 10–150 cm. Eggs are laid in masses outside the water, either on a leaf or a branch overhanging of the stream or the river. The tadpoles live in the calm water far from torrents. The period of reproduction is during the cold winter season (e.g., June and July in Mantadia). The calls of males are low and hardly audible. In the Moramanga Region, metamorphosis appears to occur from September to October. At Andranomanamponga in August 2002, the tadpoles of different development stages were observed in calm and transparent water, and these tadpoles' habitats were protected by rocks. These pools have surface water between 1–3 m^2 and a water depth between 20–100 cm. When disturbed, the tadpoles swim quickly and obliquely to hide in mud under dead leaves or retreat into rock crevices.

Distribution: Eastern and northern rainforests, including the highland of Madagascar with an elevation range between 230 to 1600 m above sea level. The species occurs as far north as the Sambirano Region (14° S) and as far south as the Anosy Mountains (25° S) (Figure 3, Table S13).

Comments: Our morphological description agrees with the *M. femoralis* shown by [29] (Figure 2, p. 247) from Antoetra, and our molecular analyses groups all our *M. femoralis* samples (Table S1) with the *M. femoralis* samples reported by [10]: AY324815 (FGMV 2001.155) AY324817 (FGMV 2002.56); and [11]: HQ610845 (ZSN 1630/2007), HQ610847 (ZSM 1643/2007), HQ610913–610914 (FGZC 271, ZCMV 370) and HQ 610916–61091618 (ZCMV 464, 937, 5874). *M. flavicrus* (with type locality "Madagascar") remains a junior synonym of *M. femoralis*.

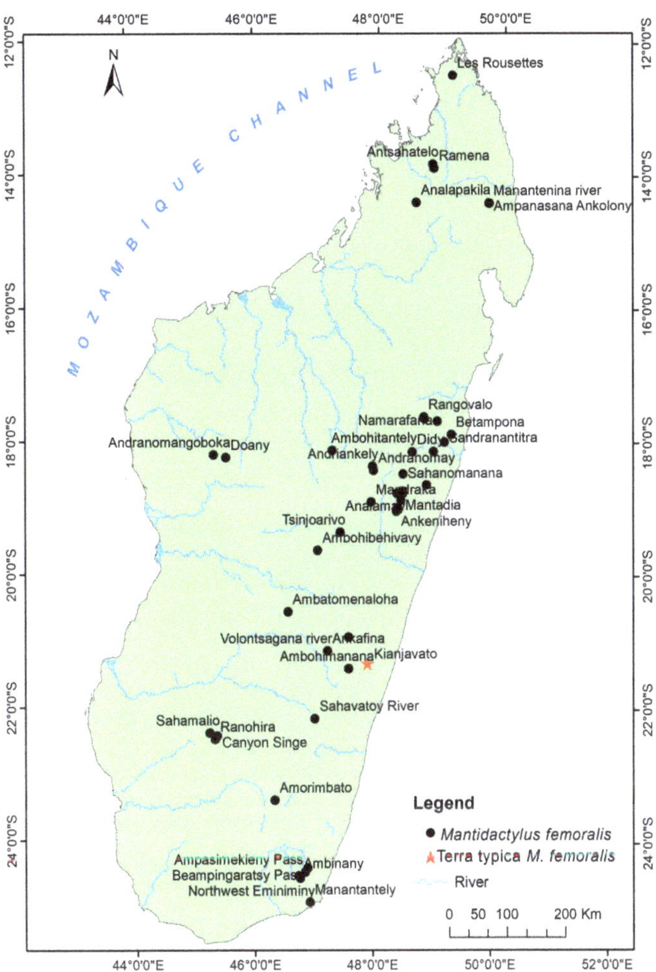

Figure 3. Distribution map of *Mantidactylus femoralis*.

3.2.2. *Mantidactylus ambreensis* [30]

Mantidactylus ambreensis [30] (p. 127). Validated as bona species by [31] (p. 67); *Mantidactylus (Ochthomantis) ambreensis:* [7] (p. 400); [3] (p. 3).

M. ambreensis was synonymized with *M. femoralis* by [22], and later recognized again as a good species by [10,31].

Holotype: MNHN 1893.241: Montagne d'Ambre, Madagascar.

Specimens examined: MNHN 1893.241 adult female: Montagne d'Ambre, Madagascar. AMNH A50521: Analalava District, Sofia Region, Madagascar, 21 January 1971. AMNH A167482* (RAX 2373) adult male: Mandrizavona, Tsaratanana Reserve, Ambanja District, Diana Region, Madagascar (13°48.043′ S 48°44.78′ E, 650 m), 12 January 2001, C. J. Raxworthy, A. Razafimanantsoa and A. Razafimanantsoa. AMNH A167485 (RAX 2468) adult male: Besahona, Tsaratanana Reserve, Ambanja District, Diana Region, Madagascar (13°54.372′ S 48°52.425′ E, 550–750 m), 25 January 2001, same collectors as AMNH A167482*. AMNH A167486 (RAX 2720): Antsahatelo, Tsaratanana Reserve, Ambanja District (13°51.588′ S 48°51.979′ E, 700 m), 7 April 2001, S. Mahaviasy, N. Rabibisoa, C. J. Raxworthy, A. Razafimanantsoa, and A. Razafimanantsoa. AMNH A167499–167500 (RAX 3289, 3330, 3343): Antsaravy, RNI Tsaratanana, Ambanja District, Diana Region, Madagascar (13°55.560′ S

48°54.353′ E, 1150 m), 20 April 2001, same collectors as AMNH A167486. AMNH A167501*–167502*–04 (RAX 6552, 6557, 6584, 6558): Irony Relict Forest Camp, Bealanana District, Sofia Region, Madagascar (14°44.937′ S 48°29.449′ E, 930 m).

Collected 1 April 2003, S. Mahaviasy, N. Rabibisoa, N. Rakotondrazafy, and C. J. Raxworthy.AMNH A 167565* (RAX 3203) juvenile: Antsaravy, RNI Tsaratanana, District Ambanja, Diana Region, Madagascar (13°55.560′ S 48°54.353′ E, 1150 m), 16 April 2001, N Rabibisoa, S. Mahaviasy, A. Razafimanantsoa, and A. Razafimanantsoa. AMNH A174618* (RAX 9589) adult male: Ambohibola forest, Tsaratanana District, Betsiboka Region, Madagascar (16°38.358′ S 47°26.165′ E, 300 m), 10 April 2006, N. Rabibisoa, and C. J. Raxworthy. UADBA 3714 (RAN 38503) adult male: Fitsahana Atomboka River, Montagne d'Ambre, Antsiranana District, Diana Region, Madagascar, (12°29.2′ S 49°10,3′ S, 1150 m), 25 December 1991, by C. J. Raxworthy, A. Raselimanana, J. B. Ramanamanjato. UADBA 5647–5650 (RAN 54050, 54105, 54130, 54054): Irony Relict Forest Camp, Bealanana District, Sofia Region, Madagascar (14°44.937′ S 48°29.449′ E, 930 m), 21–24 February 1996, C. J. Raxworthy, A. Razafimanantsoa, A. Razafimanantsoa. UADBA 7222–7224: Benavony, Ambanja District, Diana Region, Madagascar, 200 m, Mar. 21, 1994, F. Glaw, N. Rabibisoa, and O. Ramilison. UADBA 8393, 8396, 8414 (RAX 2796, 2875, 3000): Ramena River Camp, Ambanja District, Diana Region, Madagascar (13°55.071′ S 48°53.179′ E, 730–750 m), 9–13 April 2001, Mahaviasy, N. Rabibisoa, C. J. Raxworthy, A. Razafimanantsoa, and A. Razafimanantsoa. UADBA 8394, 8411, 8416 (RAX 2530, 2528, 2514): Besahona, Ambanja District, Diana Ragion (13°54.372′ S 48°52.425′ E, 550 m), 27 January 2001, A. Razafimanantsoa, and Razafimanantsoa. UADBA 8395, 8398, 8412, 8415 (RAX 2661–2658): Nirhy's cascade Camp Analabe, Tsaratanana Reserve, Ambanja District, Diana Region, Madagascar (13°51.023′ S 48°47.902′ E, 760 m), 5 April 2001, same collectors as UADBA 8393. UADBA 8397, 8413 (RAX 2557, 2401): Mandrizavona, Ambanja District, Diana Region, Madagascar (13°48.043′ S 48°44.78′ E, 450 m), 14 January and 30 January 2001, A. Razafimanantsoa, and A. Razafimanatsoa. UADBA 8399 (RAX 2554) adult female: Betaindambo, Ambanja District, Diana Region, Madagascar (13°51.932′ S 48°49.189′ E 550 m), 30 January 2001, A. Razafimanantsoa, and A. Razafimanatsoa. UADBA 8401 (RAX 2702) adult female: Antsahatelo, Tsaratanana Reserve, Ambanja District, Diana Region, Madagascar (13°51.588′ S 48°51.979′ E, 800 m), 6 April 2001, same collectors as UADBA 8393. UADBA 8408, 8410 (RAX 3195, 3236): Antsaravy Valley Camp, Tsaratanana Reserve, Ambanja District, Diana Region, Madagascar (13°55.560′ S 48°54.353′ E, 1150 m), 15 April and 17 April 2001, S. Mahaviasy, N. Rabibisoa, A. Razafimantsoa, and A. Razafimanantsoa. UADBA 9056–9057 (NR 548, 547): Les Rousettes Camp, Montagne d'Ambre National Park, Antsiranana District, Diana Region, Madagascar (12°31′ S 49°10′ E, 1000 m), 3 March 1996, N. Rabibisoa, D. Rakotomalala, and O. Ramilison. UADBA 26120, 26222–26224, 26271–26272 (RAX 6556, 6586, 6553, 6555, 6554, 6559): Irony Relict Forest Camp, District Antsohihy, Sofia Region, Madagascar (14°44.937′ S 48°29.449′ E, 930) and (14°45.140′ S 48°29.690′ E, 950 m), 1–2 April 2003, S. Mahaviasy, N. Rabibisoa, N. Rakotondrazafy, and C. J. Raxworthy. UMMZ 212435* (RAN 38501): Antomboka River, Montagne d'Ambre, Antsiranana District, Diana Region, Madagascar (12°32.3′ S 49°10′ S, 650 m), 25 December 1991, J B Ramanamanjato, A. Raselimanana, and C.J. Raxworthy.

Diagnosis: A relatively small-sized *Ochthomantis* (adult male SVL 33.3–39 mm; adult female 38.2–42 mm) with a distinct lateral white strip running along lateral surfaces of head and body, from snout tip to inguinal area. Distinguished from *M. ambony* by its larger size and from the remaining species by the presence of a sharply white lateral stripe well-defined. Character diagnostics are summarized in Tables 1 and 2.

Description of the reference specimen UADBA 8393 (RAX 2796): Adult male (SVL = 39.00 mm) in excellent state of preservation. Measurements are presented in Table S3. In dorsal view and lateral view, the snout tip is pointed. Snout tip with 2.0 mm ventral extension beyond the mouth. In the dorsal view, the head is clearly longer than large, i.e., 1.38 times longer than wide. Head length 0.44 times SVL. Canthus rostralis indistinct. Loreal area concave. Tympanum diameter 1.04 times eye. Round tympanum in contact with a supratympanic fold along their borders except in its posterior part. Supratympanic fold running posteriorly to one point of three large granules above the upper arm and shoulder girdle articulation. Dark tympanum with a small notch in its median superior area. Internarial distance 0.27 times head width. The tongue is ovoid anteriorly and bifid in its posterior part. Round nostrils with lateral aperture. Eye to nostril distance 1.97 times nostril to snout distance. Forearm length 0.48 times SVL. Hand length (including discs) 0.17 times SVL. The inner metacarpal is not obvious and the outer metacarpal has a flattened granule. Fingers without webbing. Relative finger length 1 < 2 < 4 < 3. Digits with large terminal discs (the widest part is twice the width of the basal disc). Tibiotarsal articulation between eye and nostril. Lateral metatarsal separated. Hindlimb 1.77 times SVL. Thigh length 0.97 times tibia length. Foot including tarsus 0.74 times SVL. Inner metatarsal tubercle not obvious (length 1.80 mm) at base of toe 1. Outer metatarsal tubercle absent. Toe with relative extending webbing and webbing formula: 1 (1) 2i (1) 2e (0.50), 3i (1.25) 3e (0.50) 4i (1.50) 4e (1.50) 5 (0.50). The total sum of free phalanges 7.75. The relative toe length is 1 < 2 < 3 < 5 < 4. Lateral surfaces of the body, belly, and sacral areas with very small granules. Oblong femoral glands are well-developed and its medio–proximal area with a pore surrounded by many granules, giving it a crater-like pattern. Internally, femoral gland type 3 [2,28]. In the preservative, the dorsal surface of the head, body, and limbs is dark in color. The lateral surface of the body is less dark than the dorsal surface of the body. Lips with white band colors continue along the lateral surface of the body to above hindlimb articulation. Dark iris surrounded by a white ring. The ventral surface of the body is whitish with dark marbling except in the belly region. Throat with dark-brown parallel bands. Ventral surface of the forelimb with some dark spots. Thigh with large dark spots except on the femoral gland region. Hind and forelimbs with alternate and transverse dorsal bands, dark and light gray colors.

Variation: Morphometric variation is summarized in Table S3. Sexual dimorphism is evident: males have smaller SVL (33.3–39 mm versus 38.2–42 mm), relatively longer tibia and feet, and shorter toe 3; ratio td/ed larger in males than females (0.80–1.16 vs. 0.52–0.79); femoral glands more swollen in males and smaller and more circular in females. The free phalanges on toes vary: toe 1 (0–1), internal edge of toe 2 (1), external edge of toe 2 (0–0.50), internal edge of toe 3 (1–1.50), external edge of toe 3 (0–1), internal edge of toe 4 (1–2), external edge of toe 4 (1–2), and in toe 5 (0–1). Small skin granules may be present or not on the dorsal surface of the body and above the eyes. Some individuals with granules on the sacral area.

Coloration in life: The iris has a gold ring on its outer area. A dark crossbar may be present between the eyes. There is no vertebral line on the dorsum. A white or yellow band is running along the flank. The dorsal surfaces of the head, body, and limbs may be dark green, brown, or grayish brown in color. The ventral surfaces of the body are usually mottled brown, but for some individuals, there is almost no mottling. Some individuals have a diverged pair of short longitudinal dark bands on the throat and this fades on the thorax. Large and round dark spots may also be present on the throat and thorax.

Habits: semi-aquatic forest species living next to flowing streams or rivers generally and close to rocks. This species was found between 9.00–23.00 h, but it is rather nocturnal than diurnal frogs. At night, females rested on leaves rather than on branches, and during the day, we observed them sometimes resting on the banks of the river and very rarely on the ground. For males, during the day they rested on banks and at night on leaves, and sometimes we can find them on rocks. The vertical distribution of individuals on shrubs is different between both sexes: males between 10–200 cm, whereas females between 100–200 cm. Distinguished from *M. femoralis*, this taxon prefers rivers to small streams. At Montagne d'Ambre (Station les Roussettes), it is observed near the irrigation canal. Calling males were heard in March in the afternoon from the ground along the forest brook [7]. A clutch consisting of about 100–120 eggs was found deposited on a rock edge close to a calm and shallow stream in April 2006 at Vohibola Tsaratanana and those eggs were hatched after a couple of days in a plastic bag. The egg diameter is about 2–3 mm.

Distribution: species of low and mid-altitude forests of Northwestern and Northern Madagascar, with elevational ranges between 200 to 1150 m. This species is found as far north as Montagne d'Ambre and across humid forests in the northern highlands such as Manongarivo, Tsaratanana, Andramanalana, and Sorata; and extending to the northern limit of the High Plateau at Irony River and Ambohibola Forest (Figure 4, Table S13).

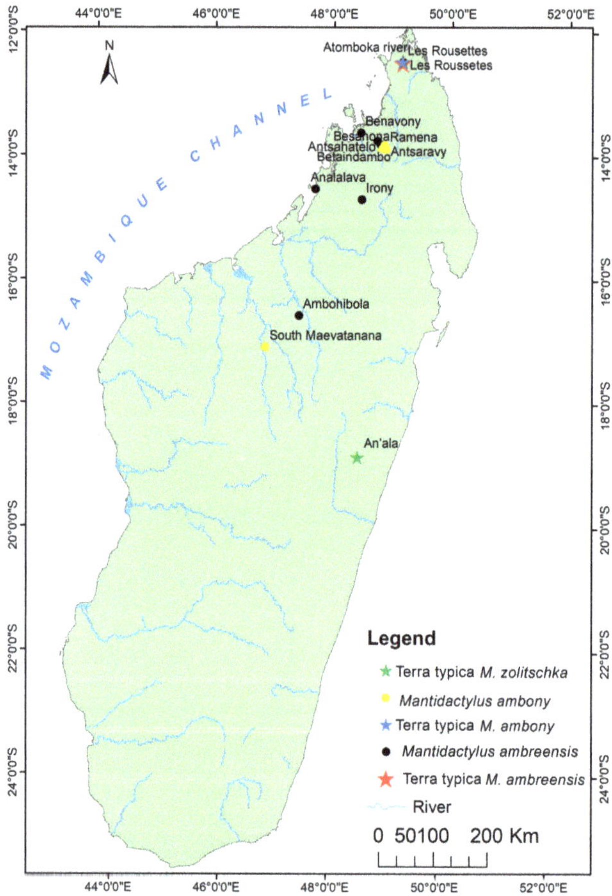

Figure 4. Distribution map of *Mantidactylus ambreensis*, *M. ambony* and *M. zolitschka*.

Comments: Our molecular analyses group all our *M. ambreensis* samples with the *M. ambreensis* samples reported by [10]: AY324822 (ZSM 492/2000); [11]: HQ610870 (FGMV 2002.1950); and by the holotype MNHN 1893.241 (sequence accession MT982173) [5].

3.2.3. *Mantidactylus ambony* [5]

Holotype: ZSM 2078/2007 (FGZC 1039): adult female, Montagne d'Ambre National Park, Antsiranana District (12.5280° S, 049.1720° E, 1050 m), 24 February 2007, F. Glaw, P. Bora, H. Enting, J. Köhler, and A. Knoll [5].

Specimens Examined: AMNH A167487 (RAX 2758): Antsahatelo, Tsaratanana Reserve, Ambanja District (13°51.588′ S 48°51,979′ E, 700 m), 7 April 2001, S. Mahaviasy, N. Rabibisoa, C. J. Raxworthy, A. Razafimanantsoa, and A. Razafimanantsoa. AMNH A167490–16795, 167497 (RAX 2816, 2872–2874, 2876, 2932, 2970): Ramena, Ambanja District, Diana Region, Madagascar (13°55,071′ S 48°53,179′ E, 730–750 m), 11 April 2001 except AMNH A167497 12 April 2001, S. Mahaviasy, N. Rabibisoa, C. J. Raxworthy, A. Razafimanantsoa and A.

Razafimanantsoa. AMNH A167498 (RAX 3289): Antsaravy, RNI Tsaratanana, Ambanja District, Diana Region, Madagascar (13°55.560′ S 48°54.353′ E, 1150 m), 20 April 2001 18 April 2001, S. Mahaviasy, N. Rabibisoa, C. J. Raxworthy, A. Razafimanantsoa, and A. Razafimanantsoa. UADBA 7225–7226: Les Roussetes Camp, Montagne d'Ambre National Park, Antsiranana District, Diana Region, Madagascar (12°31′ S 49°10′ E, 1000 m), Feb. 02, 1994, F. Glaw, N. Rabibisoa, and O. Ramilison. UADBA 8406 (RAX 2871): Ramena River Camp, Ambanja District, Diana Region, Madagascar (13°55.071′ S 48°53.179′ E, 730–750 m), 9–13 April 2001, Mahaviasy, N. Rabibisoa, C. J. Raxworthy, A. Razafimanantsoa, and A. Razafimanantsoa. UADBA 9058 (NR 564): Les Rousettes Camp, Montagne d'Ambre National Park, Antsiranana District, Diana Region, Madagascar (12°31′ S 49°10′ E, 1000 m), 3 March 1996, N. Rabibisoa, D. Rakotomalala, and O. Ramilison. UADBA 5726 (RAN 51382): 34 km South from Maevatanana, Maevatanana District, Betsiboka Region, Madagascar (17°09.092′ S 46°51.365′ E, 350 m), 24 January 1996, A. Raselimanana, C. J. Raxworthy, A. Razafimanantsoa, and A. Razafimanantsoa. UMMZ 212426* (RAN 38009): Antomboka River, Montagne d'Ambre, Antsiranana District, Diana Region, Madagascar (12°32.3′ S 49°10′ S, 1150 m), 15 November 1991, J.B. Ramanamanjato, A. Raselimanana, and C.J. Raxworthy.

Diagnosis: A small sized *Ochthomantis* (adult male SVL 30.0–31.8 mm; adult female 34.0–37.9 mm) with sharply white to yellow stripe, well-defined along the lateral head and lateral surface of body, running from snout tip to groin area. Distinguished from *M. ambreensis* by its smaller size and from the remaining species by the presence of a sharply white to yellow well-defined lateral stripe. Character diagnostics in Tables 1 and 2.

Description of reference specimen: UADBA 8406 (RAX 2871): Adult male (SVL = 31.8 mm) in excellent state of preservation. Measurements are presented in Table S4. In the dorsal view and lateral view, the snout tip relatively pointed. Snout tip with 1.7 mm ventral extension beyond mouth. In the dorsal view, the head is clearly longer than large (i.e., 1.35 times longer than wide). Head length 0.47 times SVL. Canthus rostralis indistinct. Loreal area concave. Tympanum diameter 1.19 times eye. The round tympanum is in contact with the supratympanic fold along their borders and running posteriorly above towards the upper arm and shoulder girdle articulation. Large and dark tympanum with a small notch in its median superior area. Internarial distance 0.30 times head width. Tongue ovoid anteriorly and bifid in posterior part. Round nostrils with lateral aperture. Eye to nostril distance is 1.40 times nostril to snout distance. Forearm length 0.47 times SVL. Hand length (including discs) 0.19 times SVL. The inner metacarpal is not obvious and the outer metacarpal has obvious granule. Fingers are without webbing. The relative finger length is 1 < 2 < 4 < 3. Digits with large terminal discs (widest part twice of basal disc width). Tibiotarsal articulation between eye and nostril. Lateral metatarsal separated. Hindlimb 1.69 times SVL. Thigh length 0.90 times the tibia length. Foot including tarsus 0.73 times the SVL. Inner metatarsal tubercle not obvious (length 1.35 mm) at base of toe 1. Outer metatarsal tubercle absent. Toe with less extensive webbing and webbing formula: 1 (1) 2i (1) 2e (0.50), 3i (1.25) 3e (1) 4i (2) 4e (1.50) 5 (0.50), and total sum of free phalanges 8.75. Relative toe length 1 < 2 < 3 < 5 < 4. The surfaces of the body: flank, belly, and sacral areas with very small granules and dorsal with obvious granules. Oblong femoral glands are well-developed and pores in its medio-proximal area are surrounded by many granules, giving it a crater-like pattern. Internally, femoral gland type 3 [2,28]. In preservative, the dorsal surface of the head, body, and limbs has a dark color. The lateral surface of the body is less dark than the dorsal surface of the body. Lips with white band colors run along the lateral surface of the body to the hindlimb insertion. Dark iris surrounded by a white ring. The ventral surface of the body is whitish with dark marbling except in the belly region. Throat with dark brown parallel bands. Forelimb with some dark spots on its ventral surface. Thigh with large dark spots except in the femoral gland region. Fore and hindlimbs with dorsal transverse bands, alternating dark and light gray colors.

Variation: Morphometric variation is summarized in Table S4. Sexual dimorphism is evident: males have smaller SVL (30–31.8 mm vs. 34–37.9 mm), relatively longer tibia and feet, and shorter toe 3; the ratio td/ed larger in males than females (0.93–1.33 vs. 0.59–0.79); the femoral glands are more swollen in males, but smaller and more circular in females. The free phalanges on toes vary: toe 1 (0.50–1), internal edge of toe 2 (1–1.25), external edge of toe 2 (0–0.75), internal edge of toe 3 (1.25–1.50), external edge of toe 3 (0.25–1), internal edge of toe 4 (1.50–2), external edge of toe 4 (1–2), and in toe 5 (0–1). Dorsal skin above the eyes with obvious granules. Some individuals have some granules in their sacral area.

Coloration in life: The iris has a gold ring on its outer area. A dark crossbar may be present between the eyes. No vertebral line on dorsum. A white or yellow band runs along the flank. The dorsal surfaces of the head, body, and limbs may be dark green or brown in color. The ventral surfaces are usually mottled brown. All individuals have a diverged pair of short longitudinal dark bands on the throat and fade on the thorax. Large and round dark spots were absent on the throat and thorax.

Habits: semi-aquatic forest species living next to quiet rivers with rocks. It was observed between 11.00–20.00 h. Like *M. ambreensis*, this taxon prefers rivers over small streams. At Montagne d'Ambre (Station les Roussettes), it was observed near the irrigation canals. According to [5], this is a rheophilous species and frequently terrestrial by day, sitting on the ground, on rocks, wood, lichen, or hiding under rocks. At night, often observed on perches above the water, once at 2 m height, and sitting on substrates like leaves, rocks, dead wood, and plant stems.

Distribution: Species of low and mid-altitude forests of Northwestern and Northern Madagascar, with an elevational range between 300 to 1150 m above sea level. It was observed from a relict deciduous forest, south of Maevatanana, to a humid forest at Montagne d'Ambre, extreme Northern Madagascar. All of the sites for their presence are presented in Figure 4 and Table S13.

Comments: This species occurs only at Montagne d'Ambre [5]. Even though our single molecular state this analysis, the specimens that we have recorded outside Montagne d'Ambre based on morphological characters only should be considered too, according to character diagnosis from [5]. This is why we have grouped all our *M. ambony* samples (molecular and morphological characters) with the *M. ambony* samples reported by [5]: holotype ZSM 2078/2007 (FGZC 1039).

3.2.4. *Mantidactylus mocquardi* [32]

Mantidactylus mocquardi [32]: 359 (Holotype MNHN 1929.207, according to [33]: 50, secondary homonym of *Rhacophurus mocquardi* [34]; *Mantidactylus (Mantidactylus) mocquardi*: [25]: 37; *Mantidactylus (Hylobatrachus) mocquardi*: [27]: 312; *Mantidactylus (Ochthomantis) mocquardi*: [7]: 400, [3]: 3.

Holotype: MNHN 1929.207 collected in Rogez, Moramanga District, Alaotra–Mangoro Region, Madagascar.

Specimens examined: AMNH A157111 (APR 234) adult female: Ampanasana Ankolony, Marojejy National Park, Andapa District, Sava Region, Madagascar (14°26.2′ S 49°46.5′ E, 1300 m), November 1998, A. Raselimanana and D. Rakotomalala. AMNH A157118–157119 (APR 351, 354) adult male and female: Andapimbazaha, Marojejy National Park, Andapa District, Sava Region, Madagascar (14°26′ S 49°46.7′ E, 850 m), the same date and collector as AMNH A157111. AMNH A167583 (RAX3806) adult female and AMNH A167585, 167587* (RAX 3903, 3915) adult males: Bezavona, Vohémar District, Sava Region, Madagascar (13°31.962′ S 49°51.954′ E, 350 m), 2 February and 8 February 2002, S. Mahaviasy, N. Rabibisoa, and C. J. Raxworthy. AMNH A167588 (RAX 4687) adult male: Ankitsika, Vohemar District, Sava Region, Madagascar (13°52′20.6″ S 49°47′02.7″ E, 650 m), 22 March 2002, N. Rabibisoa and S. Mahaviasy. AMNH A167589 (RAX 5298) adult female and AMNH A167597 (RAX 5297) adult male: Sorata, Vohémar District, Sava Region, Madagascar (13°41.986′ S 49°26.687′ E, 980 m), 22 April 2002, S. Mahaviasy, N. Rabibisoa, C. J. Raxworthy. AMNH A174621* (RAX 3669) adult female: Ambolokopatrika, Andapa District, Sava Region, Madagascar (14°32′18.1″ S 49°26′14.6″ E, 850 m), 29 November 2001, S. Mahaviasy, N. Rabibisoa, N. Rakotondrazafy, A. Razafimanantsoa and A. Razafimanantsoa. AMNH

A174622* (RAX 4670): Ankitsika, Vohemar District, Sava Region, Madagascar (13°52′20.6″ S 49°47′02.7″ E, 650 m), 22 March 2002, N. Rabibisoa, S. Mahaviasy, and N. Rakotondrazafy. AMNH A174628* (RAX 7524): Betampona Strict Reserve, Toamasina II District, Atsinanana Region, Madagascar (17°54.858′ S 49°12.474′ E 350 m), Feb 28, 2004, C. J. Raxworthy, N. Rabibisoa, M. Randriambahiniharime, and F. Ranjanaharisoa. AMNH A174652* (RAX 9022): Ambodiriana, Soanierana Ivongo District, Analanjirofo Region, Madagascar (16°40.469′ S 49°42,167′E, 100 m), 3 March 2006, N. Rabibisoa. UADBA 7769 (MRJ 107) adult female: Ampanasantongotra, Marojejy National Park, Andapa District, Sava Region, Madagascar (14°26′ S 49°46.5′ E, 350 m), 10 October 1994, N. Rabibisoa, J. B. Ramanamanjato, and O. Ramilison. UADBA 8118 (NR 285) adult female: Anjanaharibe–Sud Special Reserve, Andapa District, Sava Region, Madagascar (14°44.5′ S 49°26.5′ E, 1550 m), Nov. 11, 1994, N. Rabibisoa. UADBA 12312–12313 (NR 1371–1372) adult females: Sandranantitra, Toamasina District, Atsinanana Region, Madagascar (18°2.9′ S 49°5.5′ E, 450 m), 10 January 1999, J. Randrianirina and J. Razafimanantsoa. UADBA 19596 (RAX 3680) adult male: Ambolokopatrika River, Andapa District, Sava Region, Madagascar (14°32′18.1″ S 49°26′14.6″ E, 875 m), 30 November 2001, S. Mahaviasy, N. Rabibisoa, C. J. Raxworthy, A. Razafimanantsoa and A. Razafimanantsoa. UADBA 19647* (RAX 3641) adult female: Ambolokopatrika, Andapa District, Sava Region, Madagascar (14°32′18.1″ S 49°26′14.6″ E, 850 m), 29 November 2001, the same collectors as UADBA 19596. UADBA 26238 (RAX 8021) adult female, and UADBA 26287, 26290 (RAX 7539, 8036) adult males: Betampona Strict Natural Reserve, Toamasina District, Atsinanana Region, Madagascar (17°54.858′ S 49°12.474′ E, 390–450 m), 8–18 March 2004, C. J. Raxworthy, N. Rabibisoa, M. Randriambahiniharime, and F. Ranjanaharisoa. UADBA 26240, 26242 (RAN 45476, 45363) adult females, and UADBA 26298 (RAN 45370) adult male: Rangovalo, Zahamena National Park, Fenoarivo Atsinanana District, Analanjirofo Region, Madagascar (17°40.5′ S 48°45.5′ E, and 17°42′ S 48°46′ E, 850–1150 m), 28 February 1994–3 March 1994, F. Rabemananjara, J. B. Ramanamanjato, A. Raselimanana, A. Ravoninjatovo, C. J. Raxworthy, J. Razafimanantsoa, A. Razafimanantsoa, and A. Razafimanantsoa. UADBA 26283 (RAN 47954) adult female: Sahamalio, Isalo National Park, Ihorombe Region, Madagascar, (22°26.315′ S 45°15.648′ E, 700 m), 18 February 1995, J. B. Ramanamanjato, A. Raselimanana, C. J. Raxworthy, A. Razafimanantsoa, and A. Razafimanantsoa. UMMZ 212824* (RAN 37992) juvenile: Manantenina River, close to Marojejy National Park, Andapa District, Sava Region, Madagascar (14°26′ S 49°46′ E, 600 m), 16 November 1992, R. A. Nussbaum, C. J. Raxworthy, A. Razafimanantsoa, and A. Razafimanantsoa. UMMZ 21235 (RAN 39291) adult male: Ambalafary, Ambanja District, Diana Region, Madagascar (14°04′ S 48°17′ E, 250 m), 24 february 1992, C. J. Raxworthy, A. Raselimanana, J. B. Ramanamanjato, A. Razafimanantsoa, and A. Razafimanantsoa; UMMZ 212881* (RAN 42737) juvenile: Ankavanana River, Masoala National Park, Antalaha District, Sava Region, Madagascar (15°18,5′ S/50°14′ E, 70–100 m), 12 January 1993, C. J. Raxworthy, A. Razafimanantsoa, and A. Razafimanantsoa.

For other specimens examined: see Appendix B.

Diagnosis: A medium to large-sized *Ochthomantis* species (adult male 36–48 mm; adult female 44–65 mm); tibiotarsal articulation between eye and nostril but sometimes beyond nostril; body dark brown or black color (black in preservative) with few white spots scattered on the lateral body; the upper lip of mouth paler brown with dark brown spots and densely spotted in its posterior part; snout tip pointed laterally and extending > 1.75 mm beyond the lower jaw. Distinguished from all other species by blackish body coloration, white spots along the lateral body, and pale upper lip with dark brown spots. Character diagnostics are summarized in Tables 1 and 2.

Description of reference specimen UADBA 19596 (RAX 3680): adult male (SVL = 39.90 mm) in excellent state of preservation. Measurements are presented in Table S5. In the dorsal view and lateral view, the snout tip is very pointed. Snout tip with 2.30 mm straight ventral extension beyond the mouth. Head 1.52 times longer than wide. Head length 0.47 times SVL. Canthus rostralis obvious. Loreal indented. Tympanum diameter 0.80 times eye. Round tympanum with a small notch in its median superior area, clearly separated with supratympanic fold. This supratympanic fold has an umbrella-like pattern and runs to above one point between the shoulder girdle and upper arm articulation but behind one large strong granule. The anterior half part of the tympanum with light background color and the posterior part dark color. Internarial distance 0.27 times head width. Tongue ovoid anteriorly and bifid posteriorly. Non-protruding nostril with relatively close lateral aperture. Eye to nostril distance 1.59 times nostril to snout distance. Forearm length 0.47 times SVL. Hand length (including discs) 0.29 times SVL. Fingers without webbing. Outer and inner metacarpals are poorly developed. Finger relative length size 1 <2 <4 <3. Digits with a large terminal disc (the widest part twice the width of the basal disc). Tibiotarsal articulation reaching nostril. Lateral metatarsal separated. Hindlimb 1.71 times SVL. The thigh is the same length as the tibia. Foot including tarsus 0.72 times SVL. Inner metatarsal tubercle in bell-like pattern (length 1.6 mm) at base of toe 1. Outer metatarsal tubercle absent. Webbing formula 1 (0), 2i (1), 2e (0), 3i (1), 3e (0), 4i (1.5), 4e (1), 5 (0) and total sum of free phalanges 4. Relative toe length 1 <2 <3 <5 <4. Importance and repartition of body granules differently in shape and color: side and edge of the dorsal surface and above tympanum highly granulated; inguinal area, basal of flanks, and posterior of upper mouth with white evident granules; dorsum with little granules and belly finely granular. Oblong femoral glands are relatively developed with centro–distal pores surrounded by many granules, giving a crater-like pattern form. Internally, femoral glands type 3 [2,28]. In preservative, dorsum with black color. Upper lip and flanks with clean spots. The throat and the thorax with white and some silver reticulated dark brown pigments, and the belly with light yellow color. Thorax with parallel dark bands: X-like pattern on its left side and divided into two forms, spot and "+/−like" pattern on its right side. The ventral surface of the thigh is partially mottled with brown color and weak in the femoral gland. The lower part of the hindlimb is completely pigmented. The ventral side of the forelimb with clean shape and yellowish color. Evident alternate and transverse dorsal bands, shiny and dark, on the hindlimb and indistinct bands on the forelimb, but with some pinkish reticles indifferently distributed in their dorsal surfaces.

Variation: Morphometric variation is summarized in Table S5. Sexual dimorphism is evident: males have smaller SVL (36.10–48.15 mm vs. 44.30–65.40 mm) with large eyes, a higher head, a broader terminal disc, a shorter tibia, and slightly developed nostrils. Colors vary from blackish brown (e.g., holotype, UADBA 12312, 26211, 26238, 26287, 26290, AMNH A157111, A157118, A167583, A167585) to full black one (e.g., UADBA 7769, 12313, 19647, 26240, 26242, 26298, AMNH A167588–167589). The differences are also reflected in the number of granules and pigments between these two groups of specimens: the black specimens: (1) granules almost absent and body almost smooth except on flanks, (2) with or without obvious round pigment patterns, (3) upper mouth and flank without silver and white spots, (4) ventral surface with dark brown except in belly that is a clean pattern; and the blackish brown specimens: (1) holotype with dense granules, (2) flanks with evident granules, (3) some specimens with dorsal dark spots (UADBA 12313, 26238, 26240, 26290), (4) flanks, upper mouth, throat, and thorax with obvious spots, and (5) thorax and throat with dark brown color mottled by white or silver spots and belly has no spots. The border of the tympanic region is smooth in males and finely granular in females. In males, tympanum and supratympanic fold in contact with each other except for the holotype. In addition, free phalanges on toes vary: toe 1 (0–0.50) internal edge of toe 2, (0.50–1), internal edge of toe 3 (1–1.25), external edge of toe 3 (0–0.50), internal edge of toe 4 (1–2), and external edge of toe 4 (1–1.75).

Coloration in life: The iris has a golden ring on its outer area. Just the dorsum can have a black crossband in V or Y-like pattern. The dorsum is either dark brown or thoroughly black in color. The upper mouth is either dark brown or blackish. Whitish gray spots may be present or not in the inguinal region. The ventral surface shape is very heterogeneous: throat and thorax with black color and usually with numerous small white spots; belly without pigments; and throat either with two dark spots or not. The lateral surface of the body is unicolor with no evident white spots and few granules. The ventral surface of the thigh has at least some clean pattern surfaces. Hind and forelimbs have alternate transversal bands, black and brown.

Habits: Semi-aquatic rainforest species can live in open and degraded forests, especially in Northern Madagascar. This species was observed resting along the riverbanks and streams at different stages of water speeds, between 9.00 a.m to 23.00 p.m. However, it appears rather diurnal than nocturnal and prefers rocky areas to trees. On trees, it prefers resting on leaves and branches.

Distribution: Low, mid, and high-altitude rainforest species in Central Eastern and Northern Madagascar, from 100 to 1550 m elevations (Figure 5, Table S13).

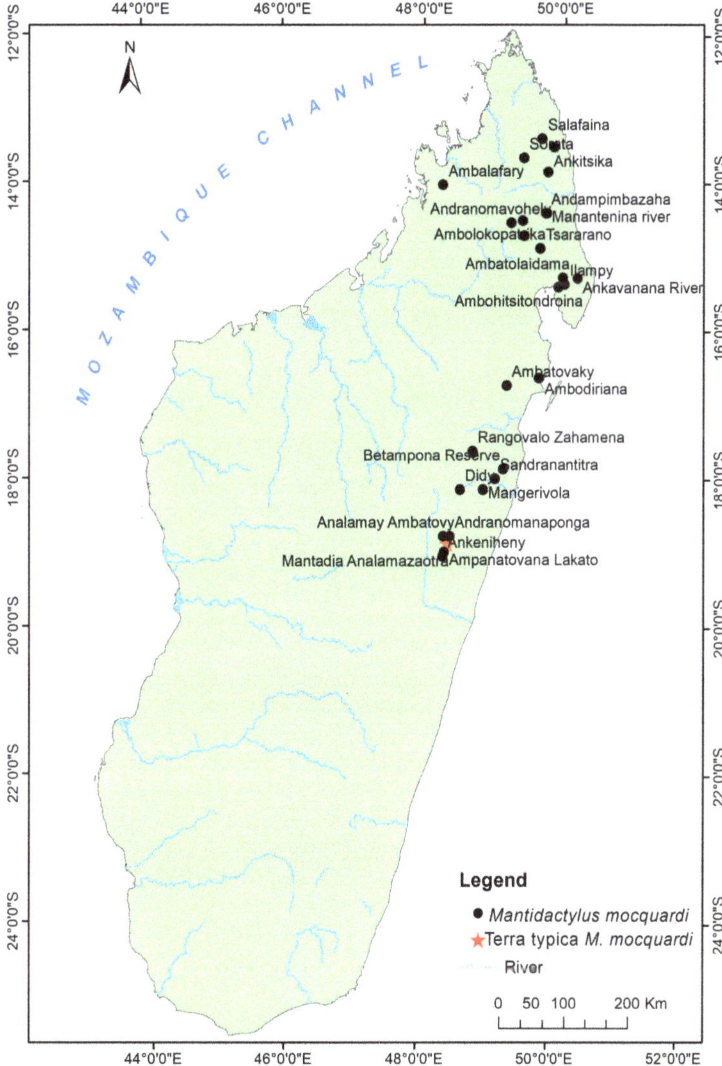

Figure 5. Distribution map of *Mantidactylus mocquardi*.

Comments: Our morphological description agrees with the *M. mocquardi* specimen shown by [29] (Figure 1, p. 249) from Andasibe, and our molecular analyses group all our *M. mocquardi* samples with the *M. mocquardi* samples reported by [10]: AF215317 (ZFMK 66668) from Ambato, Masoala; and [11]: HQ610861 (ZSM 1846/2007) and HQ610921 (ZCMV 8818) from An'Ala and Mahasoa.

3.2.5. *Mantidactylus zolitschka* [10]

Holotype: ZFMK 60110: close to An'Ala Forest (18°56′ S 48°28′ E, 840 m), 21 March 1995, F. Glaw and D. Vallan.

Paratype: ZFMK 60112–60116, ZSM 939/2000, same data as holotype. ZSM 184/2003, the same location as holotype, 2 March 2003, G. Aprea, F. Glaw, M. Puente, L. Raharivololoniaina, R. D. Randrianiaina, and M. Thomas.

Specimens examined: UADBA 6965–66, same data as holotype.

Diagnosis: A small-sized *Ochthomantis* (adult male SVL 29.6–30.6 mm, adult female SVL 37.6–37.7 mm); tibiotarsal articulation at least up to the nostril. Width of terminal disc 1.78 times basal disc. Distinguished from all other species, except *M. ambony*, by its smaller size (adult male SVL < 31 mm, female < 38 mm) and from all species by foot with less well-developed webbing (total sum of free phalanges, WS > 9). Distinguished from *M. ambony* by the absence of lateral stripe running from snout tip to inguinal region. Character diagnostics are summarized in Tables 1 and 2.

Description of reference specimen UADBA 6966: Adult male (SVL = 27.65 mm) in good state. Measurements are presented in Table S6. In the dorsal view, the body is clearly slender. In lateral and dorsal view, the snout tip is pointed. Snout tip with 1.20 mm ventral extension beyond the mouth. Head 1.40 times longer than wide. Head length 0.47 times SVL. Canthus rostralis are distinct and straight. Loreal is weakly concave. Tympanum diameter 0.83 times eye diameter. Round tympanum distinctly from the supratympanic fold. This tympanic fold runs straight first and rather curves in midway before reaching the forepart of the shoulder girdle and upper arm articulation. Internarial distance 0.24 times head width. Tongue ovoid anteriorly and distinctly bifid posteriorly. Small and round nostril without protuberant lateral aperture. Eye to nostril distance 1.44 times nostril to snout distance. Forearm length 0.52 times SVL. Hand length (including discs) 0.33 times SVL. Fingers without webbing. Relative finger length 1 < 2 < 4 < 3. Inner and outer metacarpal with tubercles. Digits with slightly enlarged terminal discs (widest part 1.78 times of basal disc width). Legs slender. Tibiotarsal articulation reaching nostril. Lateral metatarsal separated. Hindlimb 1.85 times SVL. Thigh the same length as the tibia. Foot including tarsus 0.78 times SVL. The inner metatarsal tubercle is rather small at the base of toe 1 (0.85 mm). Metatarsal with small tubercle. Webbing formula: 1(1), 2i (1.25), 2e (1), 3i (1.5), 3e (1), 4i (2), 4e (2), 5 (1) and total sum of free phalanges 9.75. Relative toe length 1 < 2 < 3 < 5 < 4. Skin is rather smooth on its upper surface and slightly granular on its flanks. Ventral side smooth one. Obvious femoral glands are in contact with the cloacal area and sharply delimited by granules with irregular tubercle-like patterns, and the presence of a central porus gives it a crater-like pattern. Internally, femoral gland type 3 [2,28]. In preservative, gray-brownish dorsal color with irregular dark and light marble shape. Upper lip and loreal area whitish color. Tympanic region dark brown color. Lower lip with alternate spots, light and dark. The lateral surface of the body is dark in color and the ventral surface of the body is light in color. The ventral surface of the body with different color pattern: the throat is whitish and becoming more yellowish on the belly. Throat with two longitudinal brown markings from the lip to the thorax and both merge at the shoulder girdle as a "Y-like pattern". One light longitudinal stripe runs from the inguinal area and fades towards forelimb insertion. Light brown forelimb and hindlimb with different numerous dark crossbands (six on hand including third finger, four on thigh, three on tibia, and five on tarsus and foot). Hindlimbs with irregular dark mottling.

Variation: Morphometric variation is summarized in Table S6. Sexual dimorphism is evident: males have smaller SVL (26.5–30.6 mm vs. 33.6–37.7 mm), and large tympanums. The free phalanges on toes vary: toe 1 (0.5–1); internal edge of toe 2 (1–1.5); external edge of toe 2 (0.5–0.75); internal edge of toe 3 (1.75–2); external edge of toe 3 (0.75–1); external edge of toe 4 (1.75–2), and in toe 5 (0.5–0.75).

Coloration in life: The iris has a golden ring on its outer area. The dorsal surface of the body has a strong gray-brownish color with a small light stripe running along it. There is a shiny yellow blotch on the inguinal region.

Habits: semi-aquatic rainforest species living close to stream, around An'Ala forest. The female ZFMK 30116 contains 49 eggs with yellow and dark brown center markings and diameter is 2 mm [10].

Distribution: Known only from type locality, An'Ala [10] (Figure 4, Table S13).

Comments: We included genetic data for this species from [11]: HQ610866 (ZSM 1768/2007) and HQ610867 (ZSM 1841/2007).

3.3. Resurrected Species

Based on our molecular and morphological results, we find strong evidence to recognize two species of *Mantidactylus (Ochthomantis)* that correspond to taxa that currently are considered junior synonyms of *M. femoralis*. After examining their type specimens and our new materials, we here recognize *Mantidactylus catalai* [35] (33 specimens) and *Mantidactylus poissoni* [36] (12 specimens) as valid species and provide new descriptions for both species below (Figure 6).

(a) (b)

Figure 6. Photos of the resurrected species of the subgenus *Ochthomantis*: (**a**). *Mantidactylus catalai*, Ampasimiekiny Pass (NR), (**b**). *Mantidactylus poissoni*, Mandraka (NR).

3.3.1. Mantidactylus catalai [35]

Mantidactylus catalai [35] (p. 203) (Holotype MNHN 1935.153, according to the original publication and [6] (p. 220); *Mantidactylus femoralis:* [22] (p. 26); *Mantidactylus (Hylobatrachus) femoralis:* [27] (p. 312); *Mantidactylus (Ochthomantis) femoralis:* [7] (p. 400), [3] (p. 3).

M. catalai has previously been considered by [22] as a synonym of *M. femoralis* but [10] had noted the considerable morphological differences between *M. catalai*, and *M. femoralis* of the southeast of Madagascar.

Holotype: MNHN 1935.153: Isaka–Ivondro, Tolagnaro District, Anosy Region, Madagascar, 700 m, 1935, M. R. Catala. Specimen in good condition.

Specimens examined: AMNH 7881–7882 A133689–133690, AMNH 18019 A168364: Fianarantsoa Ifanadiana Road, Southwest Ranomafana, Ifanadiana District, Vatovavy-Fitovinany Region, Madagascar, 900 m. AMNH A 181732*, A 181821* (RAX 10563, 10599): Beampingaratsy Pass, Anosy Montain, Tolagnaro District, Anosy Region, Madagascar, (24°28.244′ S 46°53.521′ E, 490–1140 m), 12 February–13 February 2009, S. Mahaviasy, N. Rakotondrazafy, and C. J. Raxworthy. UADBA 1419–1421, 1423 (RAN 36377, 36434, 36446,

36505): Ampasimekieny Pass, Tolagnaro District, Anosy Region, Madagascar (24°32.0′ S 46°51.0′ E, 800–950 m), 24 December–28 December 1990, J. B. Ramanamanjato, A. Raselimanana; C. J. Raxworthy, A. Razafimanantsoa, and A. Razafimanantsoa. UADBA 3706 (RAN 35091): Manatantely Forest, Tolagnaro District, Anosy Region, Madagascar (29°59.0′ S 46°55.083′ E, 125 m), 30 October 1990, same collectors as UADBA 1419. UADBA 4513–4514; 4516, 4521, 4523 (RAN 52831, 52699, 52762, 52807, 52830): Eminiminy, Andohahela National Park, Tolagnaro District, Anosy Region, Madagascar (24°35.04′ S 46°44.08′ E, 1000–1100 m), 11 November–15 November 1995, J. B. Ramanamanjato, and A. Raselimanana. UADBA 4522 (RAN 52472): Ambinany, Andohahela National Park, Tolagnaro District, Anosy Region, Madagascar (24°35.6′ S/46°44.3′ E, 820 m), 14 November 1995, J. B. Ramanamanjato, and A. Raselimanana. UADBA 9772–9774; 9782 (RAN 57002, 56723, 56937, 57006): Amorimbato Forest; Kalambatritra Special Reserve, Iakora District, South–Est Region, Madagascar (23°27.44′ S 46°20.02′ E, 1150–1300 m), 30 October–8 November 1996, J B Ramanamanjato, R. A. Nussbaum, and J. Spannring. UADBA 26403–26405 (RAN 44835, 44672, 44701): Sahavatoy and Volontsagana Rivers, Andringitra National Park, Ihorombe Region, Madagascar (22°13.667′ S 47°0.217′ E, 810–1240 m), 24 November–30 November 1993, N. Rabibisoa, A. Razafimanantsoa, and C. J. Raxworthy. UMMZ 191515–191516 (RAN 32567, 32597): Sainte Luce, Tolagnaro District, Anosy Region, Madagascar (24°45′ S 47°11′ E, 20 m), 7 October and 10 October 1989, R. A. Nussbaum, and C. J. Raxworthy. UMMZ 197662–197664 (RAN 35686, 35706–35707): Nahampoana, Tolagnaro District, Anosy Region, Madagascar (24°58′ S 46°58′ E, 75–300 m), 23 November–24 November 1990, R. A. Nussbaum, J. B. Ramanamanjato, and A. Raselimanana. UMMZ 197676 (RAN 36626): Manangotry, Tolagnaro District, Anosy Region, Madagascar (24°45′ S 46°52′ E, 850 m), 3 January 1991, J. B. Ramanamanjato, A. Raselimanana, and C. J. Raxworthy. UMMZ 212890* (RAN 44491): Iatara River, Andringitra National Park, Ivohibe District, Atsimo Atsinanana Region, Madagascar (22°13.333′ S 47°01.483′ E, 720 m), 18 November 1993, same collectors as UADBA 26403.

Diagnosis: A medium to large-sized *Ochthomantis* (adult male SVL 41–45 mm, female SVL 51–62 mm); tibiotarsal articulation between eyes and nostril (or very rarely at snout); toes fully webbed, except on toe 4 where 1–1.5 phalanges are free; no stripe line along superior lip; inguinal region with clean pattern area; snout tip very pointed in lateral view with large extension beyond mouth (1.75–3.45 mm); head wider and flattened but very sharp as "fish-like" pattern. Distinguished from other species by the following characters: *M. ambreensis* and *M. ambony* by the absence of white stripes along the lateral body; *M. femoralis*, *M. zolitschka*, and *M. danieli* n. sp. by the number of free phalanges in the internal edge of toe 4 (1–1.5) and absence of prominent pale yellow or white stripes in the inguinal region (horizontal or oblique); *M. mocquardi* by the number of free phalanges on the internal edge of toe 4 (1–1.5) and flanks without whitish spots; *M. olgae* n. sp. by the absence of obvious black granules on flanks and absence of crossbars in V- or Y-like pattern on dorsum in preservative; *M. tavaratra* n. sp. by digits with large terminal discs (widest part > 1.80 times of basal disc width), lack of prominent and pale inguinal streak, and absence of white strip on superior lip; *M. poissoni* by the absence of white spots below the eye and tibiotarsal articulation between eyes and nostrils; *M. macrotympanum* n. sp. by smaller adult male SVL (<60 mm). Character diagnostics are summarized in Tables 3 and 4.

Table 3. Morphological measurement diagnostic for the two resurrected species of the subgenus *Ochthomantis*. All measurements in mm, F: adult female, M: adult male. For morphological abbreviations see Appendix A.

Size Range	M. catalai		M. poissoni	
	M	**F**	**M**	**F**
SVL	51.9–63.3	41.1–45.4	53.4–65.3	30.7–48.2
TD	3.5–4.7	5.0–6.6	3.3–4.7	3.6–6.0
ED	6.0–8.2	5.0–6.2	6.0–7.6	5.8–6.5
EN	3.6–5.4	3.3–4.3	4.1–5.4	3.6–4.2
EST	1.6–3.5	1.8–3.2	1.6–3.3	1.5–2.8
NS	1.3–3.8	1.9–3.3	2.1–4.6	1.8–3.7
NN	4.7–5.8	3.4–4.4	4.6–5.9	4.0–5.0
HW	17.1–21.5	13.9–18.5	16.1–22.2	12.5–17.1
HL	22.2–27.5	17.8–21.1	21.2–27.8	17.7–22.2
HDL	13.0–18.1	12.0–17.9	13.5–20.2	13.0–14.6
RC	9.2–12.1	7.9–12.7	9.4–13.7	9.0–10.0
FE	27.0–31.9	20.6.24.0	29.1–35.8	24.4–24.8
TI	28.1–32.1	20.7–23.6	30.3–37.3	24.3–26.0
FT	26.8–32.0	20.1–24.5	29.1–35.2	22.7–26.6
TA	11.3–14.2	8.4–11.5	12.0–16.1	8.8–11.0
EO	8.8–11.3	6.6–7.7	9.5–11.4	7.4–10.0
EM	2.8–4.1	2.0–3.1	3.1–3.8	2.7–4.0
EHEAD	6.7–9.3	6.3–7.7	7.7–10.2	6.8–9.4
T1	1.8–3.0	1.6–2.3	1.7–3.1	1.8–2.0
TO3	7.1–8.8	5.3–6.4	7.2–11.0	7.3–7.6
TO5	8.0–9.7	5.8–7.2	8.5–11.4	8.3–8.6

Table 4. Qualitative morphological characters and biogeography diagnostics for two resurrected species of the subgenus *Ochthomantis*.

Morphological Characters	M. catalai	M. poissoni
Snout tip very pointed	Yes	No
Tibiotarsal articulation position	Between eye–nostril	Nostril–snout tip
Large tympanum	Yes	No
Large digit terminal disc	Yes	Yes
Body with striking granules	Yes	No
Body coloration	Brown	Brown
Throat with two parallel marks	Spot	Single 8-like pattern
Dorsum with crossbar	No	No
Mouth with whitish band	No	No (large pale or multiple white spot)
Yellowish or whitish shape in groin area	No	No
Ventral surface coloration	Clean pattern	Yellowish and whitish pigments
Dorsal surface with black spot	Yes	No
Toe fully webbed	No	No
Elevation (m)	20–1300	600–1450
Distribution	Southeast	Central East, Mandraka

Description of reference specimen UADBA 1419 (RAN 36377): Adult male (SVL = 42.60 mm) in good state of preservation. Measurements presented in Table S7. In the dorsal and lateral view, the snout tip is pointed. Snout tip with 2.60 mm ventral extension beyond the mouth. Head 1.40 times longer than wide. Head length 0.50 times SVL. Canthus rostralis well distinct. Loreal region with evident indentation. Tympanum diameter 1.03 times eye. Slightly round tympanum in contact with supratympanic fold, except in its posterior part, and this tympanic fold runs behind tympanum towards the small granule above shoulder girdle and forearm articulation. Tympanum with a small notch in its median superior area. Internarial distance 0.27 times head width. Tongue ovoid anteriorly and bifid posteriorly. Nostrils with distinct cutaneous fold and lateral oblique aperture. Eye to nostril distance 1.58 times nostril to snout distance. Forearm length 0.61 times SVL. Hand length (including discs) 0.42 times SVL. Fingers without webbing. Inner and outer metacarpal tubercles are evident. Relative fingers length 1 < 2 < 4 < 3. Digits with large terminal discs (widest part 2.07 times of basal disc width). Tibiotarsal articulation between the eye–nostril. Lateral metatarsal separated. Hindlimb 1.84 times SVL. The thigh is the same length as the tibia. Foot including tarsus 0.78 times SVL. The inner metatarsal tubercle is obvious along toe 1 (2.15 mm). The outer metatarsal tubercle is small and granule-shaped. Webbing formula: 1 (0.75) 2i (1) 2e (0.25), 3i (1.25) 3e (0.25) 4i (1.5) 4e (1) 5 (0) and total sum of free phalanges 6. Relative toe length 1 < 2 < 3 < 5 < 4. Body granules vary: flank with obvious granules, dorsum with granules irregularly distributed, granules of tympanum more concentrated above its superior part, and sacral area and belly finely granulated. Oblong femoral glands are relatively developed and its medio-proximal part with pores surrounded by some granules, giving it a crater-like pattern. Internally, femoral gland type 3 [2,28]. In preservative, the dorsal surface of the body is brown in color and the white longitudinal vertebral band runs from the snout tip to the anal pore. Lips with two vertical light stripes on the loreal region. Shiny line behind the eyes. The ventral surface of the body is orange–brown in color but has a clean pattern. The throat has a couple of dark brown spots (in front of the shoulder girdle). The inguinal region with white spots or not. The ventral surface of the forearm is bordered by a brown-sided but clean pattern. Fore and hindlimbs with obvious alternate crossbands, dark and shiny.

Variation: Morphometric variation is summarized in Table S7. Sexual dimorphism is evident: males have smaller SVL than females (41.10–45.40 mm versus 51.85–61.50 mm), wider head, bigger tympanum, longer fore and hindlimbs, and metatarsal tubercle not evident. The ratio of td/ed in males is larger than in females (0.89–1.08 vs. 0.50–0.69).

The femoral glands are more swollen in males and smaller and more circular in females. A majority of all specimens are brown darker color, except for the holotype and UADBAs (1419–1421, 4513, 9773, 26403–26404) which are light brown In the dorsal view, granules in a plate-like pattern are observed except for UADBAs (4523, 26403, 26405). A vertebral line is absent, except for UADBAs (1419, 4513, 4523). The throat with white spots in males, except for UDBAs (1419, 1421). The free phalanges of toes vary: toe 1 (0–1), internal edge of toe 2 (1–1.25), external edge of toe 2 (0–0.25), internal edge of toe 3 (1–1.25), external edge of toe 3 (0–0.50), internal edge of toe 4 (1–2), external edge of toe 4 (1–1.50), and in toe 5 (0–0.25).

Coloration in life: The dorsal surface of the body is dark brown. Superior lips with light dots. Round tympanum with central dark color surrounded by dark background. Male without yellowish vertebral line. Inguinal region with small spots or not. Reddish brown ventral surfaces with small spots and darker punctuations.

Habits: semi-aquatic rainforest species living in bamboo forests. This species is diurnal and/or nocturnal, observed between 9.00 a.m to 0.15 a.m, which is adapted to a "burrowing life", inside of holes and interstice of rocks, but close to the waters (stagnant water to the river but very rarely in a fast stream). It is mainly observed on rocks and sometimes on the ground, especially during the day. No individuals were seen on leaves or branches, but one individual UADBA 4521 was collected on roots.

Distribution: Known only from rainforest in Southeast Madagascar at low and medium elevations (Figure 7, Table S13):

Comments: Our morphological description agrees with the *M.* sp. aff. *mocquardi* shown by [29] (Figure 3, p. 249) from Ambatolahy, near Ranomafana, and our molecular analyses group all our *M. catalai* samples (Table S1) with the Ranomafana *M.* cf. *mocquardi* sample reported by [11]: AY324821 (FGMV 2002.173). This specimen has more recently been referred to as "Confirmed Candidate Species (CCS) sp. 47" [11]. The LSID number is E9176C8E-291B-48BD-9857-BF3D2FEF1517.

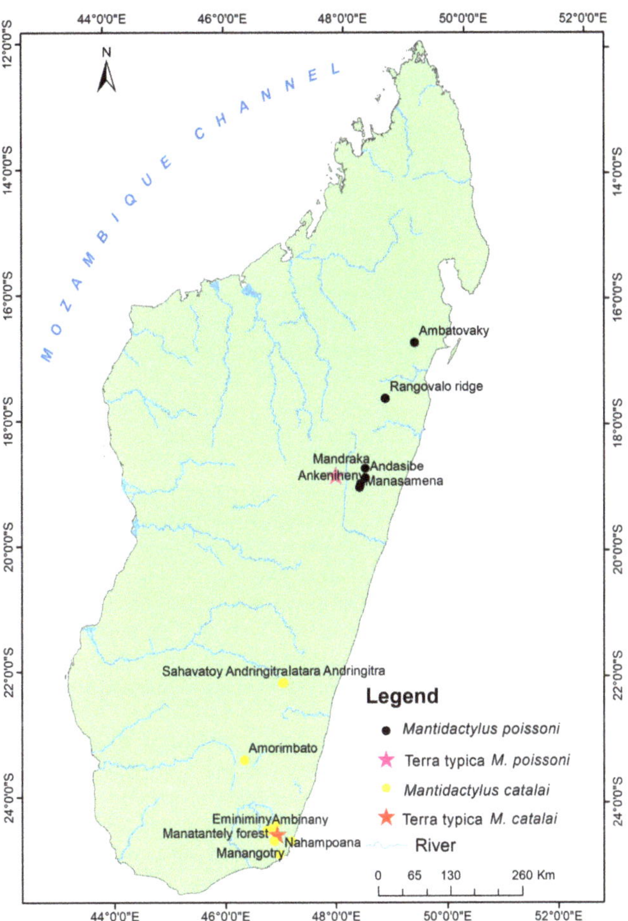

Figure 7. Maps of the distribution of *Mantidactylus catalai* and *M. poissoni*.

3.3.2. *Mantidactylus poissoni* [36]

Mantidactylus poissoni [36] (p. 178) (Holotype MNHN 1937.1) *Mantidactylus femoralis*: [22] (p. 26); *Mantidactylus (Hylobatrachus) femoralis:* [27] (p. 312); *Mantidactylus femoralis:* [37]: (p. 278); *Mantidactylus (Ochthomantis) femoralis:* [7] (p. 400), [3] (p. 3).

Holotype: MNHN 1937.1: Mandraka forest, 70 km from Antananarivo, Manjakandriana District, Analamanga Region, Madagascar, 1937, collected by M. H. Poisson. This type is in a poor state of preservation according to our observation and [22] (p. 45).

Reference specimen: AMNH A174653* (RAX 9367): Mandraka, Manjakandriana District, Analamanga Region, Madagascar, (18°54.727′ S 47°55.174′ E, 1250 m), 18 March 2006, N. Rabibisoa, J. Rafanomezantsoa, N. A. Rakotondrazafy, and P. Razafimahatratra. Same locality as the holotype.

Specimens examined: MNHN 1937.1: Mandraka Forest, 70 km from Antananarivo, Manjakandriana District, Analamanga Region, Madagascar, 1937, M. H. Poisson. AMNH A50362 adult male: Madagascar, 1971, Guibé. AMNH A174649-50 (RAX 8198–8199) adult females: Manasamena River, Lakato, Moramanga District, Alaotra Mangoro Region, Madagascar (19°02′38.2″ S 48°20′54.6″ E 950 m), 29 March 2004, N. Rabibisoa, M. Randriambahiniarime, and F. Ranjanaharisoa. AMNH A174653* (RAX 9367): Mandraka, Manjakandriana District, Analamanga Region, Madagascar (18°54.727′ S 47°55.174′ E, 1250 m), 18 March 2006, N. Rabibisoa, J. Rafanomezantsoa, N. A. Rakotondrazafy, and P. Razafimahatratra. UADBA 6876, 7125 adult females: Ankeniheny and Andasibe, Moramanga District, Alaotra Mangoro Region, Madagascar (19°05.850′ S 48°19.910′ E, 950 m, and 18°57′ S 48°26′ S, 900 m), 28 December 1994 and 15 December 1997, N. Rabibisoa, and S. Ramilison. UADBA 11899 (NR 1196) adult male: Sahaberiana, Mantadia National Park, Moramanga District, AlaotraMangoro Region, Madagascar (18°47,503′ S 48°25,572′ E, 895 m), Nov. 20, 1998, J. Rafanomezantsoa, and N. Rabibisoa. UADBA 19786 (LV77) subadult female: Ambatovaky Special Reserve, Soanierana Ivongo District, Analanjirofo Region, Madagascar (16°46.910′ S 49°14.417′ E, 600 m), 5 August 1999, by N. Rabibisoa, and S. Ramilison. UADBA 26409 (RAN 45665) adult female: Rangovalo Ridge, Zahamena National Park, Fenoarivo Atsinanana District, Analanjirofo Region, Madagascar (17°40.5′ S 48°45.5′ E, 1150 m), Mar. 4, 1994, J. B. Ramanamanjato, A. Raselimanana, C. J. Raxworthy, and A. Razafimanantsoa. UADBA 26411–26412 (RAX 8190, 8155) adult females: Manasamena River, Lakato, Moramanga District, Alaotra Mangoro Region, Madagascar (19°02.637′ S 48°20.910′ E 950 m), Mar. 29 and 27, 2004, N. Rabibisoa, M. Randriambahiniarime, and F. Ranjanaharisoa. UADBA 39000 (RAX 9368) adult male: same data as a reference specimen.

Diagnosis: A medium to large-sized *Ochthomantis* (adult male SVL 39.7–48.2 mm, adult female 53–66 mm); tibiotarsal articulation reaching at least nostril; width of digit terminal disc $\geq$ 1.70; one large white spot underneath of eye for females and numerous white spots for males. Distinguished from all other species by the following combination of characters: *M. ambreensis* and *M. ambony* by lack of white or yellow continuous line on lateral surfaces of head and body; *M. femoralis* by lack of prominent and pale inguinal patch or line and upper lip without white stripe; *M. mocquardi* by lack of white spots along lateral surface of body and surfaces of body neither black nor very dark brown color; *M. catalai* by the presence of white spots below eye and tibiotarsal articulation reaching at least nostril; *M. olgae* n. sp. by the absence of black granules on dorsal surface of head and lateral surface of body; *M. tavaratra* n. sp. and *M. danieli* n. sp. by presence of white spots below eye; *M. zolitschka* by its larger size (SVL $\geq$ 39.7 mm); and *M. macrotympanum* n. sp. by its smaller adult male size (SVL < 49 mm). Character diagnostics in Tables 3 and 4.

Description of UADBA 39000 (RAX 9368): adult male (SVL = 48 mm) in excellent state of preservation. Measurements are presented in Table S8. In the dorsal view and lateral view, the snout tip is relatively obtuse. Snout tip with 1.35 mm ventral extension beyond the mouth. Head 1.25 times longer than wide. Head length 0.45 times SVL. Canthus rostralis distinct. Loreal with groove. Tympanum diameter 0.98 times eye. Slightly round tympanum with a vivid small notch in the middle of the superior area and in contact with the supratympanic fold in its anterior part and separate in its posterior part, and this supratympanic fold runs towards before reaching the shoulder girdle and upper arm articulation. Internarial distance 0.25 times head width. Tongue ovoid anteriorly and bifid posteriorly. Round nostrils with distinct cutaneous fold and lateral aperture. Eye to nostril distance 1.81 times nostril to snout distance. Forearm length 0.49 times SVL. Hand length (including discs) 0.28 times SVL. Fingers without webbing. Inner and outer metacarpals exist. Fingers without webbing. Relative finger length 1 < 2 < 4 < 3. Digits with large terminal discs (widest part > 1.70 times of basal disc width). Tibiotarsal articulation between nostril and snout tip. Lateral metatarsal separated. Hindlimb 1.76 times SVL. The thigh is the same length as the tibia. Foot including tarsus 0.72 times SVL. The inner metatarsal tubercle is shield-shaped (length 2 mm) at the base of the toe 1. Outer metatarsal tubercle in the small granule. Webbing formula: 1 (0), 2i (1), 2e (0), 3i (1.25), 3e (0.50), 4i (1.75), 4e (1.75), 5 (0.25) and total sum of free phalanges 6.50. Relative toe length is 1 < 2 < 3 < 5 < 4. Lateral surfaces of the body and dorsum with numerous granules. Little and swollen femoral glands elongated, and its pore on the media-distal area was surrounded by many granules, giving it a crater-like pattern. Internally, femoral glands type 3 [2,28]. In preservative, the dorsal surface of the body with blackish brown color. Upper lip with shiny transverse and interrupted band oriented to eyes. The belly and ventral surface of the thigh with yellowish-white spots. Obvious white pigments on boundary surfaces of the thorax and abdomen and making them together an 8-like pattern. Thorax and throat were almost pigmented by whites with some scattered brown spots and the thorax with two brownish parallel dark bands. The inguinal region with whitish L-shaped bed pattern. Dorsal transverse bands are rather indistinct on the hind and forelimbs.

Variation: Morphometric variation is summarized in Table S8. Sexual dimorphism is evident: males have a smaller SVL (39.7–46.8 mm), thicker snout, shorter hand, larger terminal disc, and loreal less elongated. The ratio of td/ed is larger in males than in females (0.62–0.82 vs. 0.51–0.68). Tibiotarsal articulation between nostril and snout tip except UADBAs (19786, 26411, 26412) and AMNH A174650 beyond snout tip. Femoral glands are more swollen in males and smaller in females. Free phalanges on toes vary: toe 1 (0–0.75), external edge of toe 2 (0–0.25), internal edge of toe 3 (1–1.50), external edge of the toe 3 (0–0.25), internal edge of the toe 4 (1.25–1.75), external edge of the toe 4 (1–1.50), and in toe 5 (0–0.25).

Coloration in life: The iris has a golden ring on its outer area with some black spots. The dorsal surface of the body with a dark brown color has more small granules. There is a more or less round yellow spots in the inguinal region for females and a stick-like pattern for males. The ventral surface of the body is pigmented by a white color with yellowish border bands in an 8-like pattern. Hind and forelimbs with alternate dark and shiny brown transverse bands.

Habits: Semi-aquatic rainforest species living close to small streams with rock. Stream depth does not reach 1 m. All specimens were observed during the day and night between 16.00 pm to 21.30 pm. In the daytime, they were observed between 5–10 m far from the bank on the ground and at night 3–5 m far from the river, and roosting on leaves between 50–100 cm in height. UADBA 7125 was seen on rocks in the middle of the river where water is very speedy, maybe ending up there by misfortune. It is rather a ground and tree-dwelling than an aquatic frog.

Distribution: Mid and high elevations of the eastern slope forest of Madagascar (Figure 6, Table S13).

Comments: The only genetic data known for this species are the sequence from the reference specimen AMNH A174653 (RAX 9367) (Table S1). The LSID number is 57C53062-B2DF-41E5-88A3-EFE7B083DA84.

3.4. New Species Descriptions

Based on our molecular and morphological results, we find strong evidence to recognize four species of *Mantidactylus (Ochthomantis)* that correspond to taxa that cannot be assigned to any nominal species (or to names considered as junior synonyms). After examining our new materials, and developing diagnoses for each taxon, we here provide descriptions for each of these new species, such as *Mantidactylus danieli* n. sp. (54 specimens), *M. macrotympanum* n. sp. (5 specimens), *M. olgae* n. sp. (60 specimens), and *M. tavaratra* n. sp. (150 specimens) (Figure 8).

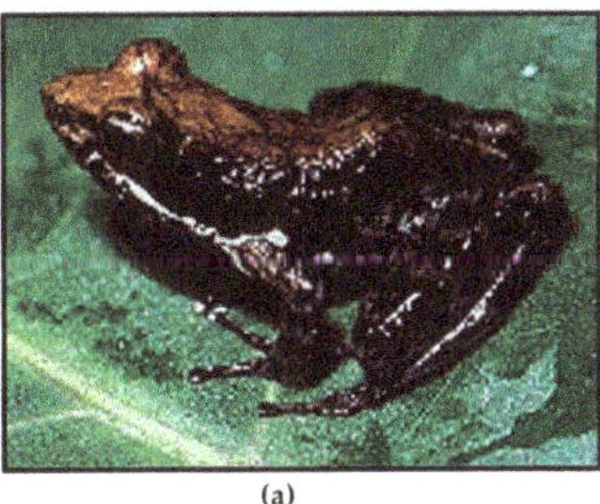

(a) (b)

Figure 8. *Cont.*

(c)

(d)

Figure 8. Photos of the new species of the subgenus *Ochthomantis*: (**a**). *Mantidactylus danieli* n. sp. (CJR), (**b**). *Mantidactylus macrotympanum* n. sp. (CJR), (**c**). *Mantidactylus olgae* n sp. (NR), (**d**). *Mantidactylus tavaratra* n. sp. (CJR).

3.4.1. *Mantidactylus danieli*, New Species

Holotype: AMNH A167590* (RAX 4268) adult female collected 22 February 2002 at Salafaina Forest, District Vohemar, Sava Region, Madagascar, 400 m, 13°26.257′ S 49°43.001′ E by S. Mahaviasy and N. Rabibisoa.

Paratypes: AMNH A167523* (RAX 6595) juvenile collected 2 April 2003 at the relict Irony Forest, Antsohihy District, Sofia Region, Madagascar, 950 m, 14°45.140′ S 48°29.690′ E by S. Mahaviasy, N. Rabibisoa, N. Rakotondrazafy and C. J. Raxworthy; AMNH A167582 (RAX 2999) adult female collected 9 April 2001 at Ramena River, Ambanja District, Diana Region, Madagascar, 750 m, 13°55.071′ S 48°53.179′ E, by S. Mahaviasy, N. Rabibisoa, C. J. Raxworthy, A Razafimanantsoa, and A. Razafimanantsoa; AMNH A 167592 (RAX 4371) adult female, AMNH A 167591 (RAX 4339) and UADBA 19595 (RAX 4372) adult males collected 24–25 February 2002, the same locality as the holotype; AMNH A 167517 (RAX 6110) and UADBA 26359 (RAX 6118) adult males, collected 11 March 2003 at Matsaborimena Trois Lacs, 1550 m, 14°19.859′ S 48°35.240′ E, Bealanana District, Sofia Region by S. Mahaviasy, N. Rakotondrazafy, and N. Rabibisoa; AMNH A181773* (RAX 10204) collected 5 April 2008 at Andramanalana, Andapa district, Sava Region, Madagascar, 850 m, 14°22.351′ S 49° 21.747′ by S. Mahaviasy and N. Rakotondrazafy; AMNH A181731* (RAX 10392) collected 16 April 2008, Tsararano, Anjanaharibe–Sud/Masoala corridor, Analanjirofo Region, Madagascar, 490 m, 14°54.667′ S 49°41.383′ E, by S. Mahaviasy and N. Rakotondrazafy; UADBA 3716 (RAN 39507) adult male, collected 9 March 1992 at Bekolosy Manongarivo, Ambanja District, Diana Region, Madagascar, 1200 m, 14°02.5′ S 48°18′ E by J-B. Ramanamanjato, A. Raselimanana, and C. J. Raxworthy; UADBA 7770 (MRJ 108) juvenile, collected 12 October 1992 at Ampanasatongotra, Marojezy National Park, Andapa District, Sava Region, Madagascar, 600 m, 14°26.2′ S 49°46.5′ E by N. Rabibisoa, J. B. Ramanamanjato, and O. Ramilison; UADBA 8382 (RAX 2737) adult female, collected 9 April 2001: at Ramena River Analabe, Tsaratanana Reserve, Ambanja District, 750 m, 13°55.071′ S 48°53.179′ E by S. Mahaviasy, N. Rabibisoa, C. J. Raxworthy, A Razafimanantsoa, and A. Razafimanantsoa; UADBA 19593 (RAX 3454) adult female, collected 4 December 2006 at Ambolokopatrika, Anjanaharibe–Sud/Marojey Corridor, Andapa District, Madagascar, 880 m, 14°32.302′ S 49°26.243′ E, the same collectors as holotype; UADBA 19594 (RAX 3785), adult male collected 15 December 2001 at Andranomavohely, Andapa District, Sava Region, Madagascar, 800 m, 14°34.165′ S 49°16.568′ E, by S. Mahaviasy, N. Rabibisoa, N. Rakotondrazafy, A. Razafimanantsoa, and A. Razafimanantsoa; UADBA 26361 (RAX 6482) adult female, collected 21 March 2003, Analapakila, Trois Lacs, 1450 m, 14°26.233′ S 48°36.696′ E, District Bealanana, Sofia Region by S. Mahaviasy, N. Rakotondrazafy and N. Rabibisoa; UADBA 26366–26368, 26373 (RD 918, 839–840, 881), collected November 2000 at Ambolokopatrika, 880 m, 14°32.302′ S 49°26.243′ E, Andapa, Sava Region by D. Rakotomalala; UADBA 26371 (RAX 3783) adult male, the same condition as the holotype; UMMZ 212827* (RAN 38186) collected 22 November 1991 at Antomboka River, Montagne d'Ambre, Antsiranana, Diana Region, 1150 m, 12°32.3′ S 49°10′ S by C J Raxworthy, J B Ramanamanjato, and A Raselimanana; UMMZ

212835 (RAN 39291) adult male: collected 24 February 1992 at Ambalafary, Manongarivo Special Reserve, Ambilobe District, Diana Region, Madagascar, 250 m, 14°04′ S 48°17′ E, by C. J. Raxworthy, A. Raselimanana, J. B. Ramanamanjato, A. Razafimanantsoa, and A. Razafimanantsoa; UMMZ 212836* (RAN 39387) collected 2 March 1992, Antsahabe River, Manongarivo Special Reserve, Antsiranana, Diana Region, 1200 m, 14°02.5′ S 48°18′ E by C. J. Raxworthy, J. B. Ramanamanjato, and A. Raselimanana.

Additional specimens examined: See Appendix B.

Diagnosis: A small to medium-sized *Mantidactylus (Ochthomantis)* species (adult SVL male 33–42 mm, female 42–59 mm), with dark brown color on the dorsal surface of the body and often with black spotted granules. The inguinal region with a soft pale yellow streak and may be partly broken up or narrow; no continuous pale stripe along the lower flank; lack of obvious white spot below the eye; moderately developed foot webbing with 1.5–2 free phalanges at the internal edge of toe 4 and WS > 7; tibiotarsal articulation extends beyond nostril; and maximum width of a terminal disc on fingers < 1.70 disc width base. *Mantidactylus danieli* can be distinguished from the following species: *M. mocquardi*, *M. catalai*, *M. olgae* n. sp., and *M. tavaratra* n. sp. by more developed webbing: WS > 7, 1.5–2 free phalanges at the internal edge of toe 4, tibiotarsal articulation beyond nostril, and presence of pale yellow inguinal streak that may be partly broken up or narrow; *M. poissoni* by lack of white spots below eye; *M. femoralis* by a maximum width of the terminal disc on fingers < 1.70 base width and by inguinal pale yellow streak may be partly broken up and narrow; *M. zolitschka* by larger adult SVL > 32 mm; *M. ambreensis* and *M. ambony* by absence of continuous pale stripe along lower flank; and *M. macrotympanum* n. sp. by smaller adult SVL (<60 mm) and smaller male tympanum/eye diameter < 0.94. Tables 5 and 6 summarize the characters' diagnosis for this new species.

Table 5. Morphological measurements of the four new species of the subgenus *Ochthomantis* are described herein. All measurements in mm, F: adult female, M: adult male. For morphological abbreviations see Appendix A.

Size Range	*M. danieli*		*M. marcotympanum*		*M. olgae*		*M. tavaratra*	
	M	F	M	F	M	F	M	F
SVL	42.3–53.0	33.9–51.9	–	59.4–62.1	45.8–51.8	33.6–42.8	42.7–63.3	35.7–48.8
TD	3.5–6.5	3.3–5.1	–	7.9–8.3	2.6–4.4	3.4–5.6	2.6–5.0	3.0–5.3
ED	4.9–9.4	4.0–5.9	–	7.9–8.3	4.9–7.4	4.1–6.3	4.7–7.8	4.2–6.0
EN	2.0–4.7	2.5–4.5	–	4.6–5.3	2.7–5.0	2.4–4.2	3.1–5.4	2.7–4.5
EST	0.4–2.6	1.0–2.0	–	2.0–2.4	2.2–3.0	1.7–2.5	0.8–3.1	1.5–2.4
NS	1.1–3.3	1.1–2.5	–	2.5–3.0	1.7–3.3	2.0–2.9	1.9–4.4	1.7–3.2
NN	3.6–5.0	2.8–4.4	–	4.6–5.7	3.0–5.2	2.6–4.9	3.6–6.6	3.3–4.4
HW	15.5–20.4	11.7–17.8	–	20.3–21.7	15.2–17.1	11.3–14.4	15.3–23.3	11.6–15.8
HL	19.0–25.1	15.5–24.4	–	25.8–26.8	18.7–22.3	14.2–20.0	18.1–28.2	16.2–21.6
HDL	11.0–15.5	9.6–15.6	–	17.3–19.8	13.0–15.9	10.1–12.4	10.8–18.4	10.8–15.6
RC	8.1–11.7	6.2–10.4	–	10.7–12.1	7.5–12.4	5.5_7.0	7.5–12.3	6.2–8.1
FE	22.1–26.1	15.8–28.8	–	29.4–34.8	23.6–27.5	16.5–21.4	23.8–34.3	18.7–24.4
TI	23.4–28.1	18.0–31.6	–	31.2–34.3	22.6–26.7	16.0–20.4	23.1–33.6	18.6–24.4
FT	23.5–29.7	19.6–28.1	–	30.5–32.6	22.4–25.8	16.4–20.6	23.1–34.3	19.6–25.1
TA	11.0–14.5	7.2–12.1	–	13.7–14.3	8.8–12.5	7.1–9.6	10.0–14.6	8.2–12.4
EO	7.5–9.5	5.8–9.7	–	10.1–11.7	6.7–8.8	5.6 7.5	10.0–9.9	6.0–8.4
EM	2.5–3.3	1.7–3.3	–	3.4–4.2	2.3–3.4	1.8–3.0	2.5–4.3	1.8–3.5
EHEAD	8.5–8.9	5.3–8.6	–	9.0–10.2	5.8–9.2	4.7–6.2	6.1–10.0	4.5–7.6
T1	1.8–2.2	0.9–1.9	–	1.9–3.0	1.3–2.4	1.0–2.2	1.4–2.9	1.2–2.2
TO3	6.1–7.7	4.3–7.4	–	8.7–9.4	4.7–6.8	3.8–5.1	5.5–9.9	4.5–7.2
TO5	7.0–8.7	4.7–8.5	–	10.1–10.9	5.9–7.8	4.6–6.1	5.8–10.6	5.0–7.8

Table 6. Qualitative morphological characters and biogeography of the four new species of the subgenus *Ochthomantis* are described herein.

Morphological Characters	*M. danieli*	*M. macrotympanum*	*M. olgae*	*M. tavaratra*
Snout tip very pointed	No	No	Yes	No
Tibiotarsal articulation position	Beyond nostrils	Beyond snout tip	Between eye–nostril	Between eye–nostril
Large tympanum	No	Yes	No	No
Large digit terminal disc	No	Yes	Yes	No
Body with striking granules	Yes	No	Yes (black)	No
Body coloration	Brown with black spots	Brown	Gray	Gray
Throat with two parallel marks	L-like pattern or spot	Bar	Stripe	Stripe
Dorsum with crossbar	No	No	V-like pattern or Y-like pattern	V-like pattern or Y-like pattern
Mouth with whitish band	Yes	Yes	No	Yes
Yellowish or whitish shape in groin area	Yes (band oblique)	White plate-like pattern	No	If present (narrow white band oblique)
Ventral surface coloration	Yellowish	Clean pattern	Clean pattern	With silvery pigments
Dorsal surface with black spot	Yes	No	Yes	Yes
Toe fully webbed	No	Yes	Yes	Slightly
Elevation (m)	350–1580	180–800	600–1700	530–2650
Distribution	North, Sambirano, northeast, northwest	Analabe, Manongarivo	Analabe, Andramanalana	North

Description of holotype: Adult female (SVL 50.6 mm) in excellent state of preservation. Measurements are presented in Table S9. In the dorsal view, the head is longer than the width. Head length 1.18 times width length. Head length 0.41 times SVL. In dorsal view, the snout tip is pointed and rounded lateral, with a 3 mm ventral extension beyond the lower lip. Canthus rostralis evident. Loreal concave. The tympanum and supratympanic fold are distinct from each other, and this supratympanic fold runs towards the front of the upper arm and shoulder girdle articulation. Round tympanum diameter 0.74 times eye diameter. Tongue ovoid anteriorly and bifid posteriorly. Internarial distance 0.21 head width. Round nostril with lateral aperture. Eye–nostril distance 1.27 times nostril–snout distance. Forearm length 0.51 times SVL. Hand length (including discs) 0.26 times SVL. Fingers without webbing. Outer and inner metacarpal with developed tubercles. Relative finger length 1 < 2 <4 < 3.

The terminal is disc relatively large (the widest part is 1.45 width of its basal disc). Tibiotarsal articulation between nostril and snout tip. Hindlimb 2.01 times SVL. Thigh length 0.98 times tibia length. Foot including tarsus 0.92 times SVL. Lateral metatarsal separated. Inner metatarsal tubercle as a skin-like pattern with three-sided (1.35 mm) on base of toe 1. Outer metatarsal tubercle present. Webbing formula: 1 (1), 2i (1), 2 (0.25), 3i (1.75), 3e (0.50), 4i (2), 4e (2), 5 (0.50), and total sum of free phalanges is 9. Relative toe length 1 < 2 < 3 < 5 < 4. The dorsal surface of the thigh, the posterior area above the eye, above the tympanum, and around the femoral glands with granule skin-like patterns. Round femoral glands poorly developed with central pores. Internally, femoral gland type 3 [2], but without small granules in the proximal area, characteristics of male femoral glands [28]. In the preservative, the dorsal surface of the body is dark brown in color and has small darker spots. Upper lip dark brown color except under tympanic region clean pattern. The throat is pale brown in color with two dark parallel bands in an L-shape. The belly is pale brown in color and is almost entirely covered with fine dark brown spots, which decrease in density posteriorly. Brown forelimbs without dark bands. Brown hindlimb with darker transversal bands on its dorsal surface and pale brown with dark-brown spots on its ventral surface. The inguinal region with pale and discontinuous oblique streak.

Variation: Morphometric variation is summarized in Table S9. Sexual dimorphism is obvious in this species: males are smaller sized with relatively larger tympanum than females, smaller SVL than females (33–42 mm vs. 42–59 mm), larger tympanum diameter than females (0.66–1.12 times eye diameter vs. 0.49–0.76 times eye diameter), shorter hindlimb, and larger terminal disc. Free phalanges on the toes vary: toe 1 (0.50–1), internal edge of toe 2 (0.50–1.50), external edge of toe 2 (0–0.75), internal edge of toe 3 (1–1.75), external edge of toe 3 (0–1), internal edge of toe 4 (1.75–2), external edge of toe 4 (1.50–2), and in toe 5 (0–0.50).

Coloration in life: The iris color is golden in the superior area and fades to the lower area of the eye. The dorsum and dorsal surfaces of the head and limbs have a dark brown color and sometimes black spots for females. The upper lip has a white band that is most developed in the posterior area. Between the eyes, there is a darker brown cross-bars in a **V**-like pattern. The ventral surfaces with dark brown spot pigments have a different background color: the belly is creamy or yellowish but the paler pattern is almost dark reddish on its anterior part, two dark reddish parallel stripes may be present on the throat and become narrower in its posterior part and decline on the thorax or is absent (UADBA 8382, 26361, 26366, 26367, 26371, 26373). The inner area of the thigh is completely covered with dark reddish-brown spots. A pale yellow oblique streak in the inguinal region is typically discontinuous with all of the specimens, except for UADBA 19595, 26359, and 26367. Transversal bands, black and darker brown were observed on the dorsal surface of the hindlimb, but indistinctly clear with some dark specimens (e.g., UADBA 26406). One adult female (UADBA 3770) has a pale vertebral line running from the snout to the anal pore.

Habits: Living in rainforests between 350–1580 m elevation and semi-aquatic species; occupying the banks of small streams and rivers. Observed between 10.30–23.00 h, although more active at night and appears to be mostly nocturnal. Unlike other *Ochthomantis* species, males rest mostly on river banks, more rarely on leaves of trees and bushes, and very rarely on rocks. Females are more tree-dwelling than rupicolous. The vertical distribution appears to be different between sexes: males between 20–100 cm, whereas females are between 50–300 cm above the ground. This species appears to prefer rivers to small fast-flowing streams and females were observed mostly in areas of slow-moving water.

Etymology: The specific name danieli is a patronym, i.e., a noun in the genitive case honoring Daniel Rakontondravony for his substantial contributions to the knowledge of the Malagasy fauna.

Distribution: Low- and mid-altitude Morthern and Sambirano Region of Madagascar (180–1200 m), from subhumid to humid forests (Figure 9, Table S13).

Comments: Our morphological description agrees with the *M.* sp. aff. *femoralis* specimen shown by [29] (pp. 178–179) from Tsaratanana. Our molecular analyses group all our *M. danieli* samples (Table S1) with previously published sequences from Montagne d'Ambre AY324818 (FGMV 2002.929) and Manongarivo AY324816 (FGMV 2002.825). *Mantidactylus* cf. *femoralis* samples reported by [10]. FGMV 2002.929 has more recently been referred to as "CCS sp. 42". The LSID number is 3A4FC339-EB42-4262-8B61-0F26A27B7505.

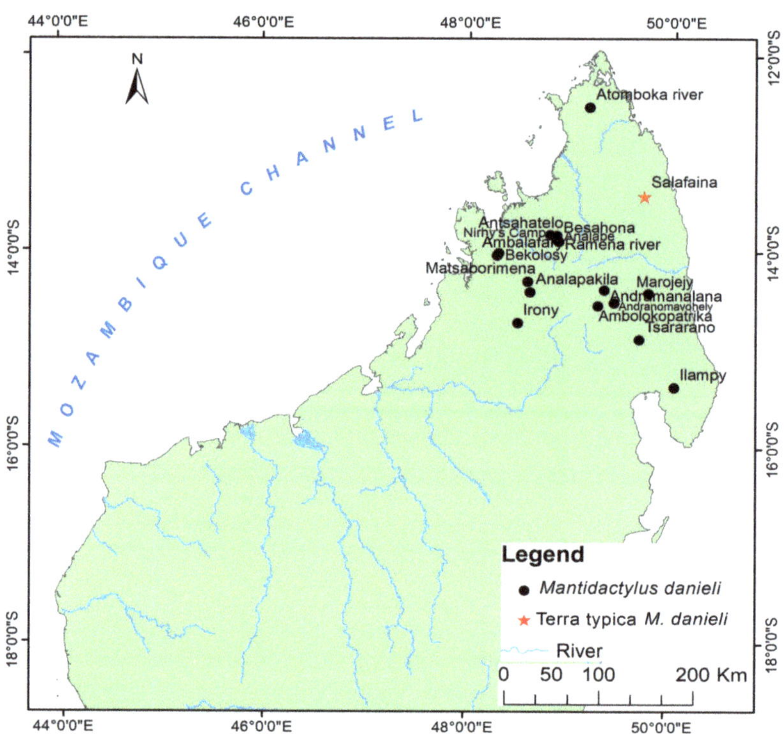

Figure 9. Maps of the distribution of *Mantidactylus danieli* n. sp.

3.4.2. *Mantidactylus macrotympanum* New Species

Holotype: AMNH A 167589* (RAX 2715): adult male at Antsahatelo, Western Slope Tsaratanana Strict Reserve, Ambanja District, Diana Region, Madagascar, 800 m, 13°51.588′ S 48°51.979′ E, 6 April 2001, N. Harilanto, S. Mahaviasy, N. Rabibisoa, C. J. Raxworthy, A. Razafimanantsoa, and A. Razafimanantsoa.

Paratypes: AMNH 167598 (RAX 2603): adult male at Antsahabe, Ramena River, Ambanja District, Diana Region, Madagascar, 180 m, 13°43.272′ S 48°39.304′ E, 4 April 2001, N. Harilanto, S. Mahaviasy, N. Rabibisoa, C. J. Raxworthy, A. Razafimanantsoa, and A. Razafimanantsoa. UMMZ 201416*–201418 (RAN 39170–39171, 39470), UMMZ 213447* (RAN 39125): adult males, Ambalafary, Manongarivo Special Reserve, Ambilobe District, Diana Region, Madagascar, 250 m, 14°04′ S 48°17′ E, 20 February 1992 and 3 March 1992, A. Raselimanana, J. B. Ramanamanjato, C. J. Raxworthy, A. Razafimanantsoa, and A. Razafimanantsoa.

Diagnosis: A large-sized *Mantidactylus (Ochthomantis)* species (adult male SVL 57–62 mm, female unknown but presumably like other *Ochthomantis* species, female larger than males > 62 mm); dorsal surfaces of head and body with dark brown color; body with skin granules especially on its posterior parts; no pale inguinal streak; large tympanum ≥ 7.6 mm; tibiotarsal articulation beyond snout tip; and webbing well-developed and total sum of free phalanges 3.75. *Mantidactylus macrotympanum* can be distinguished from all species in the subgenus *Ochthomantis* by the following characteristics: big tympanum diameter ≥ 7.6 mm, a large adult male SVL ≥ 57 mm, and presence of skin granules on lateral and dorsal surfaces of the body. Tables 5 and 6 summarize the characters' diagnosis for this new species.

Description of holotype: Adult male (SVL 61 mm) in excellent state of preservation. Measurements are presented in Table S10. In dorsal view, the head is longer than wide: head length 1.29 times head width. Head length 0.44 times SVL 0.44. In the dorsal view, the snout tip is pointed, and in the lateral view almost acuminate and curved down with 2.20 mm ventral extension beyond the lower lip. Obvious canthus rostralis and concave loreal making lip well evident. Tympanum and supratympanic fold distinct and supratympanic fold running vertically in its posterior part to one evident granule above the upper arm and shoulder girdle articulation. Round tympanum diameter 0.94 times eye diameter. Tongue ovoid anteriorly and bifid posteriorly. Internarial distance 0.23 times head width. Non-protruding nostril with lateral slanting aperture. Eye–nostril distance 1.86 times nostril–snout tip distance. Forearm length 0.50 times SVL. Hand length (including discs) 0.32 times SVL. Fingers without webbing. Outer and inner metacarpal tubercles flattened and widened. Relative finger lengths 1 < 2 <4 <3. Terminal discs are large (the widest part is 2.20 times of basal disc width). Tibiotarsal articulation largely beyond the snout tip. Hindlimb 1.89 times SVL. Thigh and tibia same length. Foot including tarsus 0.76 times SVL. Lateral metatarsal separated. Outer metatarsal tubercle absent. The inner metatarsal tubercle is not evident at the base of toe 1. Webbing formula 1 (0), 2i (0.5), 2e (0), 3i (1), 3e (0), 4i (1.25), 4e (1), 5 (0) and total sum of free phalanges 3.75. Relative toe length 1 < 2 < 3 < 5 < 4. Flanks with few prominent granules in its superior part and a white spot-like pattern posteriorly. Numerous epidermic granules in the sacral area and finely granules above the eyes. Round femoral glands are not swollen. Internally, femoral gland type 4 [2,28]. In preservative, the dorsal surface of the body is blackish and marbled with some shiny spots. Ventral surfaces of the body and superior lip homogenous pattern with a brownish color. The throat is darker with two parallel bands. Fore and hindlimbs with alternate transversal bands, shiny and dark. The inguinal region has some scattered dark spots.

Variation: Morphometric variation is summarized in Table S10. Species are only known from six male specimens. SVL between 57.9 and 62.1 mm. The ventral surface of the body is heterogeneous according character of the belly: individuals with a clean pattern belly have black spots and those can be either present or not with dark belly specimens. Flanks are shiny brownish backgrounds with white plates in the central area like the holotype, but sometimes white spots are observed for specimens that have black flank backgrounds. Three internal phalanges vary between 0.50–1 for toe 2i and 1–1.50 for toe 4i.

Coloration in life: The iris has a constant black color surrounded by white color in its superior and lower parts. The body is dark brownish. The superior lip has a black crossbar anteriorly. Above the eyes, there is a white line. The belly is darker with a black spot. Thorax has two parallel bands. The vivid brown flanks have a white plate-like. A white oblique line can be present in the inguinal area. Two alternate bands, shiny and dark, are rather distinct on the fore and hindlimbs.

Habits: This is a rainforest species living along calm streams and fast rivers with scattered rocks. Individuals were also found outside the forest, e.g., in plantations at Manongarivo. Individuals were observed between 12.00–20.00 h but they are more nocturnal than diurnal. At night, they roost and rest on branches between 10–200 cm in height and during the day, they rest on the rocks. All the known specimens observed are males.

Etymology: The specific name macrotympanum is composed of the Latin words macro (large) and tympanum (tympanum), referring to the very big tympanum of this species. The name is used as a noun in apposition.

Distribution: Known only from relict forest close to Ambalafary village, the Manongarivo Special Reserve at 250 m, and from the Ramena River (Tsaratanana Reserve) between 180–800 m (Figure 10, Table S13).

Comments: The only genetic data known for this species were obtained from AMNH A167589 (RAX 2715), UMMZ 201416 (RAN 39170), and UMMZ 213447 (RAN 39125) (Table S1). The LSID number is 957B5BBF-17E1-4175-A499-2953919D7F89.

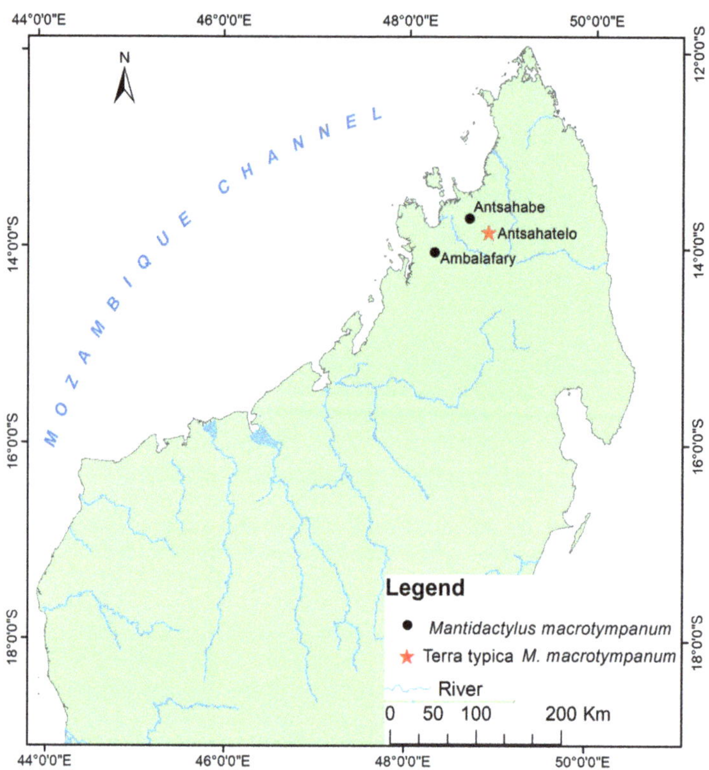

Figure 10. Map of the distribution of *Mantidactylus macrotympanum* n. sp.

3.4.3. *Mantidactylus olgae*, New Species

Holotype: AMNH A167596* (RAX 5195), subadult female collected 19 April 2002 at Sorata pic Forest, Vohemar District, Sava Region, Madagascar, 1700 m, 13°41.147′ S 49°26.511′ E, S. Mahaviasy, N. Rakotondrazafy, N. Rabibisoa, and C.J. Raxworthy.

Paratypes: AMNH A167555 (RAX 2475) adult male, and UADBA 8383, 8388 (RAX 2473, 2472) adult females collected 25 January 2001 at Besahona, Analabe forest, Ambanja District, Diana Region, Madagascar, 700 m, 13°54.372′ S 48°44.785′ E, A. Razafimanantsoa, and A. Razafimanantsoa; AMNH A167556 (RAX 2552) adult female collected 29 January 2001 at Ramena, Analabe forest, Ambanja District, Diana Region, Madagascar, 700 m, 13°42,707′ S 48°34,156′ E, same collectors as previous. AMNH A167557 (RAX 2555) adult male collected 29 January 2001 at Betaindambo, Analabe forest, Ambanja District, Diana Region, Madagascar, 600 m, 13°51.932′ S 48°49.189′ E, same collectors as previous; AMNH A167558 (RAX 2620) adult female collected 4 April 2001 at Ramena, Analabe forest, Ambanja District, Diana Region, Madagascar, 700 m, 13°42.707′ S 48°34.156′ E, N. Harilanto, S. Mahaviasy, N. Rabibisoa, C.J. Raxworthy, A. Razafimanantsoa, and A. Razafimanantsoa; AMNH A167560 (RAX 2655) adult male, and AMNH A167561–167562 (RAX 2662–63) adult females collected 5 April 2001 at Maroamalona, Analabe forest Ambanja District, Diana Region, Madagascar, 600 m, 13°51.023′ S 48°47,902′ E, same collectors as previous. AMNH A167565* (RAX 3203) juvenile, and AMNH A167567–167568 (RAX 3230, 3232) adult males, collected 15–16 April 2001 at Antsaravy Analabe forest, Ambanja District, Diana Region, Madagascar, 1150 m, 13°55.560′ S 48°54.353′ E, N. Harilanto, S. Mahaviasy, N. Rabibisoa, A. Razafimanantsoa, and A. Razafimanantsoa; AMNH A181726* (RAX 10205) collected 5 April 2008 at Andramanalana, Andapa District, Sava Region, Madagascar, 850 m, 14°24′ 24″ S 49°20′13″ E, S. Mahaviasy, and N. Rakotondrazafy; UADBA 8231, 8387, 8389–8390 (RAX 3231, 3228, 3196–3197) adult females, collected 15–16 April 2001 at Antsaravy Analabe forest, Ambanja

District, Diana Region, Madagascar, 1150 m, 13°55.560′ S 48°54.353′ E, N. Harilanto, S. Mahaviasy, N. Rabibisoa, C.J. Raxworthy, A. Razafimanantsoa, and A. Razafimanantsoa; UADBA 8404 (RAX 3312) adult male, collected 19 April 2001 at Antsaravy, Analabe forest, Ambanja District, Diana Region, Madagascar,1150 m, 13°55.560′ S 48°54.353′ E, N. Harilanto, S. Mahaviasy, N. Rabibisoa, A. Razafimanantsoa, and A. Razafimanantsoa.

Diagnosis: A small- to medium-sized *Mantidactylus (Ochthomantis)* species (adult SVL male SVL 34–40 mm, female SVL 45–52 mm) with gray color of the body; dorsal and lateral surfaces of the body with black granules; no pale yellow streak in the inguinal region; no broad pale line or white spots on the lateral surface of the body; upper lip with a pale line; the presence of crossbars in V or Y–like pattern on dorsum; snout tip pointed in lateral view; well-developed foot webbing with 1 free phalange on internal edge of toe 4 (WS < 7); tibiotarsal articulation between eye and nostril; and maximum width of terminal disc on fingers > 1.8 times of basal disc width. Distinguished from all others species by presence of obvious black skin granules both on dorsal and lateral surfaces of body and by the following characters: *M. femoralis* by one free phalange on internal edge of toe 4, no yellow patch on inguinal region, and tibiotarsal articulation between eye and nostril; *M. ambreensis* and *M. ambony* by lacking of broad pale line on lateral surface of body; *M. mocquardi* by gray body color and no white spots on lateral surface of body; *M. zolitschka* by larger male adult SVL > 33 mm and female adult > 44 mm; *M. poissoni* by lack of white spots below eye; *M. catalai* by presence of crossbars in V or Y-like pattern on dorsum; *M. danieli* n. sp. by following characters:webbing more developed (WS < 7), one free phalange at internal edge of toe 4, no pale yellow streak in inguinal region, and tibiotarsal articulation between eye and nostril; *M. macrotympanum* n. sp. by smaller male adult SVL < 60 mm and smaller male tympanum eye diameter < 0.94; and *M. tavaratra* n. sp. by larger terminal disc and maximum width of terminal discs on finger > 1.8 times of basal disc width. Character diagnostics are summarized in Tables 5 and 6.

Description of holotype: Subadult female (SVL = 40.70 mm) in excellent state of preservation. Measurements are presented in Table S11. In the dorsal view, the head is rather longer than the width: the head length is 1.23 times the head width. Head length 0.46 times SVL. In dorsal and lateral view, the snout tip is pointed with evident ventral extension 2.20 mm beyond the lower lip giving it a shell-like pattern. Canthus rostralis not evident. Loreal furrowed. The tympanum was distinct from a well-developed suratympanic fold which ran and reached towards one point before upper arm and shoulder girdle articulation. Tympanum diameter 0.56 times eye diameter. Tongue ovoid anteriorly and bifid posteriorly. Internarial distance 0.21 times head width. Round nostrils with cutaneous fold anterolateral aperture. Eye–nostril distance 1.74 times nostril–snout tip distance. Forearm length 0.55 times SVL. Hand length (including discs) 0.33 times SVL. Fingers without webbing. Outer and inner metacarpal tubercles are poorly developed. Relative fingers length 1 < 2 < 4 < 3. Larger terminal disc, i.e., widest part twice of basal disc width). Tibiotarsal articulation between eye and nostril. Hindlimb 1.98 times SVL. Thigh length 1.07 times tibia length. Foot including tarsus 0.77 times SVL. Lateral metatarsal separated. Inner metatarsal tubercle in half moon-like pattern (1.3 mm length) at base of the toe 1. Outer metatarsal with small granule. Webbing formula: 1 (0), 2i (1), 2e (0), 3i (1), 3e (0), 4i (1.25), 4e (1), 5 (0) and total sum of free phalanges 4.25. Relative toe length 1 < 2 < 3 < 5 < 4. Skin granules on the flank. Round femoral gland poorly developed with a central pore. Internally, femoral gland type 3 [2] but without small granules in its proximal area characteristics of the male femoral gland [28]. In preservatives, coloration is grayish dorsally with a Y-like pattern. Upper lip with white pigments. Throat with two dark parallel bands. Belly, ventral surface of the thigh, and forelimb clean patterns. Hind and forelimbs with gray color and alternate dorsal bands, shiny and dark, crossing them. An inguinal region without a pale inguinal streak. Flank with some black granules.

Variation: Morphometric variation is summarized in Table S11. Sexual dimorphism is evident: males have a smaller SVL than females (34.00–39.55 mm vs. 45.75–51.80 mm), males have a larger tympanum diameter than females (0.80–1.30 times eye diameter in males vs. 0.40–0.69 times eye diameter in females), and larger hindlimb; 90% of specimens have a larger thigh, between 0.91–1.18 times tibia length and the rest < 0.91; free phalanges vary: toe 1 (0–0.50), internal edge of toe 3 (1–1.25), external edge of toe 3 (0–0.50), and internal edge of toe 4 (1–1.50).

Coloration in life: The iris has a golden ring on its outer area. Body coloration is gray except for brown color, AMNH A167560 and UADBA (8391, 8477). The upper lip is white and dirty except for the holotype which is white in color. No dark parallel bands on the throat except for UADBA 8407. All of the specimens have V or Y–like patterns with dark spots on the dorsum except UADBA (8384, 8387, 8389) with X–like patterns. The ventral surface of the body is clean pattern except for juveniles which are dirty and some specimens have a throat with white pigments including the holotype and AMNH (A167557, A167564, A167568–167569, A167571, A167573–167574). The inguinal region has no pale yellow streak. The dorsal hindlimbs have darker brown transversal bands. Some specimens have a thick pale brown vertebral line.

Habits: Low- and mid-altitude rainforest species close to the edge of streams or rivers with rocky bottoms. We observed it between 10.00–22.30 h and mainly nocturnal except for juveniles. It was observed in various habitats: rock, grounds, riverbanks, and edges of water. Generally, females rest on rocks at night and some specimens overhanging on leaves between 20–50 cm above the ground. Males are more tree-dwelling than rupicola and rest on leaves of shrubs, between 10–30 cm in height. Specimens observed at all types of water, but they have a particular preference for streams or fast rivers.

Etymology: The specific name olgae honors the late Olga Ramilijaona, professor at UADBA and supervisor of the Ph.D. thesis of N.R., in recognition of her substantial mentorship.

Distribution: Species inhabits along Ramena River (Tsaratanana Strict Natural Reserve, and Analabe), Andramanalana, and Sorata in the North of Madagascar. The elevation ranges between 600–1700 m (Figure 11, Table S13).

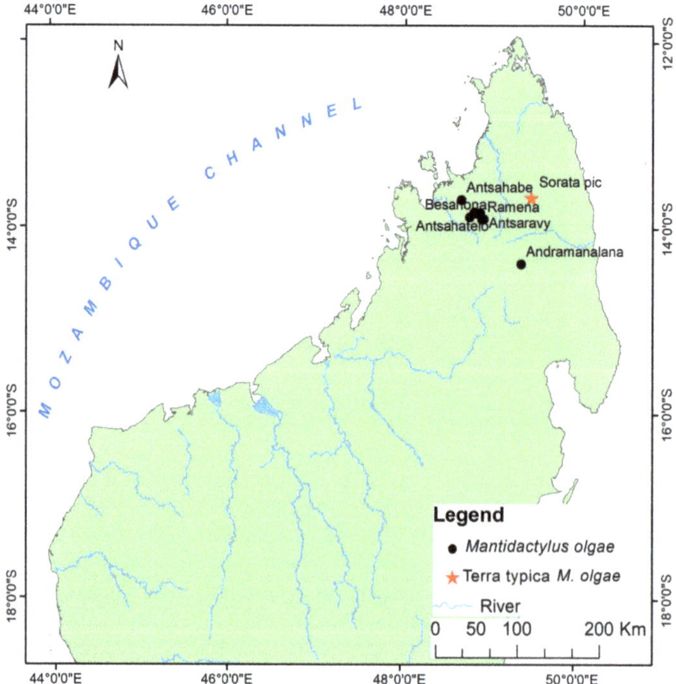

Figure 11. Map of the distribution of *Mantidactylus olgae* n. sp.

Comments: The only genetic data known for this species are presented in Table S1. Although *M. olgae* has been recorded at Andramanalana, no genetic or morphological samples have yet been confirmed from Marojejy, which is a well-sampled site about 40 km to the west. The LSID number is FB2EEFBB-92C7-460E-892A-9B5BBFF65862.

3.4.4. *Mantidactylus tavaratra*, New Species

Holotype: AMNH A 167505* (RAX 5310) adult male collected 22 April 2002 at Sorata, Vohemar District, Sava Region, Madagascar, 980 m, 13°41.986′ S 49°26.687′ E, by S. Mahaviasy, N. Rabibisoa, and C. J. Raxworthy.

Paratypes: AMNH A 167506* (RAX 5400) juvenile, AMNH A167507 (RAX 5403) adult female, and UADBA 26306 (RAX 5599) adult male collected 3 February 2003 at Matsaborimaiky Lake, Tsarananana Strict Natural Reserve, Bealanana, Sofia Region, Madagascar, 1950 m, 14°09.175′ S 48°57.431′ E, by S. Mahaviasy, N. Rabibisoa, N. Rakotondrazafy, and C. J. Raxworthy. AMNH A167510* (RAX 5769) juvenile collected 23 February 2003 at Befosa River, Tsaratanana Strict Natural Reserve, Bealanana District, Sofia Region, Madagascar, 1680 m, 14°10.455′ S 48°56.708′ E, by S. Mahaviasy, N. Rabibisoa, N. Rakotondrazafy, and C. J. Raxworthy. AMNH A167515*–167516* (RAX 6028, 6107) adult males collected 11 March 2003 at Matsaborimena, Trois Lacs, Bealanana District, Sofia Region, Madagascar,1600 m, 14°19.859′ S 48°35.240′ E, by S. Mahaviasy, N. Rabibisoa, and N. Rakotondrazafy. AMNH 167524*–167525* (RAX 6838, 6901) adult males, and AMNH A167529 (RAX 7120), UADBA 26326, 26335, 26394 (RAX 7009, 6903, 6952) adult females collected 13–17 April 2003 at Lohanandroranga River ridge, and Ambatotavaratra River, Corridor Tsaratanana Anjanaharibe–Sud, Bealanana District, Sofia Region, Madagascar, 1430–1650 m, 14°24.076′ S 49°10.253′ E, the same collectors as AMNH A167510. AMNH A167593* (RAX 5000) adult male collected 13 April, 2002 at Sorata, Vohemar District, Sava Region, Madagascar,1330 m, 13°41.147′ S 49°26.511′ E, by S. Mahaviasy, N. Rabibisoa, and C. J. Raxworthy. AMNH A181730* (RAX 10323) juvenile 10 April 2008 at Andramanalana, Andapa, Sava Region, Madagascar, 1300 m, 14°24.400′ S 49°20.21′ E, by S. Mahaviasy, and N. Rakotondrazafy. AMNH A187088*–187089* (RAX 11534–11535), and UADBA RAX 11532*–11533* juveniles collected 6 May 2010 at Marojejy, Andapa District, Sava Region, Madagascar, 1550 m, 14°32.302′ S 49°26.243′ E, by S. Mahaviasy, N. Rakotondrazafy, and C.J. Raxworthy. UADBA 19628, 19634 (RAX 4993, 4999): adult females, and UADBA 19630 (RAX 4995) adult male collected 13 April 2002 at Sorata, Vohemar District, Sava Region, Madagascar, 1300 m, 13°41.147′ S 49°26.511′ E, by S. Mahaviasy, N. Rabibisoa, and C. J. Raxworthy. UADBA 19625, 19631, 26307–26308 (RAX 4951, 4996, 5771, 5778) adult males, and UADBA 26328, 26331 (RAX 5402, 5753) adult females collected 22–23 February 2003 at Befosa River, Tsaratanana Strict Reserve, Bealanana District, Sofia Region, Madagascar, 1600 m, 14°10.455′ S 48°56.708′ E, by S. Mahaviasy, N. Rabibisoa, and N. Rakotondrazafy. UADBA 19650 (RAX 3980) adult female collected 11 February 2002 at Bezavona, Vohémar District, Sava Region, Madagascar, 350 m, 13°31.962′ S 49°51.954′ E, by S. Mahaviasy, N. Rabibisoa, and C. J. Raxworthy. UMMZ 212440* (RAN 39071) juvenile collected 16 February 1992 at Bekolosy, Manongarivo Special Reserve, Ambanja District, Diana Region, Madagascar, 1100 m, 14°02.5′ S 48°18′ E, by A Raselimanana, J. B. Ramanamanjato, and C. J. Raxworthy. UMMZ 212889* (RAN 43366) adult female collected 1 April 1993 at Befosa River, Tsaratanana Natural Reserve, Bealanana District, Sofia Region, Madagascar, 1630 m, 14°10.455′ S 48°56.708′ E, by J. B. Ramanamanjato, A. Raselimanana, C. J. Raxworthy, A. Razafimanantsoa, and A. Razafimanantsoa.

Additional specimens examined: See Appendix B.

Diagnosis: A medium to large-sized *Mantidactylus (Ochthomantis)* species (adult SVL male 35–52 mm, female 42–63 mm); brown dorsum without black granules; lateral and ventral surfaces of body well distinct by different background color, from darker to pale pattern respectively; pale white or yellow thin streak in the inguinal region; upper lip with pale stripe; webbing with 1–1.5 free phalanges at external edge of toe 4 and no free phalange at the external edge of toe 2; tibiotarsal articulation between eye and nostril; and maximum terminal disc width of fingers $\leq$ 1.8 basal disc width. Distinguished from all other species by the following characteristics: *M. ambreensis* and *M. ambony* by lack of evident and white or yellow continuous line on lateral surfaces of head and body; *M. poissoni* by lack of white spots below eye; *M. catalai* by digits with smaller terminal discs (widest part $\leq$ 1.8 of basal disc width), thin pale white or yellow oblique line in inguinal region and white stripe on upper lip; *M. femoralis* and *M. danieli* n. sp. by the presence of 1–1.5 free phalanges at the external edge of toe 4 and tibiotarsal articulation beyond nostril; *M. mocquardi* by the presence of thin pale continuous line on the upper lip, absence of white spots on the flank, and absence of gray body coloration; *M. olgae* n. sp. by flank typically with straight border from dark to pale venter coloration and maximum width of terminal disc $\leq$ 1.80 basal disc width; *M. zolitschka* by larger adult SVL > 35 mm; and *M. macrotympanum* n. sp. by smaller male adult SVL < 60 mm and smaller male tympanum-eye diameter < 0.94. Character diagnostics in Tables 5 and 6 for this new species

Description of holotype: Adult male (SVL = 37 mm) in excellent state of preservation. Measurements are presented in Table S12. In the dorsal view, the head is larger than the width: the head length is 1.34 times the head width. Head length 0.44 times SVL. In the dorsal view, the snout tip is slightly pointed and almost straight in the lateral view with a 1.55 mm ventral extension beyond the lower lip, which makes the snout tip have an obtuse appearance. Canthus rostralis obvious and loreal concave making the jaw well evident. Tympanum and supratympanic fold distinct running almost vertically towards three evident white granules in front of upper arm and shoulder girdle articulation. Round tympanum diameter 0.62 times eye diameter. Tongue ovoid anteriorly and bifid posteriorly. Internarial distance 0.30 times head width. Round nostril with non-protruding lateral aperture. Eye–nostril distance equals nostril–snout distance. Forearm length 0.46 times SVL. Hand length (including discs) 0.30 times SVL. Fingers without webbing. Outer and inner metacarpal tubercles flattened and widened. Relative finger length 1 < 2 < 4 < 3. Fingers without webbing. The terminal disc is relatively large (the widest part is 1.40 times the basal disc width). Tibiotarsal articulation between eye and nostril. Ventral surface of the forearm with three granules on its internal ridge at the base of metacarpal tubercles. Hindlimb 1.90 times SVL. The thigh length I almost equal to the tibia length. Foot including tarsus 0.77 times SVL. Lateral metatarsal separated. Outer metatarsal tubercle indistinct. Inner metatarsal tubercle flattened at base of toe 1. The webbing formula is: 1 (0.25), 2i (1), 2e (0), 3i (1.25), 3e (0.25), 4i (1.5), 4e (1.5), 5 (0), and the total sum of the free phalanges is 5.75. Relative toe length 1 < 2 < 3 < 5 < 4. The dorsal surface of the body with small granules on its edge and on the sacral region. Oblong femoral glands are relatively developed and in their centro-distal area with pores surrounded by numerous granules, giving a crater-like pattern. Internally, femoral gland type 3 [2,28]. In preservative, brown dorsum with some clean pattern or depigmented portions, especially in its posterior region. Upper lip with finely dark brown pigments posteriorly and covering with obvious round pigments anteriorly. The indistinct band runs behind the eyes and crumbles in the anterior half of the dorsum. Throat with two dark short parallel stripes. Ventral surfaces are homogeneous for different parts of the structure: throat and thorax with silver–white pigments and some reticulated dark brown pigments, belly whitish in color with some brown round spots, thigh and forearm no pigments in its part of the area. The inguinal region has fine streaks. The lower portion of the hindlimb with few scattered brown spots. Hindlimb with evident alternate bands, thick dark, and fine shiny. Forelimb with indistinct alternate bands.

Variation: Morphometric variation is summarized in Table S12. Sexual dimorphism is obvious in this species: males have smaller SVL than females (35.65–52.40 mm vs. 42.70–63.25 mm), males have a larger tympanum diameter than females (0.82–1.12 times eye diameter in males vs. 0.43–0.73 times eye diameter in females). Body coloration varies from grayish brown to reddish brown UADBA 26331 (RAX 5753) or light brown UADBA 19628, 26308 (RAX 4993, 5778). In males, half of the specimens (including the holotype) have snout tips $\leq$ 2 mm ventral extension beyond the lower lip and the other half > 2 mm. Throat with obvious dark brown parallel bands, except for some specimens with the ventral surface of the body and clean pattern: e.g., UADBA (26238, 26371, 26394). Throat and thorax with white or white silver coloration, except for UADBA (26238, 26308, 26331, and 26394). Dark dorsal crossbar in V or Y-like pattern absent, except for holotype and paratypes UADBA (19625, 19628, 19630–19631, 26308, 26394). Flanks with obvious black and white granules were observed on paratype UADBA 26391 but did not reach the roughness of *M. olgae* n. sp. Toe 3 is slightly shorter compared to toe 5 in males, whereas it is shorter in females, except UADBA 26326 (RAX 7009). Free phalanges vary: toe 1 (0–0.75), external edge of toe 2 (0–0.75), internal edge of toe 3 (0.25–1.50), external edge of toe 3 (0–0.50), internal edge of toe 4 (1–2), and external edge of toe 4 (1–1.50).

Coloration in life: The iris has a golden ring on its outer area. The body is grayish brown, light, dark brown, or sometimes reddish brown. The upper lip has an evident vivid white band or not. Between the eyes, there is a darker brown cross-bars in a rod-like pattern. The dorsal surface of the body has dark brown cross-bars in a V or Y-like pattern. The throat and thorax are silvery white or white in color. The throat is always with distinct dark brown parallel stripes. The belly has a clean pattern with smaller white spots. The inner side of the thighs has some dark reticulations. The inguinal area has a white or yellowish little line or not, but if it exists the line is narrower in shape. White brownish bands can be present on the outer side of the forelimbs and alternate transversal bands, white and brown on the inner side of the forelimbs. The dorsal surface of the hindlimbs has alternate transversal bands, light and dark browns. The black granules may be present or not on the white flank but concentrated in the posterior part of the dorsum and on the side of the forearm, but they are not very obvious like *M. olgae*.

Habits: This is a rainforest and savannah species living along the riverbank. It inhabits almost all kinds of aquatic habitats: streams, rivers, ponds, and shallow marshes except waterfalls. We observed it between 9.30–23.00 h, but mainly nocturnal except juveniles. Juveniles observed during the day. At night, they rest both on rocks and leaves, but rarely on branches and accidentally on aerial roots and *Pandanus* sp. The females prefer roosting on rocks rather than on leaves, whereas the males are indifferent to those. The vertical distribution varies between the sexes: males between 5–100 cm and females between 5–150 cm above the ground.

Etymology: The specific name tavaratra refers to the Malagasy word for "north". This name is used as a nonlatinized noun in apposition and is given in reference to the known distribution of this species in Northern Madagascar

Distribution: Northern endemic species of Madagascar, elevation varies from 80 to 2450 m elevation (Figure 12, Table S13).

Comments: Our morphological description agrees with the *M.* sp. aff. *mocquardi* shown by [29] (Figure 2, p. 249) from Marojejy. Our molecular analyses group all our *M. tavaratra* samples (Table S1) with those Manongarivo AY324819 (FGMV 2002.824) and Tsaratanana AY324820 (ZSM 643/2001), reported as *M.* cf. *mocquardi* by [10]. FGMV 2002.824 from Manongarivo has more recently been referred to as "UCS sp. 63 Tsaratanana", and this taxon also includes another specimen (FGMV 2001.114) from Tsaratanana [11]. The LSID number is BD3F2F68-88CF-4629-9A91-0196E2CF2F4A.

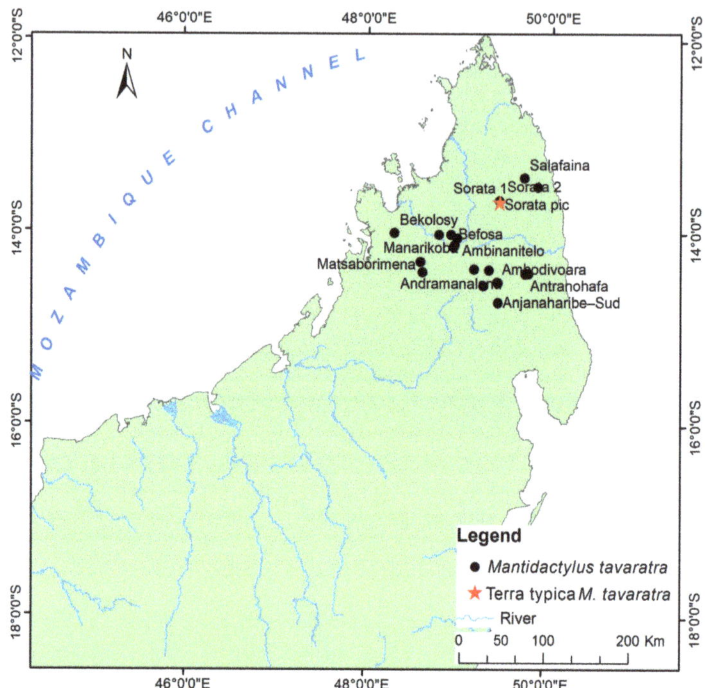

Figure 12. Map of the distribution of *Mantidactylus tavaratra*.

3.5. Identification Key

1.   Lateral head and body with a broad pale line running continuously from the snout tip to the groin area. *M. ambreensis* (SVL, adult male between 33–39 mm and adult female 38–42 mm), and *M. ambony* (SVL, adult male between 30–32 mm and adult female 34–38 mm).

 -   Lateral head and body without a broad pale line running continuously from the snout tip to the groin area. 2.

2.   Adult male SVL < 31 mm, female SVL < 38 mm, reduced foot webbing with ≥2 free phalanges on the internal edge of toe 4 and WS > 9, male tympanum diameter ≤ 7 mm. *M. zolitschka.*

 -   Adult male SVL 31–55 mm, female SVL 38–67 mm, more extensive foot webbing with ≤2 free phalanges on the internal edge of toe 4 and WS < 9, male tympanum diameter < 7 mm. 3

 -   Adult male SVL > 56 mm (female unknown, but likely ≥ 67 mm), more extensive foot webbing with 1.25 free phalanges on the internal edge of toe 4 and WS < 4, male tympanum diameter ≥ 7 mm. *M. macrotympanum.*

3.   Tibiotarsal articulation reaches between nostril and snout tip, or beyond. 4.

 -   Tibiotarsal articulation reaches between eye and nostril. 6.

4.   Upper lip with a large pale spot under the eye (female) or multiple white spots (male), groin area lacks a short pale line or spots. *M. poissoni.*

 -   Upper lip lacks a large pale spot under the eye (female) or multiple white spots (male), groin area with a short pale line or spots. 5.

5.   Large finger pads with disc width > 1.7 × disc base, upper lip with a thin pale continuous line, groin area with a short bold pale line that may be continuous or broken (in life, yellow). *M. femoralis.*

- Smaller finger pads with disc width < 1.7 × disc base, upper lip without a thin pale continuous line, groin area with a short weak pale line that is often broken or even absent. *M. danieli.*

6. Upper lip with a thin pale continuous line. 7.

    - Upper lip without a thin pale continuous line but may have small pale spots. 8.

7. Snout tip sharply angular in lateral view, flanks with an irregular border from the dark to pale venter coloration, dorsal surface of body gray in life (and often in preservation) with black granules, larger finger pads, with disc width > 1.8 × disc base. *M. olgae.*

    - Snout tip not sharply angular in lateral view, flanks typically with a straight border from the dark to pale venter coloration, dorsal surface of body brown without black granules, smaller finger pads, with disc width ≤ 1.8 × disc base. *M. tavaratra.*

8. Snout tip angular in lateral view, body dark brown (or black in preservation) with distinct white spots on flanks, groin area with a few or lacking pale spots, foot webbing with 1.5 free phalanges on the internal edge of toe 4 and WS 4. *M. mocquardi.*

    - Snout tip not angular shape in lateral view, body medium brown (never black in preservation) and lacks white spots on flanks, groin area with a prominent short pale line or spots, foot webbing with 1–1.5 free phalanges on the internal edge of toe 4 and WS 6. *M. catalai.*

4. Discussion

4.1. Taxonomy

According to this study, the subgenus *Ochthomantis* contains 11 species—five species previously considered to be valid, *M. femoralis* [8], *M. mocquardi* [32], *M. ambreensis* [30], *Mantidactylus ambony* [5], and *M. zolitschka* [10]; two resurrected species, *M. catalai* [35] and *M. poissoni* [36]; and four newly described species, *M. olgae* n. sp., *M. danieli* n. sp., *M. tavaratra* n. sp., and *M. macrotympanum* n. sp.

The increased species number is in agreement with the previous study focusing on tadpoles [11] which identified numerous candidates for species in the subgenus, several of which were validated by our study with adult specimens through morphological and cladistic analysis [28].

One species currently included in the subgenus *Ochthomantis*, *M. majori* (Figure 13), should not be included in this subgenus based on unpublished Principal Component Analysis, which places the species completely separate from the other species of the subgenus *Ochthomantis*, and on our cladistic analysis on 96 morphological patterns, habitats, and behavioral characters [28].

The divergence of *M. majori* is evidenced by its different reproductive behavior (parental care, time, and mating behavior). Based on our field observation, we confirm that males of *M. majori* guard eggs and practice parental care (Figure 13a) as [38,39] observed too; probably to protect eggs against fungi [40], predators, and desiccation. Egg deposition takes place on leaves overhanging the water (Figure 13b) in contrast to other *Ochthomantis* species where it takes place at edges of riverbanks either on twigs or on rocks, as we observed in *M. femoralis* and *M. ambreensis* (Figures 7d and 13c). Furthermore, morphological differences in *M. majori* to other *Ochthomantis* are very evident, i.e., it has very developed webbing (toes almost fully webbed, WS = 0–2.5), hindlimbs very short, internal space more developed than interorbital space, presence of the thin line vertebral, toe 3 slightly shorter than toe 5, presence of the thin white line in the side of hindlimbs and especially inguinal region free of obvious pigments (white or yellow spots; line or band patterns). However, our molecular analyses recover *M. majori* as a sister to *M. macrotympanum*, and this new subclade forms a paraphyletic group with the remnant of the subgenus *Ochthomantis* species with moderate support (posterior probability = 0.91); thus, while we do not include *M. majori* in this revision, we recommend that the placement of *M. majori* is further investigated in a study with broader taxonomic and genetic sampling.

Figure 13. (**a**) Photo of a male *Mantidactylus majori*, Andohahela (CJR), (**b**) egg deposition on leaf overhanging the water and male *M. majori* guarding eggs and practicing parental care, Vatovavy (NR), (**c**) Eggs of *Mantidactylus femoralis* deposited on a falling branch next to riverbank, Manasamena (NR), (**d**) Eggs of *M. ambreensis* laid on a rock, Ambohibola (NR).

We confirm *M. flavicrus* as a junior synonym of *M. femoralis*. The analysis of the type of *M. flavicrus*—from the description of [8], the measurement from [10], and an examination made by CJR of the type specimen of *M. flavicrus* based on our identification key—supported that they both represent the same species.

We noticed that two species in our analysis are composed of genetic subclades, *Mantidactylus mocquardi* and *M. ambreensis* (Figure 1); however, the genetic distances between the subclades are small (<3%) but may be further explored in future studies. The first of these species, *M. mocquardi*, has two subclades representative of populations from the north and south of the Sambirano River, respectively. The north of the Sambirano specimens are characterized by their black color, while those from the south of the Sambirano have a more brownish color. These subgroups also differ by additional characters: for the totally blackish specimens, (1) granules almost missing, body smoothing and homogeneous, (2) white spots or silvery in flanks and missing on superior lips, and (3) ventral face marbled with brown reticulate and white pigment except on belly that is a clean pattern in preservative; for the brownish specimens (1) granules evident on flanks, (2) white pigment very conspicuous on lower flanks continuing to lower lips, (3) some dark spots observed on back with different shape: UADBA 12313 (NR 1372), UADBA 26290 (RAX 8036), UADBA 26238 (RAX 8021), UADBA 26240 (RAN 45476), (4) ventral face completely white in living specimen, while thorax and throat with brown spots and marbled with white or silvery pigments in contrasting of clean pattern belly-shaped in preservative. Additional specimens and observation of *M. ambreensis* especially from southern populations, Ramena, and Irony, are needed to better discern morphological differences including body size.

4.2. Sexual Dimorphism and Size

Within the subgenus *Ochthomantis*, sexual dimorphism is evident in SVL and tympanum size. Males have a larger tympanum than females (0.62–1.33 vs. 0.39–0.79). The tympanum senses sound waves and detects the position and direction of animal movement [41]. Here, sexual dimorphism in tympanum size may be related to reproductive communication. The tympanum in males can play a role not only for receiving but also for emitting sound by vibration, which especially makes sense given their small vocal sacs, where their call has low intensity [6,35]. The tympanum in females only functions as a receiver, thus its smaller size. This may suggest that attraction by calls over long distances is less significant for *Ochthomantis* compared to other anuran groups that breed during the warm and rainy season in Madagascar.

Ochthomantis females are characterized by a larger body size than males. This characteristic is a mechanism to reduce the effort spent during egg laying [42] and allows for an increase in the number and size of eggs which in some dissected individuals occupied almost the whole of the ventral cavity [28]. The smaller size of males can be also explained by: 1) a higher energy spending due to their higher activity during reproduction and competition with other males.

Some other morphometric characters may be related to the ecology of each species: (1) longer hindlimbs, and thus longer tibias may be related to improved jumping capabilities and are found in species with more terrestrial behavior (*M. femoralis, and M. danieli*) than in species more specialized to living on rocks and in aquatic environments (*M. olgae, M. mocquardi, M. tavaratra, and M. catalai*), (2) *M. catalai* has pointed snout and flattened head which may be an adaptation to move within in rock fissures, (3) Terminal discs of phalanges are probably used to cling and hang on leaves or rocks. Accordingly, the more tree-dwelling and rock-dwelling species (e.g., *M. olgae, M. tavaratra*) have a larger terminal disc than those most ground-dwelling species such as *M. danieli*, (4) well-developed webbing is observed especially in the more aquatic and rock-dwelling species (*M. olgae, M. mocquardi, M. tavaratra*) compared to the terrestrial and arboreal-dwelling species (*M. femoralis, M. danieli*). Webbing is used for propulsion in the water and increases the surface to grip the rocks, 5). Generally, species with stocky bodies often have long hindlimbs. In addition, according to our observations, they appear to be more dynamic jumpers (e.g., *M. femoralis, M. danieli*).

4.3. Period of Reproduction and Reproduction

The observations made in the north and the region of Moramanga indicate that the reproduction at least of some species of *Ochthomantis* is spread throughout the year. Thus, the statement from [6] that reproduction happens during the cool season is partially true for the species of low altitude and mid-altitude in the central eastern region. The beginning of reproduction probably takes place in April during which the temperature falls to 10 °C and metamorphosis of tadpoles occurs between October and November. While for species from the high altitude in the North (>1400 m asl) as *M. tavaratra*, the reproductive season starts very early when the temperature falls to 8 °C, e.g., at Tsaratanana the tadpoles in advanced stages were observed at the end of February (stage 36). It could be hypothesized that in mid and high altitudes, the activating factor may be the decline of temperature (around 10 °C in Moramanga, and 8 °C in Tsaratanana at high altitude, 2300 m elev. asl). We also noticed that the metamorphosis from the larval stage to the juvenile stages appears to occur fast at high altitudes. This speed may be necessary to avoid predation or other factors [43], e.g., the drying up of streams at high altitudes for instance on the Tsaratanana massif.

4.4. Biogeography, Climate, and Conservation

We noticed that *Ochthomantis* are absent in the West and South of the Island. We propose three hypotheses related to their reproduction habits. The first deals with temperatures, as the factor activating mating in species of the subgenus, appears to be a decline of temperature (between 8–15 °C), and it is clear that these regions of the island have a high temperature all year long. The second is the lack of water during the dry winter season when most of the streams are dried up. Finally, the third possible factor could be the absence of consistently calm water for breeding given that during the summer season, the water velocity can be very quick after torrential rains.

Global climate change has the potential to influence the fate of *Ochthomantis* as high mountain species have difficulty rapidly adapting to new conditions or migrating to areas with suitable conditions [44]. For *Ochthomantis*, climate change may also disturb the period of reproduction, particularly in those species where a decrease in temperature is critical to the induction of reproduction. In addition, changes in water regimes in the high mountains may cause the disappearance of breeding sites due to desiccation, or to a reduction in calm stretches of water with an increased frequency of torrential rains.

Data presented during a dedicated workshop held in Madagascar on 28 January 2009 [45] projected warming across the island, with Southern Madagascar being most affected and the coast and the north showing lower projected temperature increases. Precipitation increase was projected to be centered in the northwest while drying was projected in the East.

Integrating climate change into conservation strategies for amphibians, and specifically for the subgenus *Ochthomantis*, requires more precise distribution model forecasts under alternative climate scenarios, and we anticipate that amphibians occurring at high elevations will show higher sensitivities than those from low elevations [46]. For ground-truthing these scenarios, long-term monitoring efforts are also needed [47].

5. Conclusions

This study resolves the taxonomy of various cryptic species in the subgenus *Ochthomantis* using both morphological characters and molecular data. Our analysis revealed four new species that are formally named and described herein: *M. danieli* n. sp, *M. macrotympanum* n. sp, *M. olgae* n. sp, *M. tavaratra* n. sp. Furthermore, *Mantidactylus catalai* and *M. poissoni* are revalidated from junior synonyms to good species. The subgenus *Ochthomantis* now contains 11 species, but we suspect that some specimens included here may represent additional distinct species. It is hoped that the discovery and taxonomic revision of this new cryptic biodiversity will initiate conservation activities for those species with the most restricted distributions.

Supplementary Materials: The following supporting information can be downloaded at: https://www.mdpi.com/article/0.3390/ani13172800/s1, Table S1: voucher specimens with molecular data; Tables S2–S12: species measurement data; Table S13: species collecting localities.

Author Contributions: Conceptualization by N.H.C.R. and C.J.R.; methodology, N.H.C.R. and C.J.R.; software, N.H.C.R. and R.S.W.; validation, C.J.R.; molecular and phylogenetic analysis R.S.W.; formal analysis, N.H.C.R. and C.J R.; investigation, N.H.C.R. and C.J.R.; resources, N.H.C.R. and C.J.R.; data curation, N.H.C.R. and C.J.R.; writing—original draft preparation, N.H.C.R.; writing—review and editing, C.J.R. and R.S.W.; visualization, N.H.C.R.; supervision, C.J.R.; project administration, C.J.R.; funding acquisition, C.J.R. All authors have read and agreed to the published version of the manuscript.

Funding: This research was supported by the National Science Foundation (DEB 1257610, 0641023, 0423286, 9984496, 9625873, 9322600, BSR 9024505 to C.J.R) and AMNH fellowship program to N.H.C.R.

Institutional Review Board Statement: The research was conducted in Protected Areas (MNP and NAP) and outside of PAs of Madagascar, with research permits from Ministère de l'Environnement et de Développement Durable, MEDD (previous name Ministère de l'Environnement, de l'Écologie et des Forêts, MEEF). During our fieldwork, we followed data collection procedures according to the national legislation regarding permits and international regulations regarding animal welfare. The type specimens used are from AMNH, BMNH, MNHN, and UADBA. The information reported here is the results from our data, and the findings are not reported anywhere else.

Data Availability Statement: All holotypes are housed at AMNH and paratypes are shared between AMNH and UADBA according to research permits. Specimens used for molecular analyses are also stored in both institutions. Molecular sequences reported here are available in GENBANK.

Acknowledgments: Field studies in Madagascar were made possible due to the assistance of the Ministère des Eaux et Forêts, Madagascar National Parks, and in particular by the Université d'Antananarivo, Département de Biologie Animale for administrative assistance, and granting access to their zoological collection. MNHN and BM for access to holotype specimens, and UMMZ for sending us specimens. Research and fieldwork support were provided by AMNH. We thank the many people who have contributed to this research program, especially those who participated in the fieldwork, collecting data and specimens, including: N. Harilanto, S. D. Mahaviasy, R. A. Nussbaum, F. Rabemananjara, J. Rafanomezantsoa, D. Rakotomalala, A. Rakotondrazafy, J.B. Ramanamanjato, O. Ramilison, M. Randriambahiniarime, J. Randrianirina, F. Ranjanaharisoa, A.P. Raselimanana, P. Razafimahatratra, A. Razafimanantsoa, A. Razafimanantsoa, local guides, and the Conservation Agents of the Reserves we visited. Maps were drawn with help from B. Randriamahatantsoa.

Conflicts of Interest: We declare no conflicts of interest. The funders had no role in the design of the study; in the collection, analyses, or interpretation of data; in the writing of the manuscript; or in the decision to publish the results.

Appendix A

Table A1. Explanation of the range size abbreviations.

Range Sizes	Explanation
DB	Terminal disc width, measured on dorsal view, ratio between broadest part to its basal part
ED	Eye diameter, measured horizontally between peripheral orbit
EHEAD	Head height, measured in lateral view thickness of head in tympanum region
EM	Snout height, measured in lateral view thickness of snout in nostril region
EN	Eye–nostril distance; measured as shortest distance between the center of nostril and anterior border of eye
EE	Prefrontal width, measured in dorsal view area in front of marginal orbits
EST	Snout tip extension, measured as ventrally extent of snout tip, from the shortest distance between margin of upper lip and to snout tip
FE	Thigh length, measured in dorsal view from center of knee to margin of cloaca
FT	Foot length without tarsus, measured in ventral view from base of metatarsal tubercles to tip of toe 4
HD	Hand length, measured between point of carpe and radius-ulna articulation to tip of finger n°3
HDW	Head width, measured in dorsal view, distance between tympanums
HL	Head length, measured in lateral view, from snout tip to mouth corner
NN	Internostril distance, measuring dorsally between center of each nostril
NS	Nostril-snout tip distance, measured as shortest distance between center of nostril to snout tip
RC	Radius-ulna length, measured in latero-external view as distance between base of inner metatarsal granule and center of elbow
SVL	Snout–vent length, measured in dorsal view from snout tip to cloaca
TA	Tarsus length, measured from center of tibiotarsal articulation to base of inner metatarsal tubercle
TD	Tympanum diameter, measured horizontally lateral surface of tympanum
TI	Tibia length, measured in latero-external view from center of knee to middle of tibiotarsal articulation
T1	Outer metatarsal tubercle length along toe 1, measured in lateral view from its base to its top
TO3	Toe 3 length, measured in dorsal view from base of proximal phalange to tip of toe 3
TO5	Toe 5 length, measured in dorsal view from base of proximal phalange to tip of toe 5

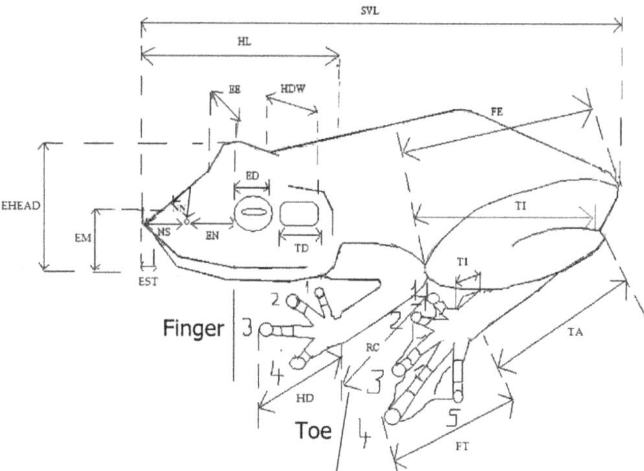

Figure A1. Showing the acronyms of the measurements of *Ochthomantis* species. The numbers 1–5 indicate the serial number from the toe 1 to toe 5 and from the finger 1 to finger 4; ED: Eye diameter, measured horizontally between peripheral orbit; EE: Prefrontal width, measured in dorsal view area in front of marginal orbits; EHEAD: Head height, measured in lateral view thickness of head in tympanum region; EM: Snout height, measured in lateral view thickness of snout in nostril region; EN: Eye–nostril distance; measured as shortest distance between the center of nostril and anterior border of eye; EST: Snout tip extension, measured as ventrally extent of snout tip, from the shortest distance between margin of upper lip and to snout tip; FE: Thigh length, measured in dorsal view from center of knee to margin of cloaca; FT: Foot length without tarsus, measured in ventral view from base of metatarsal tubercles to tip of toe 4; HD: Hand length, measured between point of carpe and radius-ulna articulation to tip of finger n°3; HDW: Head width, measured in dorsal view, distance between tympanums; HL: Head length, measured in lateral view, from snout tip to mouth corner; NS: Nostril-snout tip distance, measured as shortest distance between center of nostril to snout tip; RC: Radius-ulna length, measured in latero-external view as distance between base of inner metatarsal granule and center of elbow; TA: Tarsus length, measured from center of tibiotarsal articulation to base of inner metatarsal tubercle; TI: Tibia length, measured in latero-external view from center of knee to middle of tibiotarsal articulation; T1: Outer metatarsal tubercle length along toe 1, measured in lateral view from its base to its top; TO3: Toe 3 length, measured in dorsal view from base of proximal phalange to tip of toe 3; TO5: Toe 5 length, measured in dorsal view from base of proximal phalange to tip of toe 5.

Appendix B

Additional specimens examined

- *Mantidactylus femoralis*

Madagascar: Ambinany, Tolagnaro District, Anosy Region UADBA 4519; Ambohimanana, Tolongoina District, Vatovavy-Fitovinany Region UADBA 26397; Ambohitantely forest, Ankazobe District, Analamanga Region UADBA 26400, 39003; Amorimbato (Kalambatritra), Ivohibe District, Ihorombe Region UADBA 9772–9776, UADBA 9778; Ampasimekiany Pass, Tolagnaro District, Anosy Region UADBA 1422; Analamay (Ambatovy), Moramanga District, Alaotra–Mangoro Region UADBA 9251–9252, UADBA 9254; Andohahela, Tolagnaro District, Anosy Region UADBA 4515; Andriatantely, Brickaville District, Atsinanana Region UADBA 11974–11976; Andranomanaponga, Moramanga District, Alaotra–Mangoro Region UADBA19663–19665; Ankeniheny, Moramanga District, Alaotra–Mangoro Region UADBA 6280–81; Beambihy River/Ampanasantongotra, Marojejy National Park, Andapa District, Sava Region UADBA 7778; Canyon de Singe of Isalo, Ihosy District, Ihorombe Region UADBA 26284, 26401; Didy, Ambatondrazaka distrit,

Alaotra-Mangoro Region UADBA 7278, 11886; Manatantely, Tolagnaro District, Anosy Region UADBA 1409–1415, 1417–1418, 7517; Manantenina River, Close to Marojejy National Park, Andapa District, Sava Region UADBA 3720 Mangerivola, Brickaville District, Atsinanana Region UADBA 19691, 19693–19694; Mantadia, Moramanga District, Atsinanana Region UADBA 7241, 11887; Rangovalo ridge, Zahamena National Park, Fenoarivo-Est District, Analanjirofo Region UADBA 26398; Ranohira, Ihosy District, Ihorombe Region UADBA 6878; Sahaberiana, Moramanga District, Alaotra-Mangoro Region UADBA 11890; Sahanomanana River (Fiherenana), Moramanga District, Alaotra-Mangoro Region UADBA 7515–7516; Sandranantitra, Bickaville District, Atsinanana Region UADBA 12306, 12308; Tsinjoarivo, Distict of Ambatolampy, Vakinankaratra Region UADBA 8525, 8527, 9046, 26395–96; Vohidratiana (Vohimena), Tolongoina District, Vatovavy-Fitovinany Region UADBA 19696; Volontsagana River of Zahamena, District of Ambatondrazaka, Alaotra-Mangoro Region UADBA 26375, 26379, 26399;

- *Mantidactylus mocquardi*

Madagascar: Ambatolaidama, Masoala National Park, Maroantsetra District, Analanjirofo Region PBZT/FN 7761; Ambatovaky Special Reserve, Soanierana Ivongo District, Analanjirofo Region UADBA 17861, 19685; Ambohitsitondroina, Masoala National Park, Maroantsetra District, Analanjirofo Region PBZT/RJS 314, 342; Ambolokopatrika, Anjanaharibe–Sud/Marojezy Corridor, Andapa District, Sava Region PBZT 6641, 6659, 6836, 7042, PBZT/FN 6639, UADBA 19596, 19606, 19609–19610, 19646, 26304–26305; Ampanasana Ankolony, Marojezy National Park, Andapa District, Sava Region AMNH A157113; Ampanasatongotra, Marojezy National Park, Andapa District, Sava Region UADBA 7765–7767, 7772–7773; Ampanatovana, Lakato, Moramanga District, Alaotra-Magoro Region UADBA 26111; Analamay Ambatovy, Moramanga District, Aloatra-Mangoro Region UADBA 9253; Andasimbazaha, Marojezy National Park, Andapa District, Sava Region: AMNH A157114, A157116–15717, 157120, UADBA 7774, 7777; Andranomanaponga, Andasibe–Mantadia National Park, Moramanga District, Alaotra Mangoro Region UADBA 19680–81; Andranomavohely, Anjanaharibe–Sud/Marojezy corridor, Andapa District, Sava Region UADBA 19648–19649, 26234, 26302–26303; Anjanaharibe–Sud Special Reserve, Andapa District, Sava Region UADBA 8119, 8121, 8123–8124; Ankeniheny/Lakato, Moramanga District, Alaotra-Mangoro Region UADBA 6282–6283; Ankitsika montane, Vohemar District, Sava Region UADBA 19654–55; Betampona Strict Natural Reserve, Toamasina District, Atsinanana Region UADBA 26232–33, 26236, 26239, 26273–26275, 26278–26279, 26285–26294, 26296–26297; Bezavona montane, Vohemar District, Sava Region AMNH A167584, 167586, UADBA 19597, 19600–19602, 19650, 19652–19653; Didy, Corridor Ankeniheny-Zahamena, District Ambatondrazaka, Alaotra-Mangoro Region UADBA 10112–10119; Ilampy, Masoala National Park, Maroantsetra District, Analanjirofo Region PBZT/FAZC 10107–10108; Iofa, Ankeniheny-Zahamena corridor, Moramanga District, Alaotra-Mangoro Region UADBA 10070, 11889; Manantenina River, Marojejy National Park, Andapa District, Sava Region UADBA 3717, 3719; Mangerivola Special Reserve, Toamasina District, Atsinanana Region UADBA 17862, 19684, 19686–19690, 26300; Mantadia Analamazaotra, Andasibe-Mantadia National Park, Moramanga District, Alaotra-Mangoro Region UADBA 7312–14, 11891–11892; Rangovalo ridge/Volontsagana River, Zahamena National Park, Fenoarovo Atsinanana District, Analanjirofo Region UADBA 26225, 26228–26229, 26235, 26240–26242, 26298; Salafaina montane, Vohemar District, Sava Region UADBA 19916; Sandranantitra, Toamasina II District, Atsinanana Region UADBA 12307, 12310–12311, 12314–12315; Sorata, Vohemar District, Sava Region UADBA 19603–05, 19640; Tsararano, Anjanaharive-Sud/Masoala corridor, Maroantsetra District, Sava Region PBZT/FN 6053–6054, 6057, 6110, 6212;

- *Mantidactylus danieli* n. sp.

Madagascar: AMNH A50222; Ambolokopatrika, Betaolana–Anjanaharibe–Sud corridor, Andapa District, Sava Region UADBA 19593, 26113, 26359–60, 26362, 26364–26365, 26369–26371; Ampanasana Ankolony, Marojezy National Park, Andapa District, Sava Region AMNH A157111, 157113, UADBA 7765–7767, 7769, 7772–7773; Antsahatelo Analabe,

Ambanja District, Diana Region AMNH A167580; Atomboka River, Montagne d'Ambre National Park, Antsiranana District, Diana Region UADBA 3710–3713; Besahona Analabe, Ambanja District, Diana Region AMNH A167177; Ilampy, Masoala National Park, Maroantsetra District, Analanjirofo Region PBZT/FAZC 10053; Irony relict forest, Antsohihy District, Sofia Region UADBA 26372; Manongarivo Special Reserve, Ambanja District, Diana Region UADBA 12140; Nirhy's cascade Camp Analabe, Ambanja District, Diana Region UADBA 167578–16779; Tsararano, Anjanaharibe–Sud/Masoala corridor, Maroantsetra District, Analanjirofo Region PBZT/FN 6051, PBZT/FN 6055.

- *Mantidactylus olgae* n. sp.

Madagascar: Antsaravy valley, Tsaratanana Natural Integral Reserve, Ambanja District, Diana Region: AMNH A167565–16766, 167569–16776, UADBA 8375, 8381; 8383–8384, 8387, 8392, 8402–8404, 8407, 8409, 8477, 12773; Besahona, Tsaratanana Natural Integral Reserve, Ambanja District, Diana Region AMNH A167484, 167554; Nirhy's cascade, Analabe, Ambanja District, Diana Region AMNH A167563; Ramena River, Analabe, Ambanja District, Diana Region AMNH A167564; UADBA 8386, 8391, 8409, 12771.

- *Mantidactylus tavaratra* n. sp.

Madagascar: Ambatotavatra, Bealanana District, Sofia Region AMNH A167527–16728, UADBA 26320, 26323, 26332, 26334, 26337, 26340–26341, 26346, 26349–26350, 26357, 26393; Ambinanitelo, Tsaratanana Natural Integral Reserve, Bealanana District, Sofia Region UADBA 26356; Ambodivoara village, Andapa District, Sava Region UADBA 19682; Ambolokopatrika, Anjanaharibe–Sud-Marojejy corridor, Andapa District, Sava Region: UADBA 19607, 26345; Ampanasana Ankolony, Marojezy National Park, Andapa District, Sava Region UADBA 15773; Andasimbazaha camp, Marojezy National Park, Andapa District, Sava Region AMNH A157115, UADBA 3724, 3840; Analapakila/Trois Lacs de Bemanevika, Bealanana District, Sofia Region UADBA 26318, 26333, 26344, 26348; Andranomavohely, Anjanaharibe–Sud/Marojejy/Betaolana Corridor, Andapa District, Sava Region UADBA 19608, 26230–26231; Andranomifototra River, Marojezy National Park, Andapa District, Sava Region; AMNH A157123–15726, UADBA 15771–15772, 15774–15778; Anjanaharibe–Sud Special Reserve, Andapa District, Sava Region UADBA 8117, 8120; Antranohafa camp, Marojezy National Park, Andapa District, Sava Region AMNH A157121–15722; Antsahamanitsy, Tsaratanana Natural Integral Reserve, Bealanana, Sofia Region PBZT/RJS 102, 119; 122; Befosa River, Tsaratanana Natural Integral Reserve, Bealanana District, Sofia Region: AMNH A167511–16714, 167533, UADBA 26108, 26309, 26313, 26317, 26321, 26330, 26342–26343, 26351–26352; Bekolosy, Manongarivo Special Reserve, Ambanja, Diana Region UADBA 12154–12158; Lohan'Andranomifototra, Marojezy National Park, Andapa District, Sava Region AMNH A 157127–15729; Manarikoba, Tsaratanana Natural Integral Reserve, Bealanana, Sofia Region PBZT/FAZC 11068, 11112; Matsaborimaiky, Tsaratanana Natural Integral Reserve, Bealanana District, Sofia Region AMNH A167508–167509, UADBA 26109, 26116, 26306, 26312, 26328, 26336; Matsaborimena, "Les trois Lacs" of Bemananevika, Bealanana District, Sofia Region AMNH A167518–20, UADBA 26310, 26315–26316, 26322, 26324–26325, 26327, 26338, 26354, 26358; Salafaina montane, Vohemar District, Sava Region UADBA 19612–15; Sorata, Vohemar District, Sava Region: AMNH A167595–16797, UADBA 19617–19625, 19628–19629, 19631–19632, 19634–19639, 19642–19644, 19656; Trail to Tsaratanana pic, Tsaratanana Natural Integral Reserve, Bealanana, Sofia Region UMMZ 212833.

Appendix C

DNA sequences and alignment of the subgenus *Ochthomantis* species
> *Mantidactylus peraccae*_AY454391
TCAACGGCCGCGGTACCCTAACCGCGCGAAGGTAGCATAATCACTTGTTCTTTA
AATGGGGACTAGTATCAACGGCATCACGAGGGTCCTACTGTCTCCTCCCTCCAATCA
ATGAAACTGATCTCTCCGTGAAGAAGCGGAGATTTAACTATAAGACGAGAAGACCC
CATGGAGCTTTAAACCCTACTGTAACACG-TCGGTTTTAGGTTGGGGGTGACCGCGGA

GCACAAATCAACCTCCACGATGAACGGGACTAACCCCTTATCTAAGAAGCACACCT
CTAAGAATTAGCACACTAACATA-AAATGATCC-GACATTCGATCAACGAACCAAGT
TACCCTGGGGATAACAGCGCAATCCACTTCAAGAGCCCCTATCCACAAGTGGGTTT
ACGACCTCGATGTTGGATCAGGGTATCCTAGTGGTGCAGCANNNNNNNNNNNNNNN
NNNNNNNNNNNNNNNNNNNNNNN

> *Mantidactylus_macrotympanum* n. sp._RAN39170
NTAACGGCCACGGTA-CCTAACCGTGCGAAGGTAGCACAATCACTTGTTCTTTA
AATGAGGACTAGTATCAACGGCATCACGAGGGTTCTACTGTCTCCTCTTTCTAATCA
GTGAAACTAATCTCCCCGTGAAGAAGCGCGGATAATTATATAAGACGAGAAGACC-
CCA TGGAACTTTAAACCCAATTACAAGAATATTGGTTTTAGGTTGGGGCGACCAAG-
GAGAAT AATATAACCTCCATGATAGACG-GAATAATACCTTATCTAAGAGTCACATCT
CTAAGAATTAACACACTAATAT--AAATGACCC-ATTACTTGATCAACGAACCAA GT-
TACCCTGGGGATAACAGCGCAATCCATTTTAAGAGCCCATATCGACAAATGGGTTTA
CGACCTCGATGTTGGATCAGGATACCCAAGTGGTGTAGAAGTTACTAAAGGTTCGTT
TGTTCAACGATTAAAATCCT

> *M_macrotympanum* n. sp._RAN39125
NTAACGGCCACGGTACCCTAACCGTGCGAAGGTAGCACAATCACTTGTTCTTTA
AATGAGGACTAGTATCAACGGCATCACGAGGGTTCTACTGTCTCCTCTTTCTAATCA
GTGAAACTAATCTCCCCGTGAAGAAGCGCGGATAATTATATAAGACGAGAAGACC-
CCA TGGAACTTTAAACCCAATTACAAGAATATTGGTTTTAGGTTGGGGCGACCAAG-
GAG AATAATATAACCTCCATGATAGACG-GAATAATACCTTATCTAAGAGTCACATCT
CTAAGAATTAACACACTAATAT--AAATGACCC-ATTACTTGATCAACGAACCA AGT-
TACCCTGGGGATAACAGCGCAATCCATTTTAAGAGCCCATATCGACAAATGGGTTTA
CGACCTCGATGTTGGATCAGGATACCCAAGTGGTGTAGAAGTTACTAAAGGTTCGTT
TGTTCAACGATTANNNNNNN

> *M_macrotympanum_*n. sp. RAX2715
NTAACGGCCACGGTACCCTAACCGTGCGAAGGTAGCACAATCACTTGTTCTTTA
AATGAGGACTAGTATCAACGGCATCACGAGGGTTCTACTGTCTCCTCTTTCTAATCA
GTGAAACTAATCTCCCCGTGAAGAAGCGCGGATAATTATATAAGACGAGAAGACC-
CCA TGGAACTTTAAACCCAATTACAAGAATATTGGTTTTAGGTTGGGGCGACCAAG
GAGAATAATATAACCTCCATGATAGACG-GAATAATACCTTATCTAAGAGTCACATCT
CTAAGAATTAACACACTAATAT--AAATGACCC-ATTACTTGATCAACGAACCAAGTTA
CCCTGGGGATAACAGCGCAATCCATTTTAAGAGCCCATATCGACAAATGGGTTTAC
GACCTCGATGTTGGATCAGGATACCCAAGTGGTGTANNNNNNNNNNNNNNNNNNNN
NNNNNNNNNNNNNNNNNNNNNNN

> *M_majori_*RAX8907
NNNNNNNNNNNNNNNNNNNNNNNNNNNNNNNNNNNNNNNNNNNNNNNNNNNNNNNN
NNNNNNNNNNNNNNNNNNNNNNNNNNNNNNCGGCATCACGAGGGTTCTACTGTCTC
CTCTTTCTAATCAGCGAAATTGATCTCCCCGTGAAGAAGCGGGGATATTAATATAAG
ACGAGAAGACCCCATGGAGCTTCAAACCCAATTACAGTGTAATTAGTTTTCGGTTGG
GGCGACCACGGAGTACAATCAAACCTCCATGATGAACG-GGATTACCCCTTATCTAA
GAACAACAGCTCTAAGAAATAGAACACTAACATA-AAATGACCC—TTATATGATCA
ACGAACCAAGTTACCCTGGGGATAACAGCGCAATCCATTTCAAGAGCCCCTATCGA
CAAATGGGTTTACGACCTCGATGTTGGATCAGGATATCCTAGTGGTGCAGCAGCTAC
TAAGGGTTCGTTTGTTCAACGATTAAAATCCT

> *M_cowanii_*MK447691
NNNNNGGCCTCGGTATCCTGACCGCGCGAAGGTAGCATAATCACTTGTTCTTT
AAATGGGGACTTGTATCAACGGCATCACGAGGGTTGCACTGTCTCCTTTTTCTAATC
AGTGAAACTAATCTCCCCGTGAAGAAGCGAGGATATTACTATAAGACGAAAAGACC
CCATGGAGCTTTAAACCCAACTATAATATGCTTGGTTTTAGGTTGGGGCGACCATGG
AGTAAAAAATAACCTCCACGATGAATGGGCTTACCCCCTTATCTAAGAGCCACACCT
CTAAGAATTAAAATATTAATGTA-AAATGATCC-TTTCT-TGATCAACGAACCAAGTTA
CCCTGGGGATAACAGCGCAATCCATTTCAAGAGCCCCTATCGACAAATGGGTTTAC

GACCTCGATGTTGGATCAGGATATCCTAGTGGTGCAACAACTACTAATGGTTCGTTT
GTTCAACGATTAAAATCCT
> *M_lugubris_AY341710*

TCAACGGCCACGGTACCCTGACCGCGCGAAGGTAGCATAATCACTTGTTCTTTA
AATAAGGACTCGTATCAACGGCATCACGAGGGTTATACTGTCTCCTTTTTCTAATCAG
TGAAACTAATCTCCCCGTGAAGAAGCAGGGATACTACTATAAGACGAGAAGACCCC
ATGGAGCTTTAAACCCAACTACAGTATGCTTGGTTTTAGGTTGGGGCGACCACGGAG
TAAAAACCAACCTCCATGATGAACG-GACCTACNCCTTATCTAAGAGCCACTCCTCT
AAGAATTAGTACACTAACATA-AAATGATCC-TTTCT-TGATCAACGAACCAAGTTACC
CTGGGGATAACAGCGCAATCCATTTCAAGAGCCCATATCGACAAATGGGTTTACGA
CCTCGATGTTGGATCAGGATATCCCAGTGGTGTAGCAGCTACTAATGGTTCGTTTGTT
CAACGATTAAAATCCT
> *M_guttulatus_FJ559237*

TCAACGGCCACGGTACCCTAACCGTGCGAAGGTAGCATAATCACTTGTTCTTTA
AATGGGGACTTGTATCAACGGCACCACGAGGGCTCTACTGTCTCCTTTCTCTAATCA
GTGAAACTAATCTCCCCGTGAAGAAGCGGGGATACCACCATAAGACGAGAAGACC
CCATGGAGCTTTAAACCCAACTACAGTACACT--------------GTGACCGCGGAGCAAAA
TCAAACCTCCACAATGAACGAGAATAATCTCTTATCTAAGAGCCACACCTCTAAGA
ATTAGCACACTAACATA-AAATGATCCGATAAT-CGATCAACGAACCAAGTTACCCTG
GGGATAACAGCGCAATCCATTTCAAGAGCCCCTATCGACAAATGGGTTTACGACCT
CGATGTTGGATCAGGATATCCCAGTGGTGCAGCAGCTACTAAAGGTTCGTTTGTTCA
ACGATTAAAATCCT
> *M. femoralis* HQ610916

NNNNNNNNNNNNNNNNNNNNNNNNNNNNNNNNNNNNNNNNNNNNNNNNNNNNN
NNNNNNNNNNNNNNNNNNNNNNNNNNNNNNNNNNNNNNNNNNNNNNNNNNNTGT
CTCCTTTCTCTAATCAGTGAAACTAATCTCCCCGTGAAGAAGCGGGGATAATACTAT
AAGACGAGAAGACCCCATGGAGCTTTAAACCCAATTATAATATACTTGGTTTTAGGT
TGGGGCGACCACGGAGTAAAACCAAACCTCCATGATGTACG-GAACAACCCCTCAT
CCCAGAGCTACGACTCTAAGAATTAGCACACTACCATA-AAATGACCC-ATTAT-TGAT
CAACGAACCAAGTTACCCTGGGGATAACAGCGCAATCCATTTTNNNNNNNNNNNN
NNNNNNNNNNNNNNNNNNNNNNNNNNNNNNNNNNNNNNNNNNNNNNNNNNNNN
NNNNNNNNNNNNNNNNNNNNNNNNNNNNNNNNNNNNNNNNNNNNNNNNNNNNN
> *M_femoralis_RAX2703*

NNNNCGGCCACGGTACCCTAACCGTGCGAAGGTAGCATAATCACTTGTTCTTT
AAATGAGGACTCGTATCAACGGCATCACGAGGGTTCTACTGTCTCCTTTCTCTAATC
AGTGAAACTAATCTCCCCGTGAAGAAGCGTGGATAATATTATTAGACGAGAAGACC
CCATGGAGCTTTAAACCCAACTATAATATGCTTGGTTTTAGGTTGGGGCGACCACGG
AGTAAAACCAAACCTCCATGATGTACG-GAACAACCCCTTATCTAAGAGCCACAGC
TCTAAGAATTAGCACACTATCATA-AAATGACCC-ATTAT-TGATCAACGAACCAAGTT
ACCCTGGGGATAACAGCGCAATCCATTTTAAGAGCCCCTATCGACAAATGGGTTTAC
GACCTCGATGTTGGATCAGGATATCCCAGTGGTGCAGCAACTACTAAAGGTTCGTTT
GTTCAACGATTAAAATCCT
> *M. femoralis* AY843698

TTAACGGCCACGGTACCCTAACCGTGCGAAGGTAGCATAATCACTTGTTCTTTA
AATGAGGACTCGTATCAACGGCATCACGAGGGTTCTACTGTCTCCTTTCTCTAATCA
GTGAAACTAATCTCCCCGTGAAGAAGCGTGGATAATATTATTAGACGAGAAGACCC
CATGGAGCTTTAAACCCAACTATAATATGCTTGGTTTTAGGTTGGGGCGACCACGGA
GTAAAACCAAACCTCCATGATGTACG-GAACAACCCCTTATCTAAGAGCCACAGCTC
TAAGAATTAGCACACTATCATA-AAATGACCC-ATTAT-TGATCAACGAACCAAGTTAC
CCTGGGGATAACAGCGCAATCCATTTTAAGAGCCCCTATCGACAAATGGGTTTACGA
CCTCGATGTTGGATCAGGATATCCCAGTGGTGCAGCAACTACTAAAGGTTCGTTTGT
TCAACGATTAAAATCCT
> *M_femoralis_RAX2761*

NNNNNNNNNNNNNNNNNNNNNNNNNNNNNNNNNNNNGTAGCATAATCACTTGTT
CTTTAAATGAGGACTCGTATCAACGGCATCACGAGGGTTCTACTGTCTCCTTTCTCTA
ATCAGTGAAACTAATCTCCCCGTGAAGAAGCGTGGATAATATTATTAGACGAGAAG
ACCCCATGGAGCTTTAAACCCAACTATAATATGCTTGGTTTTAGGTTGGGGCGACCA
CGGAGTAAAACCAAACCTCCATGATGTACG-GAACAACCCCTTATCTAAGAGCCAC
AGCTCTAAGAATTAGCACACTATCATA-AAATGACCC-ATTAT-TGATCAACGAACCA
AGTTACCCTGGGGATAACAGCGCAATCCATTTTAAGAGCCCCTATCGACAAATGGGT
TTACGACCTCGATGTTGGATCAGGATATCCCAGTGGTGCAGCAACTACTAAAGGTTC
GTTTGTTCAACGATTAAAATCCT

> *M. femoralis* HQ610845

NNNACGGCCACGGTACCCTAACCGTGCGAAGGTAGCATAATCACTTGTTCTTT
AAATGGGGACTCGTATCAACGGCATCACGAGGGTTTTACTGTCTCCTTTCTCTAATC
AGTGAAACTAATCTCCCCGTGAAGAAGCGAGGATAACACTATAAGACGAGAAGAC
CCCATGGAGCTTTAAACCCAATTATAATATTCTTGGTTTTAGGTTGGGGCGACCACGG
AGTAAAACCAAACCTCCATGATGTACG-GAGCAACCCCTTATCTAAGAGCCACAACT
CTAAGAATTAGCACACTATCATA-AAATGACCC-ATTAT-TGATCAACGAACCAAGTTA
CCCTGGGGATAACAGCGCAATCCACTTTAAGAGCCCCTATCGACAAGTGGGTTTAC
GACCTCGATGTTGGATCAGGATACCCCAGTGGTGCAGCAGCTACTAAAGGTTCGTTT
GTTCAACGATCAAAATCCT

> *M_femoralis_RAX6345*

NNNNCGGCCACGGTACCCTAACCGTGCGAAGGTAGCATAATCACTTGTTCTTT
AAATGGGGACTCGTATCAACGGCATCACGAGGGTTTTACTGTCTCCTTTCTCTAATCA
GTGAAACTAATCTCCCCGTGAAGAAGCGAGGATAACACTATAAGACGAGAAGACC
CCATGGAGCTTTAAACCCAATTATAATATTCTTGGTTTTAGGTTGGGGCGACCACGGA
GTAAAACCAAACCTCCATGATGTACG-GAGCAACCCCTTATCTAAGAGCTACAACTC
TAAGAATTAGCACACTATCATA-AAATGACCC-ATTAT-TGATCAACGAACCAAGTTAC
CCTGGGGATAACAGCGCAATCCACTTTAAGAGCCCCTATCGACAAGTGGGTTTACG
ACCTCGATGTTGGATCAGGATACCCCAGTGGTGCAGCAGCTACTAAAGGTTCGTTTG
TTCAACGATCAAAATCCT

> *M_femoralis_RAX8133*

TTAACGGCCACGGTACCCTAACCGTGCGAAGGTAGCATAATCACTTGTTCTTTA
AATGGGGACTTGTATCAACGGCATCACGAGGGTTCTACTGTCTCCTTTTTCTAATCAG
TGAAACTAATCTCCCCATGAAGAAGCGGGGATAACACTATAAGACGAGAAGACCC
CATGGAGCTTTAAACCCAATTATAATATACTTGGTTTTAGGTTGGGGCGACCACGGA
GTAAAACCAAACCTCCATGATGTACG-GAACAACCCCTTATCTAAGAGCCACAGCTC
TAAGAATTAGCACACTATCATA-AAATGACCC-ATTAT-TGATCAACGAACCAAGTTAC
CCTGGGGATAACAGCGCAATCCATTTTAAGAGCCCCTATCGACAAATGGGTTTACGA
CCTCGATGTTGGATCAGGATATCCCAGTGGTGCAGCAACTACTAAAGGTTCGTTTGT
TCAACGATCAAAATCCT

> *M_femoralis_RAX7523*

NNNNNNNNNNNNNNNNNNNNNNNNNNNNNNNNNNNNNNNNNNNNNNNNNNNNN
NNNNNNNNNNNNNNNNNNNNNNNNNNNNNNNNNNNNNNNNNNGGTTCTACTGTCT
CCTTTATCTAATCAGTGAAACTAATCTCCCCGTGAAGAAGCGAGGATAACACTATAA
GACGAGAAGACCCCATGGAGCTTTAAACCCAATTATAATATACTTGGTTTTAGGTTG
GGGCGACCACGGAGTAAAACTAAACCTCCATGATGTACG-GAACAACCCCTTATCTA
AGAGCCACAGCTCTAAGAATTAGCACACTATCATA-AAATGACCC-ATTAT-TGATCAA
CGAACCAAGTTACCCTGGGGATAACAGCGCAATCCATTTTAAGAGCCCCTATCGAC
AAATGGGTTTACGACCTCGATGTTGGATCAGGATATCCCAGTGGTGCAGCAACTAC
TAAAGGTTCGTTTGTTCAACGATCAAAATCCT

> *M. femoralis* HM364726

TTAACGGCCACGGTACCCTAACCGTGCGAAGGTAGCATAATCACTTGTTCTTTA
AATGGGGACTTGTATCAACGGCATCACGAGGGTTCTACTGTCTCCTTTATCTAATCAG
TGAAACTAATCTCCCCGTGAAGAAGCGAGGATAACACTATAAGACGAGAAGACCC
CATGGAGCTTTAAACCCAATTATAATATACTTGGTTTTAGGTTGGGGCGACCACGGA

GTAAAACTAAACCTCCATGATGTACG-GAACAACCCCTTATCTAAGAGCCACAGCTC
TAAGAATTAGCACACTATCATA-AAATGACCC-ATTAT-TGATCAACGAACCAAGTTAC
CCTGGGGATAACAGCGCAATCCATTTTAAGAGCCCCTATCGACAAATGGGTTTACGA
CCTCGATGTTGGATCAGGATATCCCAGTGGTGCAGCAACTACTAAAGGTTCGTTTGT
TCAACGATCAAAATCCT

> *M. femoralis* HQ610917

NNNNNNNNNNNNNNNNNNNNNNNNNNNNNNNNNNNNNNNNNNNNNNNNNNNNN
NNNNNNNNNNNNNNNNNNNNNNNNNNNNNNNNNNNNNNNNNNNNNNNNNNNNNN
NNNNNNNNTATCTAATCAGTGAAACTAATCTCCCCGTGAAGAAGCGAGGATAACA
CTATAAGACGAGAAGACCCCATGGAGCTTTAAACCCAAATATAATATACTTGGTTTT
AGGTTGGGGCGACCACGGAGTAAAACTAAACCTCCATGATGTACG-GAACAACCCC
TTATCTAAGAGCCACAGCTCTAAGAATTAGCACACTATCATA-AAATGACCC-ATTAT-
TGATCAACGAACCAAGTTACCCTGGGGATAACAGCGCAATCCATTTTAAGAGCCCC
NNNNNNNNNNNNNNNNNNNNNNNNNNNNNNNNNNNNNNNNNNNNNNNNNNNNN
NNNNNNNNNNNNNNNNNNNNNNNNNNNNNNNNNNNNNNNNNNNNNNNNNNNNNN

> *M. femoralis* HQ610907

NNNNCGGCCACGGTACCCTAACCGTGCGAAGGTCGCATAATCACTTGTTCTTT
AAATGGGGACTCGTATCAACGGCATCACGAGGGTTTTACTGTCTCCTTTCTCTAATC
AGTGAAACTAATCTCCCCGTGAAGAAGCGAGGATAACACTATAAGACGAGAAGAC
CCCATGGAGCTTTAAACCCAATTATAATATACTTGGTTTTAGGTTGGGGCGACCACG
GAGTAAAACCAAACCTCCATGATGTACG-GAACAACCCCTTATCTAAGAGCCACAG
CTCTAAGAATTAGCACACTATCATA-AAATGACCC-ATTAT-TGATCAACGAACCAAG
TTACCCTGGGGATAACAGCGCAATCCATTTTAAGAGCCCCTATCGACAAATGGGTTT
ACGACCTCGATGTTGGATCAGGATATCCCAGTGGTGTAGCAACTACTAAAGGTTCGT
TTGTTCAACGATCAAAATCCT

> *M. femoralis* HQ610911

NNNNCGGCCACGGTACCCTAACCGTGCGAAGGTAGCATAATCACTTGTTCTTT
AAATGGGGACTCGTATCAACGGCATCACGAGGGTTTTACTGTCTCCTTTCTCTAATC
AGTGAAACTAATCTCCCCGTGAAGAAGCGAGGATAACACTATAAGACGAGAAGAC
CCCATGGAGCTTTAAACCCAATTATAATATACTTGGTTTTAGGTTGGGGCGACCACG
GAGTAAAACCAAACCTCCATGATGTACG-GAACAACCCCTTATCTAAGAGCCACAG
CTCTAAGAATTAGCACACTATCATA-AAATGACCC-ATTAT-TGATCAACGAACCAAGT
TACCCTGGGGATAACAGCGCAATCCATTTTAAGAGCCCCTATCGACAAATGGGTTTA
CGACCTCGATGTTGGATCAGGATATCCCAGTGGTGTAGCAACTACTAAAGGTTCGTT
TGTTCAACGATCAAAATCCT

> *M_femoralis_RAX9498*

NNNNNNNNNNNNNNNNNNNNNNNNNNNNGCGAAGGTAGCATAATCACTTGTT
CTTTAAATGGGGACTCGTATCAACGGCATCACGAGGGTTCTACTGTCTCCTTTCTCTA
ATCAGTGAAACTAATCTCCCCGTGAAGAAGCGAGGATAACACTATAAGACGAGAA
GACCCCATGGAGCTTTAAACCCAATTATAATATACTTGGTTTTAGGTTGGGGCGACC
ACGGAGTAAAACCAAACCTCCATGATGTACG-GAACAACCCCTTATCTAAGAGCCA
CAGCTCTAAGAACTAGCACACTATCATA-AAATGACCC-ATTAT-TGATCAACGAACC
AAGTTACCCTGGGGATAACAGCGCAATCCATTTTAAGAGCCCCTATCGACAAATGG
GTTTACGACCTCGATGTTGGATCAGGATATCCCAGTGGTGCAGCAACTACTAAAGGT
TCGTTTGTTCAACGATCAAAATCCT

> *M_femoralis_RAX7196*

NNNNNNNNNNNNNNNNNNNNNNNNNCCGTGCGAAGGTAGCATAATCACTTGTTC
TTTAAATGGGGACTCGTATCAACGGCATCACGAGGGTTCTACTGTCTCCTTTCTCTAA
TCAGTGAAACTAATCTCCCCGTGAAGAAGCGAGGATAACACTATAAGACGAGAAG
ACCCCATGGAGCTTTAAACCCAATTATAATATACTTGGTTTTAGGTTGGGGCGACCA
CGGAGTAAAACCAAACCTCCATGATGTACG-GAACAACCCCTTATCTAAGAGCCAC
AGCTCTAAGAATTAGCACACTATCATA-AAATGACCC-ATTAT-TGATCAACGAACCA
AGTTACCCTGGGGATAACAGCGCAATCCATTTTAAGAGCCCCTATCGACAAATGGG

TTTACGACCTCGATGTTGGATCAGGATATCCCAGTGGTGCAGCAACTACTAAAGGTT
CGTTTGTTCAACGATCAAAATCCT
> *M. femoralis* AY324815
TTAACGGCCACGGTACCCTAACCGTGCGAAGGTAGCATAATCACTTGTTCTTTA
AATGGGGACTCGTATCAACGGCATCACGAGGGTTCTACTGTCTCCTTTCTCTAATCA
GTGAAACTAATCTCCCCGTGAAGAAGCGAGGATAACACTATAAGACGAGAAGACC
CCATGGAGCTTTAAACCCAATTATAATATACTTGGTTTTAGGTTGGGGCGACCACGG
AGTAAAACCAAACCTCCATGATGTACG-GAACAACCCCTTATCTAAGAGCCACAGCT
CTAAGAATTAGCACACTATCATA-AAATGACCC-ATTAT-TGATCAACGAACCAAGTTA
CCCTGGGGATAACAGCGCAATCCATTTTAAGAGCCCCTATCGACAAATGGGTTTACG
ACCTCGATGTTGGATCAGGATATCCCAGTGGTGCAGCAACTACTAAAGGTTCGTTTG
TTCAACGATCAAAATCCT
> *M. femoralis* GU975183
TTAACGGCCACGGTACCCTAACCGTGCGAAGGTAGCATAATCACTTGTTCTTTA
AATGGGGACTCGTATCAACGGCATCACGAGGGTTCTACTGTCTCCTTTCTCTAATCA
GTGAAACTAATCTCCCCGTGAAGAAGCGAGGATAACACTATAAGACGAGAAGACC
CCATGGAGCTTTAAACCCAATTATAATATACTTGGTTTTAGGTTGGGGCGACCACGG
AGTAAAACCAAACCTCCATGATGTACG-GAACAACCCCTTATCTAAGAGCCACAGCT
CTAAGAATTAGCACACTATCATA-AAATGACCC-ATTAT-TGATCAACGAACCAAGTTA
CCCTGGGGATAACAGCGCAATCCATTTTAAGAGCCCCTATCGACAAATGGGTTTAC
GACCTCGATGTTGGATCAGGATATCCCAGTGGTGCAGCAACTACTAAAGGTTCGTT
TGTTCAACGATCAAAATCCT
> *M. femoralis* AY324817
TTAACGGCCACGGTACCCTAACCGTGCGAAGGTAGCATAATCACTTGTTCTTTA
AATGGGGACTCGTATCAACGGCATCACGAGGGTTCTACTGTCTCCTTTCTCTAATCA
GTGAAACTAATCTCCCCGTGAAGAAGCGAGGATAACACTATAAGACGAGAAGACC
CCATGGAGCTTTAAACCCAATTATAATATACTTGGTTTTAGGTTGGGGCGACCACGG
AGTAAAACCAAACCTCCATGATGTACG-GAACAACCCCTTATCTAAGAGCCACAGCT
CTAAGAATTAGCACACTATCATA-AAATGACCC-ATTAT-TGATCAACGAACCAAGTTA
CCCTGGGGATAACAGCGCAATCCATTTTAAGAGCCCCTATCGACAAATGGGTTTACG
ACCTCGATGTTGGATCAGGATATCCCAGTGGTGCAGCAACTACTAAAGGTTCGTTTG
TTCAACGATCAAAATCCT
> *M. femoralis*_RAX10901
TTAACGGCCACGGTACCCTAACCGTGCGAAGGTAGCATAATCACTTGTTCTTTA
AATGGGGACTCGTATCAACGGCATCACGAGGGTTCTACTGTCTCCTTTCTCTAATCA
GTGAAACTAATCTCCCCGTGAAGAAGCGAGGATAACACTATAAGACGAGAAGACC
CCATGGAGCTTTAAACCCAATTATAATATACTTGGTTTTAGGTTGGGGCGACCACGG
AGTAAAACCAAACCTCCATGATGTACG-GAACAACCCCTTATCTAAGAGCCACAGCT
CTAAGAATTAGCACACTATCATA-AAATGACCC-ATTAT-TGATCAACGAACCAAGTTA
CCCTGGGGATAACAGCGCAATCCATTTTAAGAGCCCCTATCGACAAATGGGTTTACG
ACCTCGATGTTGGATCAGGATATCCCAGTGGTGCAGCAACTACTAAAGGTTCGTTTG
TTCAACGATCAAAATCCT
> *M. femoralis* JF903917
TTAACGGCCACGGTACCCTAACCGTGCGAAGGTAGCATAATCACTTGTTCTTTA
AATGGGGACTCGTATCAACGGCATCACGAGGGTTCTACTGTCTCCTTTCTCTAATCA
GTGAAACTAATCTCCCCGTGAAGAAGCGAGGATAACACTATAAGACGAGAAGACC
CCATGGAGCTTTAAACCCAATTATAATATACTTGGTTTTAGGTTGGGGCGACCACGG
AGTAAAACCAAACCTCCATGATGTACG-GAACAACCCCTTATCTAAGAGCCACAGCT
CTAAGAATTAGCACACTATCATA-AAATGACCC-ATTAT-TGATCAACGAACCAAGTTA
CCCTGGGGATAACAGCGCAATCCATTTTAAGAGCCCCTATCGACAAATGGGTTTAC
GACCTCGATGTTGGATCAGGATATCCCAGTGGTGCAGCAACTACTAAAGGTTCGTTT
GTTCAACGATCAAAANNNN
> *M. femoralis* HQ610847

TTAACGGCCACGGTACCCTAACCGTGCGAAGGTAGCATAATCACTTGTTCTTTA
AATGGGGACTCGTATCAACGGCATCACGAGGGTTCTACTGTCTCCTTTCTCTAATCA
GTGAAACTAATCTCCCCGTGAAGAAGCGAGGATAACACTATAAGACGAGAAGACC
CCATGGAGCTTTAAACCCAATTATAATATACTTGGTTTTAGGTTGGGGCGACCACGG
AGTAAAACCAAACCTCCATGATGTACG-GAACAACCCCTTATCTAAGAGCCACAGCT
CTAAGAATTAGCACACTATCATA-AAATGACCC-ATTAT-TGATCAACGAACCAAGTTA
CCCTGGNNNNNNNNNNNNNNNNNNNNNNNNNNNNNNNNNNNNNNNNNNNNNNNN
NNNNNNNNNNNNNNNNNNNNNNNNNNNNNNNNNNNNNNNNNNNNNNNNNNNNNN
NNNNNNNNNNNNNNNNNNNNNNNNNNNNNNNNNNNNNN

> *M. femoralis* HQ610918

NNNNNNNNNNNNNNNNNNNNNNNNNNNNNNNNNNNNNNNNNNNNNNNNNNNNNN
NNNNNNNNNNNNNNNNNNNNNNNNNNNNNNNNNNNNNNNNNNNNNNNNNNNNTGT
CTCCTCTCTCTAATCAGTGAAACTAATCTCCCCGTGAAGAAGCGAGGATAATACTAT
AAGACGAGAAGACCCCATGGAGCTTTAAACCCAATTATAATATACTTGGTTTTAGGT
TGGGGCGACCACGGAGTAAAACCAAACCTCCATGATGTACG-GAACAACCCCTTAT
CCAAGAGCCACAGCTCTAAGAATTAGCACACTATCAAA-AAATGACCC-ATTAT-TGA
TCAACGAACCAAGTTACCCTGGGGATAACAGCGCAATCCATTTTAAGAGCCCCTAT
CGACNNNNNNNNNNNNNNNNNNNNNNNNNNNNNNNNNNNNNNNNNNNNNNNNNNN
NNNNNNNNNNNNNNNNNNNNNNNNNNNNNNNNNNNNNNNNNNNNNNNNNNNNN

> *M. femoralis* AY848151

NNNNCGGCCACGGTACCCTAACCGTGCGAAGGTAGCATAATCACTTGTTCTTT
AAATGGGGACTCGTATCAACGGCATCACGAGGGTTCTACTGTCTCCTTTCTCTAATC
AGTGAAACTAATCTCCCCGTGAAGAAGCGAGGATAACACTATAAGACGAGAAGAC
CCCATGGAGCTTTAAACCCAATTATAATATACTTGGTTTTAGGTTGGGGCGACCACG
GAGTAAAACCAAACCTCCATGATGTACG-GAACAACCCCTTATCCAAGAGCCACAG
CTCTAAGAATTAGCACACTATCAAA-AAATGACCC-ATTAT-TGATCAACGAACCAAG
TTACCCTGGGGATAACAGCGCAATCCATTTTAAGAGCCCCTATCGACAAATGGGTTT
ACGACCTCGATGTTGGATCAGGATATCCCAGTGGTGCAGCAACTACTAAAGGTTCG
TTTGTTCAACGATCAAAATCCT

> *M. femoralis* AY848138

NNNNCGGCCACGGTACCCTAACCGTGCGAAGGTAGCATAATCACTTGTTCTTT
AAATGGGGACTCGTATCAACGGCATCACGAGGGTTCTACTGTCTCCTTTCTCTAATC
AGTGAAACTAATCTCCCCGTGAAGAAGCGAGGATAATACTATAAGACGAGAAGAC
CCCATGGAGCTTTAAACCCAATTATAATATACTTGGTTTTAGGTTGGGGCGACCACG
GAGTAAAACCAAACCTCCATGATGTACG-GAACAACCCCTTATCCAAGAGCCACAG
CTCTAAGAATTAGCACACTATCAAA-AAATGACCC-ATTAT-TGATCAACGAACCAAG
TTACCCTGGGGATAACAGCGCAATCCATTTTAAGAGCCCCTATCGACAAATGGGTTT
ACGACCTCGATGTTGGATCAGGATATCCCAGTGGTGCAGCAACTACTAAAGGTTCG
TTTGTTCAACGATCAAAATCCT

> *M_femoralis_RAX10606*

NNNNNNNNNNNNNNNNNNNNNNNNNNNNNNNNNNNNNNNNNNNNNCATAATCACTTGT
TCTTTAAATGGGGACTCGTATCAACGGCATCACGAGGGTTCCACTGTCTCCTTTCTCT
AATCAGTGAAACTAATCTCCCCGTGAAGAAGCGGGGATAATACTATAAGACGAGA
AGACCCCATGGAGCTTTAAACCCAATTATAATATACTTGGTTTTAGGTTGGGGCGAC
CACGGAGTAAAACCAAACCTCCATGATGTACG-GAACAACCCCTCATCTAAGAGCT
ACAACTCTAAGAATTAGCACACTATCATA-AAACGACCC-ATTAT-TGATCAACGAAC
CAAGTTACCCTGGGGATAACAGCGCAATCCATTTTAAGAGCCCCTATCGACAAATG
GGTTTACGACCTCGATGTTGGATCAGGATACCCCAGTGGTGCAGCAACTACTAAAG
GTTCGTTTGTTCAACGATCAAAATCCT

> *M_femoralis_RAN35289*

TTAACGGCCACGGTACCCTAACCGTGCGAAGGTAGCATAATCACTTGTTCTTTA
AATGGGGACTCGTATCAACGGCATCACGAGGGTTCCACTGTCTCCTTTCTCTAATCA
GTGAAACTAATCTCCCCGTGAAGAAGCGGGGATAATACTATAAGACGAGAAGACCC
CATGGAGCTTTAAACCCAATTATAATATACTTGGTTTTAGGTTGGGGCGACCACGGA

GTAAAACCAAACCTCCATGATGTACG-GAACAACCCCTCATCTAAGAGCTACAACTC
TAAGAATTAGCACACTATCATA-AAACGACCC-ATTAT-TGATCAACGAACCAAGTTA
CCCTGGGGATAACAGCGCAATCCATTTTAAGAGCCCCTATCGACAAATGGGTTTAC
GACCTCGATGTTGGATCAGGATACCCCAGTGGTGCAGCAACTACTAAAGGTTCGTT
TGTTCAACGATCAAAATCCN
> HQ610913
NNNNNNNNNNNNNNNNNNNNNNNNNNNNNNNNNNNNNNNNNNNNNNNNNNNNN
NNNNNNNNNNNNNNNNNNNNNNNNNNNNNNNNNNNNNNNNNNNNNNNNNNNTGT
CTCCTTTCTCTAATCAGTGAAACTAATCTCCCCGTGAAGAAGCGGGGATAATACTAT
AAGACGAGAAGACCCCATGGAGCTTTAAACCCAATTATAATATACTTGGTTTTAGGT
TGGGGCGACCACGGAGTAAAACCAAACCTCCATGATGTACG-GAACAACCCCTCAT
CTAAGAGCTACAACTCTAAGAATTAGCACACTATCATA-AAACGACCC-ATTAT-TGAT
CAACGAACCAAGTTACCCTGGGGATAACAGCGCAATCCATTTTAAGAGCCCCNNN
NNNNNNNNNNNNNNNNNNNNNNNNNNNNNNNNNNNNNNNNNNNNNNNNNNNNN
NNNNNNNNNNNNNNNNNNNNNNNNNNNNNNNNNNNNNNNNNNNNNNNNNNNN
> HQ610914
NNNNNNNNNNNNNNNNNNNNNNNNNNNNNNNNNNNNNNNNNNNNNNNNNNNNN
NNNNNNNNNNNNNNNNNNNNNNNNNNNNNNNNNNNNNNNNNNNNNNNNNNNNN
TCTCCTTTCTCTAATCAGTGAAACTAATCTCCCCGTGAAGAAGCGGGGATAACACTA
TAAGACGAGAAGACCCCATGGAGCTTTAAACCCAATTATAATATACTTGGTTTTAGG
TTGGGGCGACCACGGAGTAAAACCAAACCTCCATGATGTACG-GAACAACCCCTCA
TCTAAGAGCTACAGCTCTAAGAATTAGCACACTATCATA-AAATGACCC-ATTAT-TGA
TCAACGAACCAAGTTACCCTGGGGATAACAGCGCAATCCATTTTAANNNNNNNNNN
NNNNNNNNNNNNNNNNNNNNNNNNNNNNNNNNNNNNNNNNNNNNNNNNNNNNN
NNNNNNNNNNNNNNNNNNNNNNNNNNNNNNNNNNNNNNNNNNNNNNNNNNNN
> *M_femoralis*_RAX8896
NNNNNNNNNNNNNNNNNNNNNNNNNNNNNGTGCGAAGGTAGCATAATCACTTGTTC
TTTAAATGGGGACTCGTATCAACGGCATCACGAGGGTTCTACTGTCTCCTTTCTCTAA
TCAGTGAAACTAATCTCCCCGTGAAGAAGCGGGGATAACACTATAAGACGAGAAG
ACCCCATGGAGCTTTAAACCCAATTATAATATACTTGGTTTTAGGTTGGGGCGACCAC
GGAGTAAAACCAAACCTCCATGATGAACG-GAACAACCCCTCATCTAAGAGCTACA
GCTCTAAGAATTAGCACACTATCATA-AAATGACCC-ATTAT-TGATCAACGAACCAAG
TTACCCTGGGGATAACAGCGCAATCCATTTTAAGAGCCCCTATCGACAAATGGGTTT
ACGACCTCGATGTTGGATCAGGATACCCCAGTGGTGCAGCAACTACTAAAGGTTCG
TTTGTTCAACGATCAAAATCCT
> *M_olgae* n. sp._RAX3203
NTAACGGCCACGGTATCCTAACCGTGCGAAGGTAGCATAATCACTTGTTCCTTA
AATAGGGACTCGTATCAACGGCATCACGAGAGTTCCACTGTCTCCTTTTTCCAATCA
GTGAAACTAATCTCCCCGTGAAGAAGCGGGGATACTTCTATAAGACGAGAAGACCC
CATGGAGCTTTAAACCCAATTATAACATAATTGGTTTTAGGTTGGGGCGACCATGGA
GTATAACCAAACCTCCACGATGAACG-GAACAACCCC--ATCTAAGAATTACACCTCT
AAGAATTAGTACACTAACATA-AAATGACCC-ATTAT-TGATCAACGAACCAAGTTAC
CCTGGGGATAACAGCGCAATCCATTTTAAGAGCCCCTATCGACAAATGGGTTTACG
ACCTCGATGTTGGATCAGGATATCCCAGTGGTGCAGCAGCTACTAAAGGCTCGTTTG
TTCAACGATTAAAATCCT
> *M_olgae*_n. sp. RAX5195
NTAACGGCCACGGTATCCTAACCGTGCGAAGGTAGCATAATCACTTGTTCCTTA
AATAGGGACTCGTATCAACGGCATCACGAGAGTTCCACTGTCTCCTTTTTCCAATCA
GTGAAACTAATCTCCCCGTGAAGAAGCGGGGATACTTCTATAAGACGAGAAGACCC
CATGGAGCTTTAAACCCAATTATAACATAATTGGTTTTAGGTTGGGGCGACCATGGA
GTATAGCCAAACCTCCACGATGAACG-GAACAACCCC--ATCTAAGAATTACACCTCT
AAGAATTAGTACACTAACATA-AAATGACCC-ATTAT-TGATCAACGAACCAAGTTAC
CCTGGGGATAACAGCGCAATCCATTTTAAGAGCCCCTATCGACAAATGGGTTTACG

ACCTCGATGTTGGATCAGGATGTCCCAGTGGTGCAGCAGCTACTAAAGGCTCGTTT
GTTCAACGATTAAAATCCT

> *M_olgae*_n. sp. RAX10205

TTAACGGCCACGGTATCCTAACCGTGCGAAGGTAGCATAATCACTTGTTCCTTA
AATAGGGACTCGTATCAACGGCATCACGAGAGTTCTACTGTCTCCTTTTTCCAATCA
GTGAAACTAATCTCCCCGTGAAGAAGCGGGGATACTTCTATAAGACGAGAAGACCC
CATGGAGCTTTAAACCCAATTATAACATAATTGGTTTTAGGTTGGGGCGACCATGGA
GTATAACCAAACCTCCACGATGAACG-GAATAACCCC--ATCTAAGAATTACACCTCT
AAGAATTAGTACACTAACATA-AAATGACCC-ATTAT-TGATCAACGAACCAAGTTAC
CCTGGGGATAACAGCGCAATCCATTTTAAGAGCCCCTATCGACAAATGGGTTTACG
ACCTCGATGTTGGATCAGGATGTCCCAGTGGTGCAGCAGCTACTAAAGGCTCGTTTG
TTCAACGATTAAAATCCT

> AY324821

TTAACGGCCACGGTACCCTAACCGTGCGAAGGTAGCATAATCACTTGTTCTTTA
AATGAGGACTCGTATCAACGGCATCACGAGGGTTCCACTGTCTCCTTTATCTAATCA
GTGAAACTAATCTCCCCGTGAAGAAGCGAGGATACTCTTATAAGACGAGAAGACCC
CATGGAGCTTTAAACCCAATTATAATATACTTGGTTTTAGGTTGGGGCGACCACGGA
GCACAACCAAACCTCCACGATGAACG-GAACAACCCCTTATCTGAGAATTACACTT
CTAAGAATTAGCACACTAACATA-AAATGACCC-ATTAT-TGATCAACGAACCAAGTT
ACCCTGGGGATAACAGCGCAATCCATTTTAAGAGCCCCTATCGACAAATGGGTTTAC
GACCTCGATGTTGGATCAGGATATCCCAGTGGTGCAGCAGCTGCTAAAGGTTCGTTT
GTTCAACGATTAAAATCCT

> *M_catalai*_RAN44491

TTAACGGCCACGGTACCCTAACCGTGCGAAGGTAGCATAATCACTTGTTCTTTA
AATGAGGACTCGTATCAACGGCATCACGAGGATTCCACTGTCTCCTTTCTCTAATCA
GTGAAACTAATCTCCCCGTGAAGAAGCGGGGATACTCTTATAAGACGAGAAGACCC
CATGGAGCTTTAAACCCAATTATAATATACTTGGTTTTAGGTTGGGGCGACCACGGA
GCACAACCAAACCTCCACGATGAACG-GAACAACCCCTTATCTGAGAATTACATTTC
TAAGAATTAGCACACTAACATA-AAATGACCC-ATTAT-TGATCAACGAACCAAGTTA
CCCTGGGGATAACAGCGCAATCCATTTTAAGAGCCCCTATCGACAAATGGGTTTAC
GACCTCGATGTTGGATCAGGATATCCCAGTGGTGCAGCAGCTACTAAAGGTTCGTTT
GTTCAACGATTAAAATCCT

> *M_catalai*_RAX10563

NNNNNNNNNNNNNNNNNNNNNNNCCGTGCGAAGGTAGCATAATCACTTGTTC
TTTAAATGAGGACTCGTATCAACGGCATCACGAGGGTTCCACTGTCTCCTTTCTCTA
ATCAGTGAAACTAATCTCCCCATGAAGAAGCGGGGATACTCTTATAAGACGAGAAG
ACCCCATGGAGCTTTAAACCCAATTATAATATACTTGGTTTTAGGTTGGGGCGACCAC
GGAGCACAACCAAACCTCCACGATGAACG-GAACAACCCCTTATCTAAGAATTACA
TTTCTAAGAATTAGCACACTAACATA-AAATGACCC-ATTAT-TGATCAACGAACCAA
GTTACCCTGGGGATAACAGCGCAATCCATTTTAAGAGCCCCTATCGACAAATGGGTT
TACGACCTCGATGTTGGATCAGGATATCCCAGTGGTGCAGCAGCTACCAAAGGTTC
GTTTGTTCAACGATTAAAATCCT

> *M_catalai*_RAX10599

NNNNNNNNNNNNNNNNNNNNNNNNNNNNNNNNNNNNNNNNNNNNNNNNNNNNNNNN
NNNNNNNNNNNNNNNNNNNNNNNNNATCAACGGCATCACGAGGGTTCCACTGTCTC
CTTTCTCTAATCAGTGAAACTAATCTCCCCATGAAGAAGCGGGGATACTCTTATAAG
ACGAGAAGACCCCATGGAGCTTTAAACCCAATTATAATATACTTGGTTTTAGGTTGG
GGCGACCACGGAGCACAACCAAACCTCCACGATGAACG-GAACAACCCCTTATCTA
AGAATTACATTTCTAAGAATTAGCACACTAACATA-AAATGACCC-ATTAT-TGATCAA
CGAACCAAGTTACCCTGGGGATAACAGCGCAATCCATTTTAAGAGCCCCTATCGAC
AAATGGGTTTACGACCTCGATGTTGGATCAGGATATCCCAGTGGTGCAGCAGCTAC
CAAAGGTTCGTTTGTTCAACGATTAAAATCCT

> *M. ambony* AY324822

TCAACGGCCACGGTACCCTAACCGTGCAAAGGTAGCATAATCACTTGTTCCTTA
AATAGGGACTCGTATCAATGGCATCACGAGGGTTATACTGTCTCCTTTCTCTAATCAG
TGAAACTAATCCCCCCGTGAAGAAGCGGGGATACCCCTATAAGACGAGAAGACCC
CATGGAGCTTTAAACCCAATTATAATATAATTGGTTTTAGGTTGGGGCGACCATGGAG
TACAATAAAACCTCCACGATGAACG-GAACAACCCCTTATCCAAGAATCACACCTCT
AAGAATTAGTACACTAACACA-AAATGACCC-ATCAA-TGATCAACGAACCAAGTTA
CCCTGGGGATAACAGCGCAATCCATTTTAAGAGCCCCTATCGACAAATGGGTTTACG
ACCTCGATGTTGGATCAGGATATCCCAGTGGCGCAGCAGCTACTAAAGGTTCGTTTG
TTCAACGATTAAAATCCT

> *M. ambony* HQ610870

NNNNNGGCCACGGTACCCTAACCGTGCAAAGGTAGCATAATCACTTGTTCCTT
AAATAGGGACTCGTATCAATGGCATCACGAGGGTTATACTGTCTCCTTTCTCTAATCA
GTGAAACTAATCCCCCCGTGAAGAAGCGGGGATACCCCTATAAGACGAGAAGACC
CCATGGAGCTTTAAACCCAATTATAATATAATTGGTTTTAGGTTGGGGCGACCATGGA
GTACAATAAAACCTCCACGATGAACG-GAACAACCCCTTATCCAAGAATCACACCT
CTAAGAATTAGTACACTAACACA-AAATGACCC-ATCAA-TGATCAACGAACCAAGT
TACCCTGGGGATAACAGCGCAATCCATTTTAAGAGCCCCTATCGACAAATGGGTTTA
CGACCTCGATGTTGGATCAGGATATCCCAGTGGCGCAGCAGCTACTAAAGGTTCGTT
TGTTCAACGATTAAAATCCT

> *M_ambony_*RAN38009

NNNNNNNNNNNNNNNNNNNNNNNNNNNNNNNNNNNNNNNNNNNNNNNNNNNTT
GTTCCTTAAATAGGGACTCGTATCAATGGCATCACGAGGGTTATACTGTCTCCTTTCT
CTAATCAGTGAAACTAATCCCCCCGTGAAGAAGCGGGGATACCCCTATAAGACGAG
AAGACCCCATGGAGCTTTAAACCCAATTATAATATAATTGGTTTTAGGTTGGGGCGA
CCATGGAGTACAATAAAACCTCCACGATGAACG-GAACAACCCCTTATCCAAGAAT
CACACCTCTAAGAATTAGTACACTAACACA-AAATGACCC---CAA-TGATCAACGAA
CCAAGTTACCCTGGGGATAACAGCGCAATCCATTTTAAGAGCCCCTATCGACAAATG
GGTTTACGACCTCGATGTTGGATCAGGATATCCCAGTGGCGCAGCAGCTACTAAAGG
TNNNNNNNNNNNNNNNNNNNNNNNNNNNNN

> *M_tavaratra_*RAX5000

NNNNNNNNNNNNNNNNNNNNNNNNNNNNNNNNNNNNNNNNNNNNNCATAATCACTTGT
TCTTTAAATAAGGACTCGTATCAACGGCATCACGAGGGTTATACTGTCTCCTTTCTCT
AATCAGTGAAACTAATCTCCCCGTGAAGAAGCGGGGATATTCCTATAAGACGAGAA
GACCCCATGGAGCTTTAAACCCAACTATAATATACTTGGTTTTAGGTTGGGGCGACC
ACGGAGTATAACCAAACCTCCATGATGAACG-GAACAA-CCCTTATCTATGAACTAC
ATATCTAAGAATTAGTATACTAACATA-AAATGACCC-ATTAT-TGATCAACGAACCAA
GTTACCCTGGGGATAACAGCGCAATCCATTTTAAGAGCCCCTATCGACAAATGGGTT
TACGACCTCGATGTTGGATCAGGATATCCCAGTGGTGCAGCAGCTACTAAAGGTTCG
TTTGTTCAACGATTAAAATCCT

> *M. danieli* n sp. AY324818

TTAACGGCCACGGTACCCTAACCGTGCAAAGGTAGCATAATCACTTGTTCTTTA
AATAGGGACTCGTATCAACGGCATCACGAGGGTTTTACTGTCTCCTTTTTCTAATCAG
TGAAACTAATCCCCCCGTGAAGAAGCGGGGATATTTTTATAAGACGAGAAGACCCC
ATGGAGCTTTAAACCCAACTATAATATACTTGGTTTTAGGTTGGGGCGACCACGGAG
TATAACCTAACCTCCACGATGTACG-GAACAACCCCTTATCTAAGAACCACACCTCT
AAGAATTAGAACACTAACATA AAATGACCC-ATTAT-TGATCAACGAACCAAGTTAC
CCTGGGGATAACAGCGCAATCCATTTTAAGAGCCCCTATCGACAAATGGGTTTACG
ACCTCGATGTTGGATCAGGATATCCCAGTGGTGCAGCAGCTACTAAAGGTTCGTTTG
TTCAACGATTAAAATCCT

> *M_danieli_*n. sp. RAN38186

NNNNNNNNNNNNNNNNNNNNNNNNNCCGTGCAAAGGTAGCATAATCACTTGTTC
TTTAAATAGGGACTCGTATCAACGGCATCACGAGGGTTTTACTGTCTCCTTTTTCTAA
TCAGTGAAACTAATCCCCCCGTGAAGAAGCGGGGATATTTTTATAAGACGAGAAGA
CCCCATGGAGCTTTAAACCCAACTATAATATACTTGGTTTTAGGTTGGGGCGACCAC

GGAGTATAACCTAACCTCCACGATGTACG-GAACAACCCCTTATCTAAGAACCACA
CCTCTAAGAATTAGAACACTAACATA-AAATGACCC-ATTAT-TGATCAACGAACCAA
GTTACCCTGGGGATAACAGCGCAATCCATTTTAAGAGCCCCTATCGACAAATGGGTT
TACGACCTCGATGTTGGATCAGGATATCCCAGTGGTGCAGCAGCTACTAAAGGTTC
GTTTGTTCAACGATTAAAATCCT

> *M_danieli_*n. sp. RAX10204

TTAACGGCCACGGTACCCTAACCGTGCAAAGGTAGCATAATCACTTGTTCTTTA
AATAAGGACTCGTATCAACGGCATCACGAGGGTTTTACTGTCTCCTTTTTCTAATCAG
TGAAACTAATCCCCCCGTGAAGAAGCGGGGATATTTTTATAAGACGAGAAGACCCC
ATGGAGCTTTAAACCCAACTATAATATATTTGGTTTTAGGTTGGGGCGACCACGGAGT
ATAACCCAACCTCCACGATGTACG-GAGCAACCCCTTATCTAAGAACCACATCTCTA
AGAATTAGAACACTAACATA-AAATGACCC-ATTAT-TGATCAACGAACCAAGTTACC
CTGGGGATAACAGCGCAATCCATTTTAAGAGCCCCTATCGACAAATGGGTTTACGA
CCTCGATGTTGGATCAGGATATCCCAGTGGTGCAGCAGCTACTAAAGGTTCGTTTGT
TCAACGATTAAAATCCT

> *M_danieli_*n. sp. RAX4268

TTAACGGCCACGGTACCCTAACCGTGCAAAGGTAGCATAATCACTTGTTCTTTA
AATAAGGACTCGTATCAACGGCATCACGAGGGTTTTACTGTCTCCTTTTTCTAATCAG
TGAAACTAATCCCCCCGTGAAGAAGCGGGGATATTTTTATAAGACGAGAAGACCCC
ATGGAGCTTTAAACCCAACTATAATATACTTGGTTTTAGGTTGGGGCGACCACGGAG
TATAACCTAACCTCCACGATGTACG-GAGCAACCCCTTATCTAAGAACCACATCTCTA
AGAATTAGAACACTAACATA-AAATGACCC-ATTAT-TGATCAACGAACCAAGTTACC
CTGGGGATAACAGCGCAATCCATTTTAAGAGCCCCTATCGACAAATGGGTTTACGAC
CTCGATGTTGGATCAGGATATCCCAGTGGTGCAGCAGCTACTAAAGGTTCGTTTGTT
CAACGATTAAAATCCT

> *M_danieli_*n. sp. RAX10392

NNNNNNNNNNNNNNNNNNNNNNNNCCGTGCAAAGGTAGCATAATCACTTGTTC
TCTAAATAGGGACTCGTATCAATGGCATCACGAGGGTCCTACTGTCTCCTTTTTCTAA
TCAGTGAAACTAATCTCCCCGTGAAGAAGCGGGAATATTTTTATAAGACGAGAAGA
CCCCATGGAGCTTTAAACCCAACTATAATATATTTGGTTTTAGGTTGGGGCGACCACG
GAGTAAAGCCTAACCTCCACGATGAACG-AAATAACCCCTTATCCAAGAACCACAT
ATCTAAGAATTAGTAAACTAACATATAAATGACCC--TTAT-TGATCAACGAACCAAGT
TACCCTGGGGATAACAGCGCAATCCATTTTAAGAGCCCCTATCGACAAATGGGTTTA
CGACCTCGATGTTGGATCAGGATATCCCAGTGGTGCAGCAGCTACTAAAGGTTCGTT
TGTTCAACGATTAAAATCCT

> *M_danieli_*n. sp. RAX6595

TCAACGGCCACGGTACCCTAACCGTGCAAAGGTAGCATAATCACTTGTTCTCTA
AATAGGGACTCGTATCAACGGCATCACGAGGGTTCTACTGTCTCCTTTTTCTAATCA
GTGAAACTAATCTCCCCGTGAAGAAGCGGGAATATTTTTATAAGACGAGAAGACCC
CATGGAGCTTTAAACCCAATTATAATATATTTGGTTTTAGGTTGGGGCGACCACGGA
GTATAACCTAACCTCCATGATGAACG-AAACAACCCCTTATCTAAGAACCACATATCT
AGGAATTAGTATACTAACATA-AAATGACCC-ATTAT-TGATCAACGAACCAAGTTACC
CTGGGGATAACAGCGCAATCCATTTTAAGAGCCCCTATCGACAAATGGGTTTACGA
CCTCGATGTTGGATCAGGATATCCCAGTGGTGCAGCAGCTACTAAAGGTTCGTTTGT
TCAACGATTAAAATCCT

> *M. danieli* n. sp. AY324816

TCAACGGCCACGGTACCCTAACCGTGCAAAGGTAGCATAATCACTTGTTCTCTA
AATAGGGACTCGTATCAACGGCATCACGAGGGTTTTACTGTCTCCTTTTTCTAATCAG
TGAAACTAATCTCCCCGTGAAGAAGCGGGGATATTTTTATAAGACGAGAAGACCCC
ATGGAGCTTTAAACCCAATTATAATATATTTGGTTTTAGGTTGGGGCGACCACGGAGT
ATAACCTAACCTCCATGATGAATG-AAACAACCCCTTATCTAAGAACCACACATCTA
AGAATTAGTATACTAACATA-AAATGACCC-ATTAT-TGATCAACGAACCAAGTTACCC
TGGGGATAACAGCGCAATCCATTTTAAGAGCCCCTATCGACAAATGGGTTTACGAC

CTCGATGTTGGATCAGGATATCCCAGTGGTGCAGCAGCTACTAAAGGTTCGTTTGTT
CAACGATTAAAATCCT
> *M_danieli_*n. sp. RAN39387
NNNNNNNNNNNNNNNNNNNNNNNNNNNNNNNNNNNNNNNNNNNNNNNNNNNN
NNTTCTCTAAATAGGGACTCGTATCAACGGCATCACGAGGGTTTTACTGTCTCCTTTT
TCTAATCAGTGAAACTAATCTCCCCGTGAAGAAGCGGGGATATTTTTATAAGACGAG
AAGACCCCATGGAGCTTTAAACCCAATTATAATATATTTGGTTTTAGGTTGGGGCGAC
CACGGAGTATAACCTAACCTCCATGATGAACG-AAACAACCCC-TATCTAAGAACCA
CATATCTAAGAATTAGTATACTAACATA-AAATGACCC-ATTAT-TGATCAACGAACCA
AGTTACCCTGGGGATAACAGCGCAATCCATTTTAAGAGCCCCTATCGACAAATGGG
TTTACGACCTCGATGTTGGATCAGGATATCCCAGTGGTGCAGCAGCTACTAAAGGTT
CGTTTGTTCAACGATTAAAATCCT
> *M. mocquardi* AY848150
NNNNCGGCCACGGTACCCTAACCGTGCAAAGGTAGCATAATCACTTGTTCTTT
AAATAAGGACTCGTATCAACGGCATCACGAGGGTTCTACTGTCTCCTTTTTTTAATCA
GTGAAACTAATCTCCCCGTGAAGAAGCGGGGATACTCCTATAAGACGAGAAGACCC
CATGGAGCTTTAAACCCAACTATAATATACTTGGTTTTAGGTTGGGGCGACCACGGA
GTATAACTAAACCTCCACAATGAACG-AAACAA-CCCTTATCTAAGAATTACACTTCT
AAGAATTAGTACACTAACATA-AAATGACCC-ATTAT-TGATCAACGAACCAAGTTAC
CCTGGGGATAACAGCGCAATCCATTTTAAGAGCCCTTATCGACAAATGGGTTTACGA
CCTCGATGTTGGATCAGGATATCCCAGTGGTGCAGCAGCTACTAAAGGTTCGTTTGT
TCAACGATTAAAATCCT
> *M_mocquardi_*RAN42737
TAAACGGCCACGGTACCCTAACCGTGCAAAGGTAGCATAATCACTTGTTCTTTA
AATAAGGACTCGTATCAACGGCATCACGAGGGTTCTACTGTCTCCTTTTTTTAATCAG
TGAAACTAATCTCCCCGTGAAGAAGCGGGGATACTCCTATAAGACGAGAAGACCCC
ATGGAGCTTTAAACCCAACTATAATATACTTGGTTTTAGGTTGGGGCGACCACGGAG
TATAACTAAACCTCCACAATGAACG-AAACAA-CCCTTATCTAAGAATTACACTTCTA
AGAATTAGTACACTAACATA-AAATGACCC-ATTAT-TGATCAACGAACCAAGTTACC
CTGGGGATAACAGCGCAATCCATTTTAAGAGCCCTTATCGACAAATGGGTTTACGA
CCTCGATGTTGGATCAGGATATCCCAGTGGTGCAGCAGCTACTAAAGGTTCGTTTGT
TCAACGATTAAAATCCT
> *M. mocquardi* AF215317
TAAACGGCCACGGTACCCTAACCGTGCAAAGGTAGCATAATCACTTGTTCTTTA
AATAAGGACTCGTATCAACGGCATCACGAGGGTTCTACTGTCTCCTTTTTTTAATCAG
TGAAACTAATCTCCCCGTGAAGAAGCGGGGATACTCCTATAAGACGAGAAGACCCC
ATGGAGCTTTAAACCCAACTATAATATACTTGGTTTTAGGTTGGGGCGACCACGGAG
TATAACTAAACCTCCACAATGAACG-AAACAA-CCCTTATCTAAGAATTACACTTCTA
AGAATTAGTACACTAACATA-AAATGACCC-ATTAT-TGATCAACGAACCAAGTTACC
CTGGGGATAACAGCGCAATCCATTTTAAGAGCCCTTATCGACAAATGGGTTTACGAC
CTCGATGTTGGATCAGGATATCCCAGTGGTGCAGCAGCTACTAAAGGTTCGTTTGTT
CAACGATTAAAATCCT
> *M. mocquardi* HQ610921
NNNNNNNNNNNNNNNNNNNNNNNNNNNNNNNNNNNNNNNNNNNNNNNNNNNN
NNNNNNNNNNNNNNNNNNNNNNNNNNNNNNNNNNNNNNNNNNNNNNNNNNNG
TCTCCTTTTTTTAATCAGTGAAACTAATCTCCCCGTGAAGAAGCGGGGATATTCTTAT
AAGACGAGAAGACCCCATGGAGCTTTAAACCCAACTATAATATACTTGGTTTTAGGT
TGGGGCGACCACGGAGTATAATTAAACCTCCACAATGAACG-AAACAA-CCCTTATC
TAAGAGTCACACCTCTAAGAATTAGCACACTAACATA-AAATGACCC-ATTAT-TGATC
AACGAACCAAGTTACCCTGGGGATAACAGCGCAATCCATTTTAAGAGCCCCTATCG
ACAAATNNNNNNNNNNNNNNNNNNNNNNNNNNNNNNNNNNNNNNNNNNNNNNNNN
NNNNNNNNNNNNNNNNNNNNNNNNNNNNNNNNNNNNNNNNNNNNNN
> *M_mocquardi_*RAX7524

NNNNNNNNNNNNNNNNNNNNNNNNNNNNNNNNNNNNNNNNNNNNNNNNNNNNNNNNNNNNN
NNNNNNNNNNNNNNNNNNNNNNNNNNNNNCGGCATCACGAGGGTTCTACTGTCTC
CTTTTTTTAATCAGTGAAACTAATCTCCCCGTGAAGAAGCGGGGATATTCTTATAAG
ACGAGAAGACCCCATGGAGCTTTAAACCCAATTATAATATACTTGGTTTTAGGTTGG
GGCGACCACGGAGTATAACTAAACCTCCACAATGAACG-AAACAA-CCCTTATCTGA
GAATTACACCTCTAAGAATTAGTATACTAACATA-AAATGACCC---TAT-TGATCAACG
AACCAAGTTACCCTGGGGATAACAGCGCAATCCATTTTAAGAGCCCCTATCGACAA
ATGGGTTTACGACCTCGATGTTGGATCANNNNNNNNNNNNNNNNNNNNNNNNNNNN
NNNNNNNNNNNNNNNNNNNNNNNNNNNNNNNNNN

> *M_mocquardi*_RAX9022

NNNNNNNNNNNNNNNNNNNNNNNNNNNNNNNNNAAGGTAGCATAATCACTTGTT
CTTTAAATAGGGACTCGTATCAACGGCATCACGAGGGTTCTACTGTCTCCTTTTTTTA
ATCAGTGAAACTAATCTCCCCGTGAAGAAGCGGGGATATTCTTATAAGACGAGAAG
ACCCCATGGAGCTTTAAACCCAACTATAATATACTTGGTTTTAGGTTGGGGCGACCA
CGGAGTATAACTAAACCTCCACAATGAACG-AAACAA-CCCTTATCTAAGATTTACAC
CTCTAAGAATTAGCACACTAACATA-AAATGACCC-ATTAT-TGATCAACGAACCAAG
TTACCCTGGGGATAACAGCGCAATCCATTTTAAGAGCCCCTATCGACAAATGGGTTT
ACGACCTCGATGTTGGATCAGGATATCCTAGTGGTGCAGCAGCTACTAAAGGTTCGT
TTGTTCAACGATTAAAATCCT

> *M. mocquardi* JX570508

NNNNNNNNNNNNNNNNNNNNNNNNNNNNNNNNNCAAAGGTAGCATAATCACTTGTT
CTTTAAATAAGGACTCGTATCAACGGCATCACGAGGGTTCTACTGTCTCCTTTTTTTA
ATCAGTGAAACTAATCTCCCCGTGAAGAAGCGGGGATATTCTTATAAGACGAGAAG
ACCCCATGGAGCTTTAAACCCAACTATAATATACTTGGTTTTAGGTTGGGGCGACCA
CGGAGTATAACTAAACCTCCACAATGAACG-AAACAA-CCCTTATCTGAGAATTACA
CCTCTAAGAATTAGCACACTAACATA-AAATGACCC-ATTAT-TGATCAACGAACCAA
GTTACCCTGGGGATAACAGCGCAATCCATTTTAAGAGCCCCTATCGACAAATGGGTT
TACGACCTCGATGTTGGATCAGGATATCCTAGTGGTGCAGCAGCTACTAAAGGTTCG
TTTGTTCAACGATTAAAATCCT

> *M. mocquardi* HQ610861

TTAACGGCCACGGTACCCTAACCGTGCAAAGGTAGCATAATCACTTGTTCTTTA
AATAAGGACTCGTATCAACGGCATCACGAGGGTTCTACTGTCTCCTTTTTTTAATCAG
TGAAACTAATCTCCCCGTGAAGAAGCGGGGATATTCTTATAAGACGAGAAGACCC-
CAT GGAGCTTTAAACCCAACTATAATATATTTGGTTTTAGGTTGGGGCGACCACGGA
GTATAACTAAACCTCCACAATGAACG-AAACAA-CCCTTATCTGAGAATTACACCTCT
AAGAATTAGTACACTAACATA-AAATGACCC-ATTAT-TGATCAACGAACCAAGTTAC
CCTGGGGATAACAGCGCAATCCATTTTAAGAGCCCCTATCGACAAATGGGTTTACG
ACCTCGATGTTGGATCAGGATATCCTAGTGGTGCAGCAGCTACTAAAGGTTCGTTTG
TTCAACGATTAAAATCCT

> *M. mocquardi*_RAX3669

TTAACGGCCACGGTACCCTAACCGTGCAAAGGTAGCATAATCACTTGTTCTTTA
AATAAGGACTTGTATCAACGGCATCACGAGGGTTCCACTGTCTCCTTTTTCTAATCA
GTGAAACTAATCTCCCCGTGAAGAAGCGGGGATACTTCTATAAGACGAGAAGACCC
CATGGAGCTTTAAACCCAACTATAATATATTTGGTTTTAGGTTGGGGCGACCACGGA
GTATAACTAAACCTCCATGATGAACG-AAACAA-CCCTTATCTAAGAGTTACACCTCT
AAGAATTAGTACACTAACACA-AAATGACCC-ATTAT-TGATCAACGAACCAAGTTAC
CCTGGGGATAACAGCGCAATCCATTTTAAGAGCTCCTATCGACAAATGGGTTTACG
ACCTCGATGTTGGATCAGGATATCCCAGTGGTGCAGCAGCTACTAAAGGTTCGTTTG
TTCAACGATTAAAATCCT

> *M. mocquardi*_RAN37992

TTAACGGCCACGGTACCCTAACCGTGCAAAGGTAGCATAATCACTTGTTCTTTA
AATAAGGACTTGTATCAACGGCATCACGAGGGTTCCACTGTCTCCTTTTTCTAATCA
GTGAAACTAATCTCCCCGTGAAGAAGCGGGGATACTTCTATAAGACGAGAAGACCC
CATGGAGCTTTAAACCCAACTATAATATATTTGGTTTTAGGTTGGGGCGACCACGGA

GTATAACTAAACCTCCATGATGAACG-AAACAA-CCCTTATCTAAGAGTTACACCTCT
AAGAATTAGTACACTAACACA-AAATGACCC-ATTAT-TGATCAACGAACCAAGTTAC
CCTGGGGATAACAGCGCAATCCATTTTAAGAGCTCCTATCGACAAATGGGTTTACGA
CCTCGATGTTGGATCAGGATATCCCAGTGGTGCAGCAGCTACTAAAGGTTCGTTTGT
TCAACGATTAAAATCCT
> *M. mocquardi*_RAX3641
NNNNNNNNNNNNNNNNNNNNNNNNNCGTGCAAAGGTAGCATAATCACTTGTTC
TTTAAATAAGGACTTGTATCAACGGCATCACGAGGGTTCCACTGTCTCCTTTTTCTAA
TCAGTGAAACTAATCTCCCCGTGAAGAAGCGGGGATACTTCTATAAGACGAGAAGA
CCCCATGGAGCTTTAAACCCAACTATAATATATTTGGTTTTAGGTTGGGGCGACCAC
GGAGTATAACTAAACCTCCATGATGAACG-AAACAA-CCCTTATCTAAGAGTTACACC
TCTAAGAATTAGTACACTAACACA-AAATGACCC-ATTAT-TGATCAACGAACCAAGT
TACCCTGGGGATAACAGCGCAATCCATTTTAAGAGCTCCTATCGACAAATGGGTTTA
CGACCTCGATGTTGGATCAGGATATCCCAGTGGTGCAGCAGCTACTAAAGGTTCGTT
TGTTCAACGATTAAAATCCT
> *M. mocquardi*_RAX4670
NNNNNNNNNNNNNNNNNNNNNNNNNNNNNNNNNNNNNNNNNNNNNAATCACTTG
TTCTTTAAATAAGGACTTGTATCAACGGCATCACGAGGGTTCTACTGTCTCCTTTTTC
CAATCAGTGAAACTAATCTCCCCGTGAAGAAGCGGGGATACTTCTATAAGACGAGA
AGACCCCATGGAGCTTTAAACCCAACTATAATATACTTGGTTTTAGGTTGGGGCGAC
CACGGAGTATAACTAAACCTCCACGATGAACG-AAACAA-CCCTTATCTAAGAGTTA
CACCTCTAAGAATTAGCACACTAACATA-AAATGACCC-ATTAT-TGATCAACGAACC
AAGTTACCCTGGGGATAACAGCGCAATCCATTTCAAGAGCCCCTATCGACAAATGG
GTTTACGACCTCGATGTTGGATCAGGATATCCCAGTGGTGCAGCAGCTACTAAAGGT
TCGTTTGTTCAACGATTAAAATCCT
> *M. mocquardi*_RAX3915
NNNNNNNNNNNNNNNNNNNNNNNNNCCGTGCAAAGGTAGCATAATCACTTGTTC
TTTAAATAAGGACTTGTATCAACGGCATCACGAGGGTTCTACTGTCTCCTTTTTCCAA
TCAGTGAAACTAATCTCCCCGTGAAGAAGCGGGGATACTTCTATAAGACGAGAAGA
CCCCATGGAGCTTTAAACCCAACTATAATATACTTGGTTTTAGGTTGGGGCGACCAC
GGAGTATAACTAAACCTCCACGATGAACG-AAACAA-CCCTTATCTAAGAGTCACAC
CTCTAAGAATTAGCACACTAACATA-AAATGACCC-ATTAT-TGATCAACGAACCAAG
TTACCCTGGGGATAACAGCGCAATCCATTTCAAGAGCCCCTATCGACAAATGGGTT
TACGACCTCGATGTTGGATCAGGATATCCCAGTGGTGCAGCAGCTACTAAAGGTTCG
TTTGTTCNNNNNNNNNNNNNNNNN
> *M_ambreensis*_RAX6552
TCAACGGCCACGGTACCCTAACCGTGCAAAGGTAGCATAATCATTTGTTCTCTA
AATAGGGACTCGTATCAACGGCATCACGAGGGTTCCACTGTCTCCTTTCTCTAATCA
GTGAAACTAATCTCCCCGTGAAGAAGCGGGGATATTTTTATAAGACGAGAAGACCC
CATGGAGCTTTAAACCCAATTATAATATAATTGGTTTTAGGTTGGGGCGACCACGGA
GTATAATTTAACCTCCACGATGAACG-GAACAACCCCTTATCTAAGAGTTACACCTCT
AAGAATTAGCACACTAACACA-AAATGACCC---TAA-TGATCAACGAACCAAGTTAC
CCTGGGGATAACAGCGCAATCCATTTTAAGAGCCCCTATCGACAAATGGGTTTACG
ACCTCGATGTTGGATCAGGATATCCCAGTGGTGCAGCAGCTACTAAAGGTTCGTTTG
TTCAACGATCAAAATCCT
> *M_ambreensis*_RAX9589
NNNNNNNNNNNNNNNNNNNNNNNNNNNNNNNNNNNNNNNNNNNNNNAATCATTTG
TTCTCTAAATAGGGACTCGTATCAACGGCATCACGAGGGTTCCACTGTCTCCTTTCT
CTAATCAGTGAAACTAATCTCCCCGTGAAGAAGCGGGGATATTTTTATAAGACGAG
AAGACCCCATGGAGCTTTAAACCCAATTATAATATAATTGGTTTTAGGTTGGGGCGA
CCACGGAGTATAATTTAACCTCCACGATGAACG-GAACAACCCCTTATCTAAGAGTT
ACACCTCTAAGAATTAGCACACTAACACA-AAATGACCC---TAA-TGATCAACGAAC
CAAGTTACCCTGGGGATAACAGCGCAATCCATTTTAAGAGCCCCTATCGACAAATG

GGTTTACGACCTCGATGTTGGATCAGGATATCCCAGTGGTGCAGCAGCTACTANNN
NNNNNNNNNNNNNNNNNNNNNNNNNNNNNN
> *M_ambreensis_RAX2373*
NNNNNNNNNNNNNNNNNNNNNNNNNNNNNNNNNNNNNNNNNNNNNAATCATTTG
TTCTCTAAATAGGGACTCGTATCAACGGCATCACGAGGGTTCTACTGTCTCCTTTCTC
TAATCAGTGAAACTAATCTCCCCGTGAAGAAGCGGGGATATCTTTATAAGACGAGA
AGACCCCATGGAGCTTTAAACCCAATTATAATATAATTGGTTTTAGGTTGGGGCGAC
CACGGAGTATAATTAAACCTCCACGATGAACG-GAACAACCCCTTATCTAAGAGTCA
CACCTCTAAGAATTAGTACACTAACACA-AAATGACCC---TAA-TGATCAACGAACCA
AGTTACCCTGGGGATAACAGCGCAATCCATTTTAAGAGCCCCTATCGACAAATGGGT
TTACGACCTCGATGTTGGATCAGGATATCCCAGTGGTGCAGCAGCTACTAAAGGTTC
GTTTGTTCAACGATCAAAATCCT
> *M_ambreensis_RAN38501*
TCAACGGCCACGGTACCCTAACCGTGCAAAGGTAGCATAATCATTTGTTCTCTA
AATAGGGACTCGTATCAACGGCATCACGAGGGTTCTACTGTCTCCTTTCTCTAATCA
GTGAAACTAATCTCCCCGTGAAGAAGCGGGGATATCTTTATAAGACGAGAAGACCC
CATGGAGCTTTAAACCCAATTATAATATAATTGGTTTTAGGTTGGGGCGACCATGGA
GTATAATTAAACCTCCACGATGAACG-GAACAACCCCTTATCTAAGAGTCACACCTC
TAAGAATTAGTACACTAACATA-AAATGACCC---TAA-TGATCAACGAACCAAGTTAC
CCTGGGGATAACAGCGCAATCCATTTTAAGAGCCCCTATCGACAAATGGGTTTACG
ACCTCGATGTTGGATCAGGATATCCCAGTGGTGCAGCAGCTACNNNNNNNNNNN
NNNNNNNNNNNNNNNNNNNNN
> *M_ambreensis* AY848129
NNNNNNNNNNNNNNNNNNNNNNNNNNNNNNNNNNNNNNNNNNNNNNNNNNNN
NNNNNNNNNNNNNNNNNNNNNNNNNNNNNNCGGCATCACGAGGGTTCTACTGTCT
CCTTTCTCTAATCAGTGAAACTAATCTCCCCGTGAAGAAGCGGGGATATCTTTATAA
GACGAGAAGACCCCATGGAGCTTTAAACCCAATTATAATATAATTGGTTTTAGGTTG
GGGCGACCATGGAGTATAATTAAACCTCCACGATGAACG-GAACAACCCCTTATCTA
AGAGTCACACCTCTAAGAATTAGTACACTAACATA-AAATGACCC-ATTAA-TGATCA
ACGAACCAAGTTACCCTGGGGATAACAGCGCAATCCATTTTAAGAGCCCCTATCGA
CAAATGGGTTTACGACCTCGATGTTGGATCAGGATATCCCAGTGGTGCAGCAGCTAC
TAAAGGTTCGTTTGTTCAACGATCAAAATCCT
> *M_poissoni_RAX9367*
NCAACGGCCACGGTACCCTAACCGTGCGAAGGTAGCATAATCACTTGTTCTTTA
AATAGGGACTTGTATCAACGGCACCACGAGGGTTCTACTGTCTCCTTTCTCTAATCA
GTGAAACTAATCTCCCCGTGAAGAAGCGGGGATATTCCTATAAGACGAGAAGACCC
CATGGAGCTTTAAACCCAATTATAATATACTTGGTTTTAGGTTGGGGCGACCACGGA
GTATAACCAAACCTCCACGATGAACG-GAACAATCCCTTATCTAAGAGTCACACCTC
TAAGAATTAGTACACTAACATA-AAATGACCC-ATTAT-TGATCAACGAACCAAGTTAC
CCTGGGGATAACAGCGCAATCCATTTTAAGAGCCCCTATCGACAAATGGGTTTACG
ACCTCGATGTTGGATCAGGATATCCCAGTGGTGCAGCAGCTACTAAAGGTTCGTTTG
TTCAACGATTAAAATCCT
> *M. zolitschka* HQ610866
NNNNCGGCCACGGTACCCTAACCGTGCGAAGGTAGCATAATCACTTGTTCTTT
AAATAGGGACTCGTATCAACGGCACCACGAGGGTTCCACTGTCTCCTTTCTCTAATC
AGTGAAACTAATCTCCCCGTGAAGAAGCGGGGATATTCTATAAGACGAGAAGACC
CCATGGAGCTTTAAACCCAATTATAATATACTTGGTTTTAGGTTGGGGCGACCGCGG
AGTACAACCTAACCTCCACGATGAACG-GAACAACCCCTTATCTAAGAACCACACC
TCTAAGAATTAGTACACTACCATA-AAATGACCC-ATTAT-TGATCAACGAACCAAGTT
ACCCTGGGGATAACAGCGCAATCCATTTTAAGAGCCCCTATCGACAAATGGGTTTAC
GACCTCGATGTTGGATCANNNNNNNNNNNNNNNNNNNNNNNNNNNNNNNNNNNN
NNNNNNNNNNNNNNNNNNNNNNNN
> *M. zolitschka* HQ610867

TCAACGGCCACGGTACCCTAACCGTGCGAAGGTAGCATAATCACTTGTTCTTTA
AATAGGGACTCGTATCAACGGCACCACGAGGGTTCCACTGTCTCCTTTCTCTAATCA
GTGAAACTAATCTCCCCGTGAAGAAGCGGGGATATTTCTATAAGACGAGAAGACCC
CATGGAGCTTTAAACCCAATTATAATATACTTGGTTTTAGGTTGGGGCGACCGCGGA
GTACAACCTAACCTCCACGATGAACG-GAACAACCCCTTATCTAAGAACCACACCT
CTAAGAATTAGTACACTACCATA-AAATGACCC-ATTAT-TGATCAACGAACCAAGTTA
CCCTGGGGATAACAGCGCAATCCATTTTAAGAGCCCCTATCGACAAATGGGTTTAC
GACCTCGATGTTGGATCAGGATATCCCAGTGGTGCAGCAGCTACTAAAGGTTCGTT
TGTTCAACGATTAAAATCCT

> *M_tavaratra*_n. sp. RAX11535

TTAACGGCCACGGTACCCTAACCGTGCGAAGGTAGCATAATCACTTGTTCTTTA
AATAGGGACTTGTATCAACGGCATCACGAGGGTTGCACTGTCTCCTTTCTCTAATCA
GTGAAACTAATCTCCCCGTGAAGAAGCGGGGATACTCCTATAAGACGAGAAGACCC
CATGGAGCTTTAAACCCAACTATAATATACTTGGTTTTAGGTTGGGGCGACCACGGA
GTATAACCAAACCTCCACGATGAACG-GAATAACCCCCCATCTAAGAGCCACACCTC
TAAGAATTAGCATACTAACATA-AAATGACCC-ATTAT-TGATCAACGAACCAAGTTAC
CCTGGGGATAACAGCGCAATCCATTTTAAGAGCCCCTATCGACAAATGGGTTTACGA
CCTCGATGTTGGATCAGGATATCCCAGTGGTGCAGCAGCTACTAAAGGTTCGTTTGT
TCAACGATTAAAATCCT

> *M_tavaratra*_n. sp._RAX11534

TTAACGGCCACGGTACCCTAACCGTGCGAAGGTAGCATAATCACTTGTTCTTTA
AATAGGGACTTGTATCAACGGCATCACGAGGGTTGCACTGTCTCCTTTCTCTAATCA
GTGAAACTAATCTCCCCGTGAAGAAGCGGGGATACTCCTATAAGACGAGAAGACCC
CATGGAGCTTTAAACCCAACTATAATATACTTGGTTTTAGGTTGGGGCGACCACGGA
GTATAACCAAACCTCCACGATGAACG-GAATAACCCCCCATCTAAGAGCCACACCTC
TAAGAATTAGCATACTAACATA-AAATGACCC-ATTAT-TGATCAACGAACCAAGTTAC
CCTGGGGATAACAGCGCAATCCATTTTAAGAGCCCCTATCGACAAATGGGTTTACG
ACCTCGATGTTGGATCAGGATATCCCAGTGGTGCAGCAGCTACTAAAGGTTCGTTTG
TTCAACGATTAAAATCCT

> *M_tavaratra*_n. sp._RAX11533

TTAACGGCCACGGTACCCTAACCGTGCGAAGGTAGCATAATCACTTGTTCTTTA
AATAGGGACTTGTATCAACGGCATCACGAGGGTTGCACTGTCTCCTTTCTCTAATCA
GTGAAACTAATCTCCCCGTGAAGAAGCGGGGATACTCCTATAAGACGAGAAGACCC
CATGGAGCTTTAAACCCAACTATAATATACTTGGTTTTAGGTTGGGGCGACCACGGA
GTATAACCAAACCTCCACGATGAACG-GAATAACCCCCCATCTAAGAGCCACACCTC
TAAGAATTAGCATACTAACATA-AAATGACCC-ATTAT-TGATCAACGAACCAAGTTAC
CCTGGGGATAACAGCGCAATCCATTTTAAGAGCCCCTATCGACAAATGGGTTTACG
ACCTCGATGTTGGATCAGGATATCCCAGTGGTGCAGCAGCTACTAAAGGTTCGTTTG
TTCAACGATTAAAATCCT

> *M_tavaratra*_n. sp._RAX11532

NNNNNNNNNNNNNNNNNNNNNNNNGTGCGAAGGTAGCATAATCACTTGTTC
TTTAAATAGGGACTTGTATCAACGGCATCACGAGGGTTGCACTGTCTCCTTTCTCTAA
TCAGTGAAACTAATCTCCCCGTGAAGAAGCGGGGATACTCCTATAAGACGAGAAGA
CCCCATGGAGCTTTAAACCCAACTATAATATACTTGGTTTTAGGTTGGGGCGACCAC
GGAGTATAACCAAACCTCCACGATGAACG-GAATAACCCCCCATCTAAGAGCCACA
CCTCTAAGAATTAGCATACTAACATA-AAATGACCC-ATTAT-TGATCAACGAACCAA
GTTACCCTGGGGATAACAGCGCAATCCATTTTAAGAGCCCCTATCGACAAATGGGT
TTACGACCTCGATGTTGGATCAGGATATCCCAGTGGTGCAGCAGCTACTAAAGGTTC
GTTTGTTCAACGATTAAAATCCT

> *M_tavaratra*_n. sp._RAX10323

NNNNNNNNNNNNNNNNNNNNNNTAACCGTGCAAAGGTAGCATAATCACTTGTTCT
TTAAATAGGGACTAGTATCAACGGCACCACGAGGGTTACACTGTCTCCTCTCTCCAA
TCAGTGAAACTAATCTCCCCGTGAAAAAGCGGGAATACTCCTATAAGACGAGAAGA
CCCCATGGAGCTTTAAACCCAACTATAATATATTTGGTTTTAGGTTGGGGCGACCAC

GGAGTATAGCCAAACCTCCACGATGAACG-GAACAACCCCCTATCTAAGAGCCACA
CCTCTAAGAATTAATATACTAACATA-AAATGACCC-ATTAT-TGATCAACGAACCAAG
TTACCCTGGGGATAACAGCGCAATCCATTTTAAGAGCCCCTATCGACAAATGGGTTT
ACGACCTCGATGTTGGATCAGGATATCCCAGTGGTGCAGCAGCTACTAAAGGTTCG
TTTGTTCAACGATTAAAATCCT

> *M_tavaratra*_n. sp._RAN39071

TTAACGGCCACGGTACCCTAACCGTGCAAAGGTAGCATAATCACTTGTTCTTTA
AATAAGGACTAGTATCAACGGCACCACGAGGGTTACACTGTCTCCTCTCTCCAATCA
GTGAAACTAATCTCCCCGTGAAAAAGCGGGAATACTCCTATAAGACGAGAAGACCC
CATGGAGCTTTAAACCCAACTATAATATATTTGGTTTTAGGTTGGGGCGACCACGGA
GTATAGCCAAACCTCCACGATGAACG-GAACAACCCCCTATCTAAGAGCCACACCT
CTAAGAATTAATATACTAACATA-AAATGACCC---TAT-TGATCAACGAACCAAGTTAC
CCTGGGGATAACAGCGCAATCCATTTTAAGAGCCCCTATCGACAAATGGGTTTACGA
CCTCGATGTTGGATCAGGATATCCCAGTGGTGCAGCAGCTACTAAAGGTTCGTTTGT
TCAACGATTAAAATCCT

> *M_tavaratra*_n. sp. AY324819

TTAACGGCCACGGTACCCTAACCGTGCAAAGGTAGCATAATCACTTGTTCTTTA
AATAAGGACTAGTATCAACGGCACCACGAGGGTTACACTGTCTCCTCTCTCCAATCA
GTGAAACTAATCTCCCCGTGAAAAAGCGGGAATACTCCTATAAGACGAGAAGACC
CCATGGAGCTTTAAACCCAACTATAATATATTTGGTTTTAGGTTGGGGCGACCACGG
AGTATAGCCAAACCTCCACGATGAACG-GAACAACCCCCTATCTAAGAGCCACACC
TCTAAGAATTAATATACTAACATA-AAATGACCC-ATTAT-TGATCAACGAACCAAGTT
ACCCTGGGGATAACAGCGCAATCCATTTTAAGAGCCCCTATCGACAAATGGGTTTAC
GACCTCGATGTTGGATCAGGATATCCCAGTGGTGCAGCAGCTACTAAAGGTTCGTTT
GTTCAACGATTAAAATCCT

> *M_tavaratra*_n. sp._RAX6901

TTAACGGCCACGGTACCCTAACCGTGCAAAGGTAGCATAATCACTTGTTCTTTA
AATAAGGACTAGTATCAACGGCACCACGAGGGTTACACTGTCTCCTCTCTCCAATC
AGTGAAACTAATCTCCCCGTGAAAAAGCGGGAATACTCCTATAAGACGAGAAGAC
CCCATGGAGCTTTAAACCCAACTATAATATATTTGGTTTTAGGTTGGGGCGACCACG
GAGTATAGCCAAACCTCCACGATGAACG-GAACAACCCCCTATCTAAGAGCCACAC
CTCTAAGAATTAATATACTAACATA-AAATGACCC-ATTAT-TGATCAACGAACCAAGT
TACCCTGGGGATAACAGCGCAATCCATTTTAAGAGCCCCTATCGACAAATGGGTTTA
CGACCTCGATGTTGGATCAGGATATCCCAGTGGTGCAGCAGCTACTAAAGGTTCGTT
TGTTCAACGATTAAAATCCT

> *M_tavaratra*_n. sp._RAX5769

TTAACGGCCACGGTACCCTAACCGTGCAAAGGTAGCATAATCACTTGTTCTTTA
AATAAGGACTAGTATCAACGGCACCACGAGGGTTACACTGTCTCCTCTCTCCAATCA
GTGAAACTAATCTCCCCGTGAAAAAGCGGGAATACTCCTATAAGACGAGAAGACC
CCATGGAGCTTTAAACCCAACTATAATATATTTGGTTTTAGGTTGGGGCGACCACGG
AGTATAGCCAAACCTCCACGATGAACG-GAACAACCCCCTATCTAAGAGCCACACC
TCTAAGAATTAATATACTAACATA-AAATGACCC-ATTAT-TGATCAACGAACCAAGTT
ACCCTGGGGATAACAGCGCAATCCATTTTAAGAGCCCCTATCGACAAATGGGTTTAC
GACCTCGATGTTGGATCAGGATATCCCAGTGGTGCAGCAGCTACTAAAGGTTCGTT
TGTTCAACGA TTAAAATCCT

> *M_tavaratra*_n. sp._RAX6028

NNNNNNNNNNNNNNNACCCTAACCGTGCAAAGGTAGCATAATCACTTGTTCTT
TAAATAAGGACTAGTATCAACGGCACCACGAGGGTTACACTGTCTCCTCTCTCCAAT
CAGTGAAACTAATCTCCCCGTGAAAAAGCGGGAATACTCCTATAAGACGAGAAGA
CCCCATGGAGCTTTAAACCCAACTATAATATATTTGGTTTTAGGTTGGGGCGACCACG
GAGTATAGCCAAACCTCCACGATGAACG-GAACAACCCCCTATCTAAGAGCCACAC
CTCTAAGAATTAATATACTAACATA-AAATGACCC-ATTAT-TGATCAACGAACCAAGT
TACCCTGGGGATAACAGCGCAATCCATTTTAAGAGCCCCTATCGACAAATGGGTTTA

CGACCTCGATGTTGGATCAGGATATCCCAGTGGTGCAGCAGCTACTAAAGGTTCGTT
TGTTCAACGATTAAAATCCT
> *M_tavaratra_*n. sp._RAX6107
NNNNNNNNNNNNNNNNNNNNNNNNNNNNNNNNNNNNNNNNNNNNNAATCACTTG
TTCTTTAAATAAGGACTAGTATCAACGGCACCACGAGGGTTACACTGTCTCCTCTCT
CCAATCAGTGAAACTAATCTCCCCGTGAAAAAGCGGGAATACTCCTATAAGACGAG
AAGACCCCATGGAGCTTTAAACCCAACTATAATATATTTGGTTTTAGGTTGGGGCGAC
CACGGAGTATAGCCAAACCTCCACGATGAACG-GAACAACCCCCTATCTAAGAGCC
ACACCTCTAAGAATTAATATACTAACATA-AAATGACCC-ATTAT-TGATCAACGAACC
AAGTTACCCTGGGGATAACAGCGCAATCCATTTTAAGAGCCCCTATCGACAAATGG
GTTTACGACCTCGATGTTGGATCAGGATATCCCAGTGGTGCAGCAGCTACTAAAGGT
TCGTTTGTTCAACGATTAAAATCCT
> *M_tavaratra_*n. sp._RAX6838
TTAACGGCCACGGTACCCTAACCGTGCAAAGGTAGCATAATCACTTGTTCTTTA
AATAAGGACTAGTATCAACGGCACCACGAGGGTTACACTGTCTCCTCTCTCCAATCA
GTGAAACTAATCTCCCCGTGAAAAAGCGGGGATACTCCTATAAGACGAGAAGACCC
CATGGAGCTTTAAACCCAACTATAATATATTTGGTTTTAGGTTGGGGCGACCACGGA
GTATAGCCAAACCTCCACGATGAACG-GAACAACCCCCTATCTAAGAGCCACACCT
CTAAGAATTAATATACTAACATA-AAATGACCC-ATTAT-TGATCAACGAACCAAGTTA
CCCTGGGGATAACAGCGCAATCCATTTTAAGAGCCCCTATCGACAAATGGGTTTAC
GACCTCGATGTTGGATCAGGATATCCCAGTGGTGCAGCAGCTACTAAAGGTTCGTTT
GTTCAACGATTAAAATCCT
> *M_tavaratra_*n. sp. AY324820
TTAACGGCCACGGTACCCTAACCGTGCAAAGGTAGCATAATCACTTGTTCTTTA
AATAAGGACTAGTATCAACGGCACCACGAGGGTTACACTGTCTCCTCTCTCCAATCA
GTGAAACTAATCTCCCCGTGAAAAAGCGGGGATACTCCTATAAGACGAGAAGACCC
CATGGAGCTTTAAACCCAACTATAATATATTTGGTTTTAGGTTGGGGCGACCACGGA
GTATAACCAAACCTCCACGATGAACG-GAACAACCCCCTATCTAAGAGCCACACCT
CTAAGAATTAATATACTAACATA-AAATGACCC-ATTAT-TGATCAACGAACCAAGTTA
CCCTGGGGATAACAGCGCAATCCATTTTAAGAGCCCCTATCGACAAATGGGTTTAC
GACCTCGATGTTGGATCAGGATATCCCAGTGGTGCAGCAGCTACTAAAGGTTCGTT
TGTTCAACGATTAAAATCCT
> *M_tavaratra_*n. sp._RAN43366
TTAACGGCCACGGTACCCTAACCGTGCAAAGGTAGCATAATCACTTGTTCTTTA
AATAAGGACTAGTATCAACGGCACCACGAGGGTTACACTGTCTCCTCTCTCCAATCA
GTGAAACTAATCTCCCCGTGAAAAAGCGGGGATACTCCTATAAGACGAGAAGACCC
CATGGAGCTTTAAACCCAACTATAATATATTTGGTTTTAGGTTGGGGCGACCACGGA
GTATAACCAAACCTCCACGATGAACG-GAACAACCCCCTATCTAAGAGCCACACCT
CTAAGAATTAATATACTAACATA-AAATGACCC-ATTAT-TGATCAACGAACCAAGTTA
CCCTGGGGATAACAGCGCAATCCATTTTAAGAGCCCCTATCGACAAATGGGTTTACG
ACCTCGATGTTGGATCAGGATATCCCAGTGGTGCAGCAGCTACTAAAGGTTCGTTTG
TTCAACGATTAAAATCCT
> *M_tavaratra_*n. sp._RAX5310
NNNNNNNNNNNNNNNNNNNNNNNNNNNNNGTGCAAAGGTAGCATAATCACTTGTTC
TTTAAATAAGGACTAGTATCAACGGCACCACGAGGGTTACACTGTCTCCTCTCTCCA
ATCAGTGAAACTAATCTCCCCGTGAAAAAGCGGGGATACTCCTATAAGACGAGAAG
ACCCCATGGAGCTTTAAACCCAACTATAATATATTTGGTTTTAGGTTGGGGCGACCAC
GGAGTATAACCAAACCTCCACGATGAACG-GAACAACCCCCTATCTAAGAGCCACA
CCTCTAAGAATTAATATACTAACATA-AAATGACCC-ATTAT-TGATCAACGAACCAAG
TTACCCTGGGGATAACAGCGCAATCCATTTTAAGAGCCCCTATCGACAAATGGGTTT
ACGACCTCGATGTTGGATCAGGATATCCCAGTGGTGCAGCAGCTACTAAAGGTTCGT
TTGTTCAACGATTAAAATCCT
> *M_tavaratra_*n. sp._RAX5400

NNNNNNNNNNNNNNNNNNNNNNNNNNNNNNNNNNNNNNNNNNNNNNNNNNNNN
NNNNNNNNNNNNNNNNNNNNNNNNNNNNNNNNNNCGGCACCACGAGGGTTACACTGTCT
CCTCTCTCCAATCAGTGAAACTAATCTCCCCGTGAAAAAGCGGGGATACTCCTATAA
GACGAGAAGACCCCATGGAGCTTTAAACCCAACTATAATATATTTGGTTTTAGGTTG
GGGCGACCACGGAGTATAACCAAACCTCCACGATGAACG-GAACAACCCCCTATCT
AAGAGCCACACCTCTAAGAATTAATATACTAACATA-AAATGACCC-ATTAT-TGATCA
ACGAACCAAGTTACCCTGGGGATAACAGCGCAATCCATTTTAAGAGCCCCTATCGA
CAAATGGGTTTACGACCTCGATGTTGGATCAGGATATCCCAGTGGTGCAGCAGCTAC
TAAAGGTTCGTTTGTTCAACGATTAAAATCCT

References

1. Amphibiaweb Amphibian in the World. Available online: https://amphibiaweb.org (accessed on 30 July 2023).
2. Glaw, F.; Vences, M. Phylogeny and genus level classification of mantellids frogs (Amphibia, Anura). *Org. Divers. Evol.* **2006**, *6*, 236–253. [CrossRef]
3. Glaw, F.; Vences, M. Complete classification of species in the family Mantellidae Laurent, 1946. *Org. Divers Evol.* **2006**, *6*, 236–253. [CrossRef]
4. Frost, D.R. Amphibian Species of the World, Version 6. Available online: http://research.amnh.org/vz/herpetology/amphibia/index.php//Amphibia/Anura/Mantellidae/Mantellinae/Mantidactylus (accessed on 3 August 2018).
5. Scherz, M.D.; Rasolonjatovo, S.M.; Köhler, J.; Rancilhac, L.; Rakotoarison, A.; Raselimanana, A.P.; Ohler, A.; Preick, M.; Hofreiter, M.; Glaw, F.; et al. "Barcode fishing" for archival DNA from historical type material overcomes taxonomic hurdles, enabling the description of a new frog species. *Sci. Rep.* **2020**, *10*, 19109. [CrossRef]
6. Blommers–Schlösser, R.M.A.; Blanc, C.P. Amphibien. In *Faune de Madagascar*; Partie I.; MNHN: Paris, France, 1991; pp. 1–379.
7. Glaw, F.; Vences, M. *A Fieldguide to the Amphibians and Reptiles of Madagascar*, 2nd ed.; Vences, M., Glaw, F., Verlags, G.b.R., Eds.; Moss Druck: Leverkusen, Germany, 1994; pp. 1–480. ISBN 3-929449-01-3.
8. Boulenger, G.A. Catalogue of the Batrachia Salientia s. Ecaudata. In *BHL Collections*, 2nd ed.; Taylor and Francis: London, UK, 1882; p. 495. [CrossRef]
9. Boulenger, G.A. On a genus of frog peculiar to Madagascar. In *Annals and Magazine of Natural History*; Série 6; Ann.Mag.nat.Hist: Toronto, Canada, 1896; pp. 420–421. [CrossRef]
10. Glaw, F.; Vences, M. A preliminary review of cryptic diversity in frogs of the subgenus *Ochthomantis* based on mt DNA sequence and morphology. *Spixiana* **2004**, *27*, 83–91.
11. Randrianiaina, R.D.; Strauß, A.; Glos, J.; Glaw, F.; Vences, M. Diversity, external morphology and 'reverse taxonomy' in the specialized tadpoles of Malagasy river bank frogs of the subgenus *Ochthomantis* (genus *Mantidactylus*). *Contrib. Zool.* **2011**, *80*, 17–65. [CrossRef]
12. Wollenberg, K.C.; Vieites, D.R.; Glaw, F.; Vences, M. Speciation in little: The role of range and body size in the diversification of Malagasy mantellid frogs. *Artic. Rev.* **2011**, *11*, 217. [CrossRef]
13. Leviton, A.E.; Gibbs, R.H., Jr.; Heal, E.; Dawson, C.E. Standards in Herpetology and Ichthylogy: Part I. Standard Symbolic Codes for International Resource Collections in Herpetology and Icthyology. *Copeia* **1985**, *1985*, 802–832.
14. Hayek, L.C.; Heyer, W.R.; Gascon, C. Frog morphometrics: A cautionary tale. *Alytes* **2001**, *18*, 153–177.
15. Savage, J.M.; Heyer, W.R. Digital webbing formulae for anurans: A refinement. *Herp. Rev.* **1997**, *28*, 131.
16. Blommers–Schlösser, R.M.A. *Biosystematics of the Malagasy frogs: Mantellinae (Ranidae)*, 1st ed.; Beaufortia: Amsterdam, Pays-Bas, 1979; pp. 1–77.
17. Fang, Y.Z.; Yang, S.; Wu, G. Free Radicals, Antioxidants, and Nutrition. *Nutrition* **2002**, *18*, 872–879. [CrossRef]
18. Glenn, T.C.; Staton, J.L.; Vu, A.T.; Davis, L.M.; Bremer, J.R.A.; Rhodes, W.E.; Brisbin, I.L., Jr.; Sawyer, R.H. Low Mitochondrial DNA Variation among American Alligators and Non-Coding Region in Crocodilians. *J. Exp. Zool. Mol. Dev. Evol.* **2002**, *294*, 312–324. [CrossRef] [PubMed]
19. Rambaut, A.; Drummond, A.J. *Tracer: A Program for Analysing Results from Bayesian MCMC Programs such as BEAST & MrBayes*, 5th ed.; University of Oxford: Oxford, UK, 2003; p. 1. Available online: https://beast.bio.ed.ac.uk/tracer (accessed on 15 November 2022).
20. de Queiroz, K. The General Lineage Concept of Species, Species Criteria, and the Process of Speciation. In *Endless Forms: Species and Speciation*; Howard, D.J., Berlocher, S.H., Eds.; Oxford University Press: Oxford, UK, 1998; pp. 57–75.
21. Vieites, D.R.; Wollenberg, K.C.; Andreone, F.; Köhler, J.; Glaw, F.; Vences, M. Vast undersestimation of Madagascar's biodiversity evidenced by an integrative amphibian inventory. *Proc. Natl. Acad. Sci. USA* **2009**, *106*, 8267–8272. [CrossRef] [PubMed]
22. Guibé, J. *Les Batraciens de Madagascar*; Bonn. Zool. Monogr.; Smithsonian Libraries: Bonn, Germany; Washington, DC, USA, 1978; Volume 11, pp. 1–140. ISBN 0302-671X.
23. Boulenger, G.A. Descriptions of new reptiles and batrachians from Madagascar. In *Annals and Magazine of Natural History*; London Taylor and Francis, Ltd.: London, UK, 1888; Volume 4, pp. 244–248.
24. Parker, H.W. New and rare reptiles and Batrachians from Madagascar. In *Annals and Magazine of Natural History, Series 9*; London Taylor and Francis, Ltd.: London, UK, 1925; Volume 16, pp. 390–394.

25. Ahl, E. *Amphibia Anura III: Polypedatidae*; Das Tierreich; Walter de Gruyter & Company: Lieferung, Germany, 1931; Volume 55, pp. 1–477. Available online: https://books.google.mg/books/about/Amphibia_Anura_III_Polypedatidae.html?id=M3 HYtgAACAAJ&redir_esc=y (accessed on 25 March 2023).

26. Guibé, J. Etude comparée des espèces lugubris, femoralis et cowani appartenant au genre *Mantidactylus* (Batraciens). *Bull Mus. Natl. Hist. Nat.* **1948**, *20*, 235–238. Available online: https://archive.org/details/biostor-234457 (accessed on 3 February 2007).

27. Dubois, A. Notes sur la classification des Ranidae (Amphibiens anoures). *Bull Mens. Soc. Linn.* **1992**, *61*, 305–352. [CrossRef]

28. Rabibisoa, N.H.C. Systématique et Biogéographie d'un groupe d'Amphibiens des cours d'eau Malagasy: Cas du genre Mantidactylus sous–genre Ochthomantis. Thèse de Doctorat, Université d'Antananarivo, Antananarivo, Madagascar, 8 June 2008.

29. Glaw, F.; Vences, M. *Field Guide to the Amphibians and Reptiles of Madagascar*, 3rd ed.; Vences, M., Glaw, F., Eds.; Verlags, GbR: Köln, Germany, 2007; pp. 1–495. ISBN 978-3-929449-04-4.

30. Mocquard, F. Sur une collection de reptiles recueillis à Madagascar par MM. Alluaud et Belly. *Bull Soc. Philomath* **1895**, *7*, 93–136.

31. Raxworthy, C.J.; Nussbaum, R.A. A rainforest survey of amphibians, reptiles and small mammals at Montagne d'Ambre, Madagascar. *Biol. Conserv.* **1994**, *69*, 65–73. [CrossRef]

32. Angel, M.F. Matériaux de la mission Petit G. à Madagascar. Description de trois batraciens nouveaux appartenant aux genres *Mantidactylus* et *Gephyromantis*. *Bull Mus. Natl. Hist. Nat.* **1928**, *1*, 358–362.

33. Guibé, J. *Catalogue des Types d'Amphibiens du Muséum National d'Histoire Naturelle*; Imprimérie Nationale: Paris, France, 1950.

34. Boulenger, G.A. Descriptions of two new frogs obtained in Madagascar by Dr. Forsyth Major. *Ann. MagnatHist* **1896**, *18*, 420–421. [CrossRef]

35. Angel, M.F. Batraciens nouveaux de Madagascar récoltés par M. R. Cartala. *Bull. Soc. Zool.* **1935**, *60*, 202–207.

36. Angel, M.F. Une grenouille nouvelle de Madagascar appartenant au genre Mantidactylus. *Bull Mus. Natl. Hist. Nat.* **1937**, *2*, 178–179.

37. Glaw, F.; Vences, M. *A Fieldguide to the Amphibians and Reptiles of Madagascar*, 1st ed.; Moos Druck: Köln, Germany, 1992; pp. 1–478. ISBN 3-929449-01-03.

38. Vences, M.; De La Riva, I. *Mantidactylus majori* (NCN). Male egg guarding. *Herpetol. Rev.* **2005**, *36*, 435–436.

39. Porcel, X.; Dubos, N.; Noël, J.; Lava, H.; Velo, J.H.; Georges; Melo, M.; Rosa, G.; Andreone, F.; Crottini, A. Male parental care in Malagasy stream-dwelling frogs of the *Mantidactylus femoralis* group (Anura: Mantellidae: *Ochthomantis*). *Herpetol. Notes* **2022**, *15*, 55–61.

40. Lehtinen, R.M. Parental care and reproduction in two species of Mantidactylus (Anura: Mantellidae). *J. Herpetol.* **2003**, *37*, 766–768. [CrossRef]

41. Wever, R.A. Internal interactions within the human circadian system: The masking effect. *Experientia* **1985**, *47*, 333–342. [CrossRef]

42. Zug, R.G.; Vitt, L.J.; Caldwell, J.P. *Herpetology: An Introductory Biology of Amphibians and Reptiles*, 2nd ed.; Academic Press: London, UK, 2001; pp. 1–645. ISBN 0-12-782622-X.

43. Werner, E.E. Amphibian metamorphosis: Growth rate, predator risk, and the optimal size at transformation. *Am. Nat.* **1986**, *128*, 319–341. [CrossRef]

44. Raxworthy, C.J.; Pearson, R.G.; Rabibisoa, N.; Rakotondrazafy, A.M.; Ramanamanjato, J.B.; Raselimanana, A.P.; Wu, S.; Nussbaum, R.A.; Stone, D.A. Extinction vulnerability of tropical montane endemism from warming and upslope displacement: A preliminary appraisal for the highest massif in Madagascar. *Glob. Chang. Biol.* **2008**, *14*, 1703–1720. [CrossRef]

45. CI; USAID; WWF; Mac Arthur; MEEFT. Assessing the impacts of climate change on Madagascar's biodiversity and livelihoods. In Proceedings of the Madagascar Climate Change and Livelihoods Workshop, Palais de Sport Antananarivo, Antananarivo, Madagascar, 28 January 2008; pp. 1–113.

46. Burney, D.A. Climate change and Fire ecology as factors in the quaternary biogeography of Madagascar. In Proceedings of the Actes du colloque International Biogéographie de Madagascar, ORSTOM, Paris, France, 26–28 September 1996; pp. 49–58.

47. Andreone, F.; Randriamahazo, H. *Sahonagasy Action Plan: Conservation Programs for the Amphibians of Madagascar*; Andreone, F., Randriamahazo, H., Rabibisoa, N.H.C., Eds.; A Conservation Strategy for the Amphibians of Madagascar (ACSAM); MRSN, CI, IUCN; MRSN: Torino, Italy; CI: Washington, DC, USA; IUCN: Gland, Switzerland, 2008; ISBN 978-88-86041-83-6.

Article

Spatial Ecology of an Arboreal Iguana (*Oplurus cyclurus*) in a Treeless Landscape

Fulvio Licata [1,2], Paolo Eusebio Bergò [3], Devin Edmonds [4,5], Franco Andreone [3] and Gonçalo M. Rosa [6,7,*]

1   Centro de Investigação em Biodiversidade e Recursos Genéticos (CIBIO), InBIO Laboratório Associado, Universidade do Porto, Campus de Vairão, 4485-661 Vairão, Portugal
2   Biopolis Program in Genomics, Biodiversity and Land Planning, CIBIO, Campus de Vairão, 4485-661 Vairão, Portugal
3   Museo Regionale di Scienze Naturali, 10123 Torino, Italy; franco.andreone@gmail.com (F.A.)
4   Illinois Natural History Survey, Prairie Research Institute, Champaign, IL 61820, USA
5   Department of Natural Resources and Environmental Sciences, University of Illinois Urbana-Champaign, Champaign, IL 61801, USA
6   Institute of Zoology, Zoological Society of London, London NW1 4RY, UK
7   Centre for Ecology, Evolution and Environmental Changes (cE3c) & Global Change and Sustainability Institute (CHANGE), Faculdade de Ciências da Universidade de Lisboa, Campo Grande, 1749-016 Lisboa, Portugal
*   Correspondence: goncalo.m.rosa@gmail.com

Simple Summary: The spiny-tailed lizard *Oplurus cyclurus* is a widespread endemic iguanian occurring in southern and western Madagascar dry areas. This species is mostly arboreal, and little is known about its spatial ecology. We conducted a radio tracking study on 19 individuals of a population with saxicolous habits, inhabiting an open, treeless savannah in the Isalo sandstone massif (central-southern Madagascar). Tracked lizards had a small home range size (95% isopleth = 247.8 m^2) and showed high site and burrow fidelity. The activity pattern was unimodal, increasing along the day and with juveniles more active than adults in unfavourable weather conditions. Basking occurred mostly near the burrow entrance. Despite high burrow fidelity, lizards changed shelters regularly (approx. once a week), but there was no obvious relation between lizards' body and/or tail size and the width and depth of selected burrows. We argue that the saxicolous habits of this population may entail local behavioural adaptations.

Citation: Licata, F.; Eusebio Bergò, P.; Edmonds, D.; Andreone, F.; Rosa, G.M. Spatial Ecology of an Arboreal Iguana (*Oplurus cyclurus*) in a Treeless Landscape. *Animals* **2023**, *13*, 3198. https://doi.org/10.3390/ani13203198

Academic Editors: Clive J. C. Phillips and Xiang Ji

Received: 3 July 2023
Revised: 12 September 2023
Accepted: 10 October 2023
Published: 13 October 2023

Abstract: Understanding the spatial ecology of species has important implications for conservation, as it helps identify suitable habitats and minimum requirements for biodiversity monitoring and management. The spiny-tailed lizard *Oplurus cyclurus* is a widespread endemic iguanid occurring in dry areas of southern and western Madagascar. While the species is known to be mostly arboreal, populations of the Isalo sandstone massif suggest local adaptation to a less forested savannah and a more exposed habitat. We radio-tracked 19 spiny-tailed lizards to investigate the species' rock-dwelling behaviour and spatial ecology at Isalo National Park. Tracked individuals showed high site and burrow fidelity, and a basking behaviour mostly tied to the accessibility of their burrow, the time of day, and their life stage. Activity peaked during the sunniest hours, while juveniles were more active than adults with unfavourable weather conditions. Despite high burrow fidelity, lizards used shelters non-exclusively, regularly changing (approx. once a week) with neighbouring burrows (average distance between burrows = 13.6 m). However, there was no obvious relation between lizards' body and/or tail size and the width and depth of selected burrows. Dynamic Brownian Bridge Movement Models estimated frequented areas over 247.8 m^2 (95% isopleth), where territorial overlap is common. Our results challenge the notion that burrow-site fidelity is the sole driving factor behind space utilization in the studied population. We argue that the apparently unusual saxicolous habits imposed by habitat features (the absence of trees) may lead to local behavioural adjustments influencing antipredatory and foraging strategies, as well as intraspecific interactions.

Keywords: Sauria; ecology; Isalo; Madagascar; radio-telemetry; home range; site fidelity

1. Introduction

With the herpetofauna of Madagascar facing rapid ongoing habitat loss, the description of new reptile species is often a compulsory action so that species-based conservation measures can promptly be enacted. In particular, the IUCN Red List of Threatened Species requires species to be named to propose conservation categories and apply safeguard measures [1–3]. While much time and effort has been spent on taxonomy, outside a handful of popular species common in the pet trade (e.g., *Furcifer pardalis* [4]), the ethological and ecological requirements of most reptile species in Madagascar remain largely unknown. Indeed, compared to birds and mammals, reptile behaviour overall has been neglected by researchers [5], and for lizards, we are missing critical habitat use data for many species [6].

Iguanians are among the most diverse and species-rich groups of lizards, which occur in a vast range of habitats and have developed a plethora of life history strategies [7]. The family Opluridae contains eight species native to Madagascar, one of which also occurs in the Comoros [8]. Oplurid lizards are the only iguanians known outside the Americas and Pacific Islands [7]. Only a few studies have been conducted on *Oplurus* ecology and behaviour, with the majority concentrated on the better-known *O. cuvieri*. Research has revealed that this species primarily utilizes an ambush strategy to capture prey [9,10] and exhibits notable territorial behaviour and site fidelity [8], which is also observed in other *Oplurus* [11,12]. However, most advancements have been made within captive settings [13,14], leaving the understanding of *Oplurus* spatial ecology in the wild relatively unexplored. Oplurids include mostly saxicolous species (*O. grandidieri*, *O. fierinensis*, *O. saxicola*, and *O. quadrimaculatus*) and the arboreal sister species *O. cuvieri* and *O. cyclurus* [8]. The latter is known to have higher body activity temperatures than its sister species, making it more likely to inhabit warmer environments [15]. Furthermore, contrary to *O. cuvieri*, *O. cyclurus* is known to have at least in part terrestrial habits, as it sometimes catches its prey on the ground [15].

During fieldwork carried out at the Isalo Massif in central-southern Madagascar, we found a population of *O. cyclurus* with fully saxicolous habits. Here, instead of perching on tree branches, *Oplurus* lizards basked on piles of stones, large rocks, or close to burrows that were used as shelters in dry open grassland habitats, often near canyons. Investigating the spatial ecology of populations with different, unusual life habits may help to disentangle the role of habitat in determining behavioural, physiological, or life history changes. Furthermore, contextualising habitat features and the spatial ecology of species may also contribute to a better understanding of their ecological requirements. Thus, studies in these fields represent an exciting and much-needed area of research to advance our understanding of eco-evolutionary processes, providing useful insights into species' ecological requirements for conservation purposes.

Saxicolous habitats are structurally simpler than arboreal ones, and may therefore impose different challenges to species by increasing exposure to predation, altering basking, foraging, and sheltering strategies, while affecting intraspecific interactions [6]. In this study, we hypothesized that saxicolous habits could drive a peculiar utilization of the space in the arboreal *O. cyclurus*. Using radiotracking, we investigated the spatial ecology of this species in an unusual habitat by focusing on (1) home range size, (2) activity patterns, and (3) burrow use.

2. Materials and Methods

2.1. Study Site and Period

Isalo is a large sandstone massif within the Ihorombe Region of central-southern Madagascar, ranging between 510–1268 m a.s.l. A network of canyons cuts through the stone matrix, in particular in the eastern and north-western sectors [16]. The climate of Isalo is sub-desertic and dry tropical, but some canyons are on the limit of the humid eastern and dry western biomes, hosting enclaves of humid forests. Around 850–1200 mm of rain falls every year, with 90% of the precipitation occurring between November and March [17]. A few rivers are permanent, whereas many others are seasonal. Temperatures vary greatly between monthly means: 17 °C in June and 25 °C in February [18].

The northernmost part of the massif is currently managed as a national park by MNP (Madagascar National Park) and is one of the largest protected areas of Madagascar, with 815 km^2 of extension [19]. Extensive areas of the massif are covered by bare rock or savannahs, the latter being maintained by human actions through the centuries. Annual fires are set to manage the savannah to allow grazing for zebu cattle, which are bred in high numbers by the Bara ethnic group in this region [20]. Grass savannah is interspersed with a mosaic of tree savannah mostly represented by fire-resistant species (such as the palm *Bismarckia nobilis* and *Uapaca bojeri*), which sometimes form forest patches. In open areas, particularly on steep slopes or exposed ridges, vegetation is sporadic, dwarf and mainly xerophytic.

Data collection took place during a four-week period in January and February 2011, when increased rainfall tended to heighten foraging activity in Malagasy oplurids [9]. The study was conducted at a site locally known as Malaso (Figure 1; −22.59146°, 45.35744°; elevation 899 m), within the boundary of the national park. Malaso mostly consists of a wide rocky area surrounded by a largely treeless savannah and holds a narrow canyon (extending for about 260 m) below ground level, cut by a small stream. The internal parts of the canyon are deep, around five meters, and several small pools form along the stream bed. Canyons such as this assure a peculiar microclimate and micro-habitat that is consistently wetter and more humid than in the outside savannahs.

Figure 1. Study system: Study site (Malaso) in Isalo Massif, central-southern Madagascar (**bottom**); individual of *Oplurus cyclurus* with a radio-transmitter harnessed to its waist (**top**).

2.2. Radio-Tracking, Distance Measurements, and Environmental Variables

We used snares to capture lizards. Each individual was then weighed with an electronic scale (precision at 0.1 g) and their total length (TL) and tail length (TaL) was measured with a digital calliper to the nearest 0.1 mm. Once measured, we attached a radio-transmitter with a unique frequency harnessed to their waist. We used two kinds of transmitting devices: (1) the ATS A2414 glue-on-transmitter (Advanced Telemetry Systems, Inc., Isanti, MN, USA), and (2) the Holohil BD-2N transmitter (Holohil Systems Ltd., Ottawa, ON, Canada), both with a weight of approx. 0.3 g. We followed Tramontano [21] and affixed the transmitters externally with a cotton thread. The antenna was external, about ten mm long, and left trailing behind the hind legs (Figure 1). Individuals' positions were obtained by homing-in [22], with a Biotrack SIKA receiver and a folding three-element Biotrack antenna.

Search and detection of the radio-tagged individuals was performed once a day. Signals ranged from a maximum distance of 75–120 m. Once located, we marked the position with a flag and recorded GPS coordinates. We then moved the flag and recorded new locations every session. We also recorded if a detected individual was "active" (basking) or "non-active" (hiding in the burrow), and then we measured the distance of the displacement of each individual in relation to the former location. Thus, we categorized locations as "displacement" if the lizards were found at a distance ≥ 50 mm from the former location. The distance between two consecutive contacts (despite the orientation) was measured linearly using a ribbon meter and in the subsequent analysis considered as a simple value (distance). We defined "basking distance" as the Euclidean distance of the basking location to the burrow frequented by the individual on the same day.

Cloud coverage was assessed daily and categorized into three levels: 'clear' (less than 30% sky coverage), 'partially cloudy' (30–70% sky coverage), and 'cloudy' (over 70% sky coverage). We obtained the hourly temperature and the wind speed at 1.2 m above ground level for the centroid of the Minimum Convex Polygons (MCP) including all lizard locations using the R packages *microclima* [23] and *NicheMapR* [24], which allow for producing realistic estimates of microclimate at fine (<30 m) spatial and temporal scales anywhere on Earth [25]. This method allows for obtaining terrain and sub-daily atmospheric forcing data from the National Centers for Environmental Prediction (NCEP) (using the R packages *elevatr* and *RNCEP* [26,27]), which are downscaled and interpolated through a microclimate modelling approach which account for terrain and shade adjustments [25].

2.3. Home Range Estimates

We used Dynamic Brownian Bridge Movement Models (dBBMMs) to calculate the home range of the lizards. This class of models provides occurrence distribution probabilities based on animal movement paths [28], accounting for temporal autocorrelation and incorporating errors associated with the radiotracking device used. Moreover, dBBMMs have been successfully used to estimate home range in reptiles with very different behaviours [29]. We set the window size and margin, which regulate the Brownian motion variance parameter, at 7 and 3 subsequent locations, respectively, to allow calculations for all individuals with more than 7 observations ($n = 10$). When time gaps between observations did not allow the computation of dBBMMs, we estimated the home range using MCP and Kernel Density Estimation (KDE) using the *href bandwidth* selection algorithm. To compute dBBMMs, we used the R package *move* [30], while for MCP and KDE we used the R package *adehabitatHR* [31].

2.4. Activity Pattern

We used binomial Generalized Linear Mixed Models (GLMMs) to test the relationship between activity status and time of observation (minutes from sunrise), including its quadratic term to assess non-linear relationships, the difference between adults and juveniles (i.e., TL more and less than 100 mm, respectively), and the effect of cloud coverage (no clouds, partially cloudy, and cloudy), temperature, and wind speed.

2.5. Burrow Use

We obtained the number of burrows used, the number of times individuals changed burrows, and the number of burrows revisited during the tracking period. To assess the temporal use of burrows, we assumed that consecutive observations in the same burrow did not include undetected displacements to other burrows, while single observations in a burrow were considered as a single day of burrow use. Furthermore, we measured the burrows' width and depth and used Linear Mixed Models (LMMs) to test whether these burrows' features (ln-transformed to reduce skewness) were related to TL and TaL of individuals.

2.6. Basking Behaviour

We used LMMs to test the relationship between the basking distance (ln-transformed to reduce skewness) and time of observation (minutes from sunrise), with its quadratic term, while taking into account the effect of the ontogenetic stages (adults vs. juveniles), cloud coverage (no clouds, partially cloudy, and cloudy), temperature, and wind speed.

2.7. Modelling Procedure

In all mixed models, we considered "individual" as a random effect to account for possible interindividual variation in the response variables. Prior to running the models, we assessed potential collinearity issues [32] between independent variables using (VIF) and its generalised version (i.e., gVIF [33]), which indicated a lack of multicollinearity issues (i.e., VIF and gVIF always <2). We built models using all combinations of explanatory variables and ranked them by corrected Akaike Information Criterion (AIC$_c$) [34]. We used the 'nesting rule' to select the final set of candidate models, excluding all models which had a simpler nested model with lower AIC$_c$ [35]. For each model, we calculated the AIC$_c$ weight, which indicates the relative likelihood of a model given the data and the candidate set of models. Models were run in the *R* environment [36], using the package *lme4* for mixed effect models [37], and the package *MuMIn* for computing models with all combinations of explanatory variables (function *'dredge'* [38]). For LMMs, test statistics were obtained using the *lmerTest* package [39], while for binomial GLMMs we calculated test statistics using the likelihood-ratio test (LRT) [34]. Data visualization was performed using *ggplot2* [40].

3. Results

We made 234 field observations on 19 *Oplurus* individuals, found over an area of 1.9 ha (100% MCP of all observations), which corresponds to 9.7 individuals per hectare. On average, we recorded 12 observations per individual (SD = 8; median = 8; range = 1–46; Table S1). The average time between observations was 21.6 h (SD = 16). Seventeen individuals were monitored for an average of 11 days each (SD = 7; median = 9; range = 1–30; see Table S1), while two were followed for less than 24 h (Table S1).

3.1. Home Range Estimates

The average home range size for nine individuals was 1036.6 m^2 (SD = 1511.4; range = 36.0–4650.7) for the 99% dBBMM isopleths, 247.8 m^2 (SD = 277.9; range = 11.7–762.1) for the 95% isopleths, and 40.9 m^2 (SD = 48.8; range = 4–145.9) for the 75% isopleths (see Table S1). The size of the 99% isopleths (ln-transformed to reduce skewness) was correlated with the duration of the tracking period (Pearson's correlation r = 0.8, df = 7, p = 0.002), but it was not correlated with the body size of the individuals (Pearson's correlation r = 0.3, df = 7, p = 0.491).

Six individuals had overlapping home ranges, resulting in 12 cases of overlapped home ranges. In four cases, three lizards had the entire home range included in the 99% isopleth of other individuals, with two individuals even sharing (at different times) the same burrow. In the remaining cases (n = 8), home ranges overlapped by an average of 12% (SD = 21; range = 0.9–63.3) (Figure 2). Lastly, we failed to estimate the home range size of one individual (i.e., 011) due to excessive time gaps between observations. However, the 100% Minimum Convex Polygon for this individual was 98.5 m^2, while KDE returned 95% fixed kernels of 602.8 m^2 (75% KDE = 315.3 m^2; 99% KDE = 851.1 m^2).

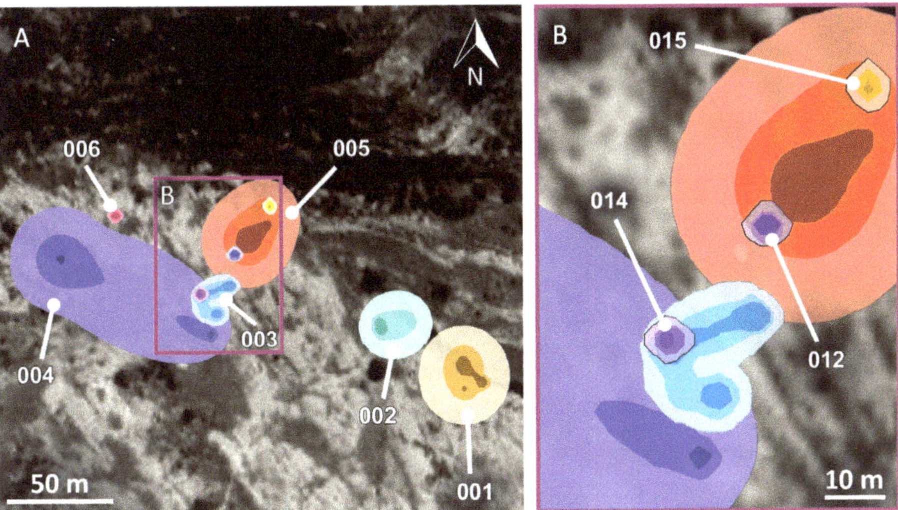

Figure 2. (**A**) Estimated home ranges of nine individuals of *Oplurus cyclurus* in Malaso, Isalo National Park (Madagascar) for which dBBMMs could be computed; (**B**) Individuals exhibiting overlapping home ranges.

3.2. Activity Status

The best model explaining variation in activity status had no competing models (i.e., models with $\Delta AIC_c < 2$; Table S2), and indicated that the activity pattern followed a quadratic trend (binomial GLMM, LRT; $\chi^2 = 9.2$, df = 2, $p = 0.009$) with a peak of activity during solar noon (Figure 3A,B). Furthermore, weather conditions significantly predicted the activity of individuals, which were more active with higher temperatures ($\chi^2 = 5.9$, df = 1, $p = 0.015$) and less active during cloudy days ($\chi^2 = 12.1$, df = 2, $p = 0.002$) (Table S2). Lastly, juveniles showed a wider activity window than adults ($\chi^2 = 7.4$, df = 1, $p = 0.006$; Figure 3A). Past 17:00, all individuals were sheltering or in burrows (Figure 3B). No other *Oplurus* were observed past that time, even if there was still relative sunlight.

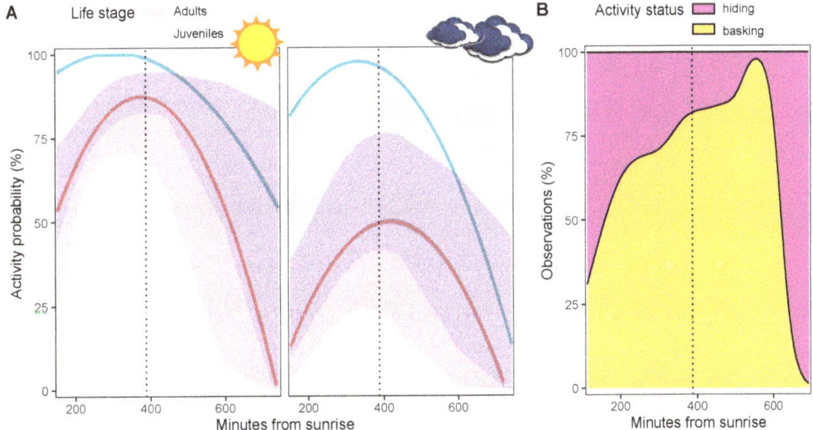

Figure 3. Activity pattern of 19 tracked individuals of *Oplurus cyclurus* in Malaso, Isalo National Park (Madagascar). (**A**) Predicted probability of *Oplurus* lizard activity in relation to the minutes from the sunrise during sunny days (**left**) and cloudy days (**right**); (**B**) Frequency distribution of observations of active and non-active (see legend) *Oplurus* lizards in relation to the minutes from the sunrise. In all graphs, the dotted vertical line indicates the solar noon.

3.3. Burrow Use

Oplurus cyclurus individuals were commonly observed seeking refuge in narrow rocky spaces, specifically utilizing crevices and natural burrows in rock formations. Burrows were on average 30.4 cm wide (range = 19.6–58.6 cm), 154.7 cm deep (range = 16.5–280 cm) and, excluding the out-and-back movements to the same burrow, the mean distance between consecutive burrows was 13.6 m (mean = 14.5; SD = 16.9; range = 4.1–79.2) (Figure 4A). Burrow features were not related to any of the morphological traits considered (i.e., total and tail lengths; Table S2).

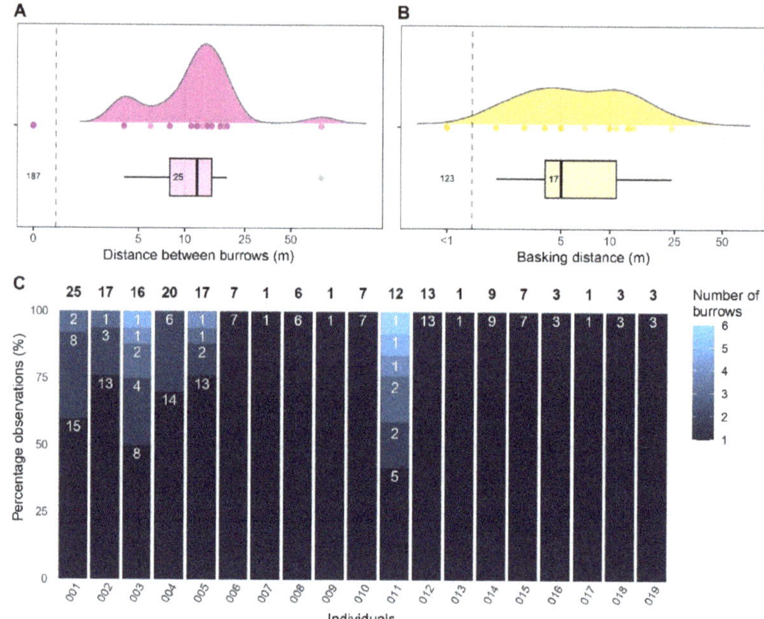

Figure 4. Burrow use and basking distances of 19 tracked individuals of *Oplurus cyclurus* in Malaso, Isalo National Park (Madagascar). (**A**) Frequency distribution of distances between consecutive burrows, and (**B**) frequency distribution of basking distances from the last frequented burrow. Box and density plots are plotted excluding non-movements and movements up to one meter, respectively. The number of observations used to create the boxplot is indicated within the boxplot; (**C**) Observations of burrow use in the tracked *Oplurus* lizards. Numbers on the top of the bars represent the total number of days of observation in a burrow, while numbers inside the bars refer to the number of days in which the individual was observed in the same burrow.

The number of burrows used per individual ranged from one to six (Figure 4C; Table S1) and was positively correlated with the number of days of the tracking period (Pearson's correlation r = 0.7, df = 17, p = 0.001). Considering those individuals tracked for more than four days (n = 12), the average number of burrows frequented per week was 1.1 (SD = 0.6; range = 0.5–2.9). Excluding the first observations of the tracking period (n = 17) and the observations of individuals monitored for less than 24 h (n = 3), in 87.8% of cases (n = 186), the individuals used the same burrow of the prior observation. Six individuals changed burrows on twenty-six occasions (i.e., 12.2% of observations), nine times of which they revisited a previously occupied burrow. The number of revisited burrows represented, on average, 32% of the total burrows frequented by these individuals (SD = 30; range = 0–67). However, considering the differences in the duration of the tracking period between individuals, this probably represents an underestimate of the actual burrow-site fidelity, as longer tracking periods resulted in more revisited burrows (Pearson's correlation r = 0.7, df = 17, p = 0.001).

Oplurus lizards spent on average 5.9 consecutive days in the same burrow (SD = 3.9; range = 1–14), corresponding to 83.1% of the tracking period (SD = 26.6; range = 29.4–100). In cumulative terms, the amount of time (i.e., number of days) spent in a burrow by a single individual was on average 88.1% of the tracking period (SD = 19.4; range = 42–100), corresponding to 6.8 days (SD = 4.8; range = 1–15) (Figure 4C).

3.4. Basking Behaviour

We recorded 140 observations of basking lizards. The median distance between the basking spot and the respective burrow was 0.2 m (SD = 5.9 m; range = 0.0–24.5 m; Table S1). However, excluding the basking distances of over one meter, the median value was 4.7 m, and only five times (3.5% of observations) were individuals recorded at more than ten meters from their reference burrow (Figure 4B). Furthermore, the maximum distance of basking was not correlated with the size of the 99% dBBMM isopleth (Pearson's correlation r = −0.1, df = 7, p = 0.917). Lastly, the best LMM had no competing models (Table S2) and indicated a linear increase in the basking distance with the time of the day (t = 2.6, p = 0.010).

4. Discussion

Overall, *Oplurus* lizards showed small home range sizes, not related to body size, and high burrow-site fidelity, with individuals occupying the same burrow for up to two consecutive weeks. Nonetheless, individuals made use of multiple burrows (up to six in our study), and even shared burrows with other conspecifics at different times. The activity pattern was unimodal, with activity and basking distance from the burrow increasing throughout the day until sunset. We argue that the saxicolous habits imposed by habitat features (the absence of trees) may result in local behavioural adjustments influencing antipredatory and foraging strategies, as well as intraspecific interactions.

4.1. Space Use

Lizards' home range size is believed to result from the interplay of intrinsic (e.g., behavioural and physiological) and extrinsic factors (e.g., habitat features and resource availability) [41]. Arboreal and saxicolous lizards are predicted to have smaller home ranges than terrestrial (open ground) species [41]. The home range size of *O. cyclurus* is much smaller than the terrestrial leopard lizards *Gambelia sila* and *G. wislizenii*, in which home range can cover areas of multiple hectares [42,43], or other terrestrial iguanian lizards of comparable size (e.g., *Crotaphytus reticulatus* [44]). Conversely, our estimates largely overlap with other saxicolous (e.g., *Tropidurus hispidus* [45]) or arboreal species (e.g., *Sceloporus undulatus* [46]), including the sister species *O. cuvieri* [47] where males' home ranges are on average 1315 m^2 [48]. However, the possible influence of individual sex on home range size [41] was not assessed in our study, leaving the potential of a skewed estimated average home range size open.

Lizard body size is positively related to home range size and likely reflects the energetic requirements of the species [41]. *Oplurus cuvieri* is bigger than *O. cyclurus* [49], which could explain the larger home range size (approx. 30% of difference) (Table S1). Alternatively, the different methods used in other spatial ecology studies (i.e., home range size estimators, radio transmitter models, and attachment techniques) may have hampered a direct comparison of results. In the past, the most widely supported method for estimating home range size and making comparisons across studies was MCP [50]. Nonetheless, MCP is found to substantially overestimate home range areas [51], contrary to dBBMMs which currently represent the best trade-off in describing space utilization in reptiles [29].

Foraging strategies also determine the home range size of lizards, with active foragers using larger areas than ambushers or herbivorous species [7]. These differences can become evident even within the same genus of lizard, such as in the case of *G. sila* and *G. wislizenii*, with the latter having a consistently larger home range as a result of the active foraging strategy [42]. *Oplurus cuvieri* and *O. cyclurus* (and oplurids in general) are sit-and-wait predators [8,9], and their foraging strategy might therefore be reflected in their similar home range sizes. Relatedly, our estimate of home range size refers to the rainy season, when dietary needs and food availability are probably higher for this species; therefore, our estimates of home range size likely reflect the maximum across seasons.

Lastly, attachment techniques and radio transmitter mass can sometimes affect the locomotor performances of lizards [52]. The low ratio between body weight and radio-transmitter weight in our study should be a positive factor, but we cannot exclude that the kind of external affixation and external antenna, while not being as invasive as implantation, could alter the animals' displacements, influence predation episodes, or reduce hiding capacity. However, our fieldwork showed that the technique of external tagging was effective, as we did not record any lost or dead individual nor any predation event.

4.2. Activity Pattern and Burrow Use

The lizards showed a unimodal activity pattern, which is common in species with high thermal tolerance, like *O. cyclurus* [15]. However, our study period was during a favourable season for the species, and it is not uncommon for activity patterns to shift to bimodality in hotter periods of the year [53].

The higher activity rates in younger individuals can be explained by differences in heating rates between age classes (smaller individuals have higher surface-to-volume ratios; e.g., [52,54]). Conversely, the basking strategy was similar between juveniles and adults, which mostly basked near their burrow entrance. In lizards, distance to the refuge is often adjusted as a trade-off between perceived predatory risk and body temperature and locomotor performance [52,55]. The tracked *Oplurus* increased the basking distance throughout the day, likely as a result of increased body temperatures and lower perceived predation risk. Habitat structural complexity and seasonality may affect predation risk in lizards [56,57]. Our study area (a rocky soil in a fairly treeless savannah) has a structurally simple habitat in which predation efficiency is predicted to be high; therefore, if the *O. cyclurus* were overexposed to predators at our site, they might have adopted a basking strategy aimed at reducing predation risk.

Our results show high burrow-site fidelity in *O. cyclurus*; however, individuals also changed burrows frequently, using alternative refuges for up to 32% of the tracking period (see Results; Figure 4C). This behaviour suggests they did not rely exclusively on specific burrows. On the one hand, in sit-and-wait forager lizards, frequent burrow change may increase feeding opportunities, and reduce predation risk and parasite load [58]. On the other hand, habitat features and shelter availability may also play a key role in determining spatial utilization and burrow use in lizards, which may also reflect on population structure [59]. For instance, the arboreal *O. cuvieri* principally makes use of a single shelter (a tree hollow [48]) and attains densities of 3.5 lizards/ha, which is nearly three times lower than our study population (i.e., 9.7 lizards/ha). Saxicolous habitats may harbour a higher number of potential shelters, which could result in different selective pressures on *Oplurus* populations. These may include alteration in territoriality, antipredatory, and foraging strategies, but also have repercussions on parasite load or persistence, ultimately reflecting on local adaptive fitness responses.

5. Conclusions

Despite its Least Concern status, *O. cyclurus* is facing local decline attributed to significant habitat pressure [60]. Understanding the spatial ecology of this species provides crucial insights into habitat use, territoriality, and foraging strategies, enabling targeted conservation efforts. On the other hand, emphasizing behavioural and ecological aspects of species under unconventional environmental conditions may prove valuable in supporting theory about local adaptation.

Supplementary Materials: The following supporting information can be downloaded at: https://www.mdpi.com/article/10.3390/ani13203198/s1, Table S1: Table of radio-tracked individuals of *Oplurus* *cyclurus* in Isalo Massif (central-southern Madagascar); Table S2: List of Generalized and Linear Mixed models explaining the activity status, the basking behaviour and the burrow use of *Oplurus cyclurus* in Isalo Massif (central-southern Madagascar).

Author Contributions: Conceptualization, P.E.B. and G.M.R.; Methodology, P.E.B., D.E. and G.M.R.; Validation, F.L.; Formal analysis, F.L. and G.M.R.; Investigation, P.E.B., D.E. and G.M.R.; Resources, F.A.; Data curation, F.L. and G.M.R.; Writing—original draft, F.L. and G.M.R.; Writing—review & editing, F.L., D.E. and G.M.R.; Supervision, F.A. and G.M.R.; Funding acquisition, F.A. All authors have read and agreed to the published version of the manuscript.

Funding: The work was supported by the Mohamed bin Zayed Species Conservation Fund (grant number 0925453), EDGE (Zoological Society of London), Amphibian Specialist Group, Conservation International, Gondwana Conservation and Research, Madagascar Fauna Group, Museo Regionale di Scienze Naturali, Zoo Zürich, and Nando Peretti Foundation.

Institutional Review Board Statement: All fieldwork conformed to the highest standards of animal welfare for animal capture and handling following the "Guidelines for the ethical use of animals in applied ethology studies" [61]. Methods and protocols were performed in accordance with the relevant local regulations and licensing after review by Commission ad'hoc Faune et Flore (CAFF)/Comité d'Orientation de la Recherche Environnementale (CORE).

Informed Consent Statement: Not applicable.

Data Availability Statement: Any computer codes used to generate results reported in the manuscript, as well as raw data that support the findings of this study, are available on request from the corresponding author, without undue reservation.

Acknowledgments: We are indebted to the Parc Botanique et Zoologique de Tsimbazaza, University of Antananarivo, and the Direction des Eaux et Forêts for permits issued to conduct research at the protected area of Isalo (permit n. 191/09/MEF/SG/DGF/DCB.SAP/SLRSE of 22.IX.2009, 222/09/MEF/SG/DGF/DCB.SAP/SLRSE of 12.X.2009, and 357/10/MEF/SG/DGF/DCB.SAP/SCB of 17.XII.2010). The team of Madagascar National Parks (MNP) at Ranohira-Isalo was also very collaborative and helpful. The people of MICET helped with logistics and assistance. We are also grateful to our guide Anicet, as well as N. H. C. Rabibisoa, T. J. Razafindrabe, J. E. Randrianirina, N. Rabibisoa, and I. Lau for assistance in the field and useful talks.

Conflicts of Interest: The authors declare no conflict of interest.

References

1. Stuart, S.N.; Chanson, J.S.; Cox, N.A.; Young, B.E.; Rodrigues, A.S.L.; Fischman, D.L.; Waller, R.W. Status and Trends of Amphibian Declines and Extinctions Worldwide. *Science* **2004**, *306*, 1783–1786. [CrossRef] [PubMed]
2. Andreone, F.; Cadle, J.E.; Cox, N.; Glaw, F.; Nussbaum, R.A.; Raxworthy, C.J.; Stuart, S.N.; Vallan, D.; Vences, M. Species Review of Amphibian Extinction Risks in Madagascar: Conclusions from the Global Amphibian Assessment. *Conserv. Biol.* **2005**, *19*, 1790–1802. [CrossRef]
3. Andreone, F.; Carpenter, A.I.; Cox, N.; du Preez, L.; Freeman, K.; Furrer, S.; Garcia, G.; Glaw, F.; Glos, J.; Knox, D.; et al. The Challenge of Conserving Amphibian Megadiversity in Madagascar. *PLoS Biol.* **2008**, *6*, e118. [CrossRef] [PubMed]
4. Gehring, P.-S.; Lutzmann, N.; Furrer, S.; Sossinka, R. Habitat Preferences and Activity Patterns of *Furcifer pardalis* (CUVIER, 1829) in the Masoala Rain Forest Hall of the Zurich Zoo. *Salamandra* **2008**, *44*, 129–140.
5. Doody, J.S.; Burghardt, G.M.; Dinets, V. Breaking the Social–Non-Social Dichotomy: A Role for Reptiles in Vertebrate Social Behavior Research? *Ethology* **2013**, *119*, 95–103. [CrossRef]
6. Smith, G.R.; Ballinger, R.E. The Ecological Consequences of Habitat and Microhabitat Use in Lizards: A Review. *Contemp. Herpetol.* **2001**, *3*, 1–28. [CrossRef]
7. Vitt, L.J.; Caldwell, J.P. *Herpetology: An Introductory Biology of Amphibians and Reptiles*; Academic Press: Cambridge, UK, 2013.
8. Cadle, J.E.; Chan, L.; Miralles, A. Opluridae, Malagasy Iguanas, Androngovato. In *The New Natural History of Madagascar*; Goodman, S.M., Andrianarimisa, A., Armstrong, A.H., Cooke, A., De Wit, M., Ganzhorn, J.U., Gautier, L., Eds.; Princeton University Press: Princeton, NJ, USA, 2022; pp. 1502–1505, ISBN 978-0-691-22262-2.
9. Mori, A.; Randriamahazo, H.J.A.R. Foraging Mode of a Madagascan Iguanian Lizard, *Oplurus cuvieri cuvieri*. *Afr. J. Ecol.* **2002**, *40*, 61–64. [CrossRef]
10. Rosa, G.M.; Rakotozafy, S. Opportunistic but Voracious: Madagascan Spiny-Tailed Iguana, *Oplurus cuvieri* (Reptilia: Opluridae) Predation upon a Small Mammal. *Herpetol. Notes* **2013**, *6*, 159–161.
11. Brillet, C. Contribution à l'étude des relations entre individus chez cinq espèces d'iguanes malgaches du genre *Oplurus*. *Rev. D'ecologie Terre Vie* **1982**, *36*, 79–148. [CrossRef]

12. Lobón-Rovira, J.; Belluardo, F.; Crottini, A.; Rosa, G.M. Territorial Behaviour in *Oplurus grandidieri* (Mocquard, 1900), a Rock Dwelling Lizard from Madagascar. *Herpetol. Notes* **2019**, *12*, 783–786.

13. Dickinson, H.C.; Fa, J.E. Ultraviolet Light and Heat Source Selection in Captive Spiny-Tailed Iguanas (*Oplurus cuvieri*). *Zoo Biol.* **1997**, *16*, 391–401. [CrossRef]

14. Gibson, R.C.; Buley, K.R. Captive Management and Breeding of Madagascar Spiny Iguanas *Oplurus cuvieri Cuvieri* Gray, 1831. *Dodo* **1996**, *32*, 137–143.

15. Randriamahazo, H.J.A.R. Activity Temperatures in *Oplurus cyclurus*, *Oplurus cuvieri* and *Zonosaurus laticaudatus* and Resting Metabolic Rates in the Latter Two Species. *Amphib.-Reptil.* **1998**, *19*, 215–220. [CrossRef]

16. Rakotonomenjanahary, O.M.; Hawkins, A.F.A. Le Projet 'Zicoma' Ou 'Zones d'Importance Pour La Conservation Des Oiseaux a Madagascar'. *Ostrich* **2000**, *71*, 168–171. [CrossRef]

17. Cocca, W.; Rosa, G.; Andreone, F.; Aprea, G.; Eusebio Bergò, P.; Mattioli, F.; Mercurio, M.; Randrianirina, J.; Rosado, D.; Vences, M.; et al. The Herpetofauna (*Amphibia, Crocodylia, Squamata, Testudines*) of the Isalo Massif, Southwest Madagascar: Combining Morphological, Molecular and Museum Data. *Salamandra* **2018**, *54*, 178–200.

18. Mercurio, V.; Aprea, G.; Crottini, A.; Mattioli, F.; Randrianirina, J.E.; Razafindrabe, T.J.; Andreone, F. The Amphibians of Isalo Massif, Southern-Central Madagascar: High Frog Diversity in an Apparently Hostile Dry Habitat. *Monogr. Del. Mus. Reg. Sci. Nat. Torino* **2008**, *XLI*, 5–58.

19. ANGAP (Association Nationale pour la Gestion des Aires Protegées). *Plan de Gestion du Réseau National des Aires Protégées de Madagascar*; ANGAP and Ministère de l'Environnement: Antananarivo, Madagascar, 2003.

20. Kull, C.A. The "Degraded" Tapia Woodlands of Highland Madagascar: Rural Economy, Fire Ecology, and Forest Conservation. *J. Cult. Geogr.* **2002**, *19*, 95–128. [CrossRef]

21. Tramontano, R. Continuous Radiotracking of the Common Frog, *Rana temporaria*. In *Herpetologia Bonnensis*; SEH: Bonn, Germany, 1997; pp. 359–365.

22. White, G.C.; Garrott, R.A. *Analysis of Wildlife Radio-Tracking Data*; Academic Press: Cambridge, UK, 1990; ISBN 978-0-08-092657-5.

23. Maclean, I.M.D.; Mosedale, J.R.; Bennie, J.J. Microclima: An r Package for Modelling Meso- and Microclimate. *Methods Ecol. Evol.* **2019**, *10*, 280–290. [CrossRef]

24. Kearney, M.R.; Porter, W.P. NicheMapR—An R Package for Biophysical Modelling: The Microclimate Model. *Ecography* **2017**, *40*, 664–674. [CrossRef]

25. Kearney, M.R.; Gillingham, P.K.; Bramer, I.; Duffy, J.P.; Maclean, I.M.D. A Method for Computing Hourly, Historical, Terrain-Corrected Microclimate Anywhere on Earth. *Methods Ecol. Evol.* **2020**, *11*, 38–43. [CrossRef]

26. Hollister, J.; Shah, T.; Robitaille, A.L.; Beck, M.W.; Johnson, M. Elevatr: Access Elevation Data from Various APIs. Version 0.4.5. 2023. Available online: https://CRAN.R-project.org/package=elevatr (accessed on 6 September 2023).

27. Kemp, M.U.; van Loon, E.E.; Shamoun-Baranes, J.; Bouten, W. RNCEP: Obtain, Organize, and Visualize NCEP Weather Data. Version 1.0.10. 2020. Available online: https://CRAN.R-project.org/package=RNCEP (accessed on 6 September 2023).

28. Kranstauber, B.; Kays, R.; LaPoint, S.D.; Wikelski, M.; Safi, K. A Dynamic Brownian Bridge Movement Model to Estimate Utilization Distributions for Heterogeneous Animal Movement. *J. Anim. Ecol.* **2012**, *81*, 738–746. [CrossRef] [PubMed]

29. Silva, I.; Crane, M.; Marshall, B.M.; Strine, C.T. Reptiles on the Wrong Track? Moving beyond Traditional Estimators with Dynamic Brownian Bridge Movement Models. *Mov. Ecol.* **2020**, *8*, 43. [CrossRef]

30. Kranstauber, B.; Smolla, M.; Scharf, A.K. Move: Visualizing and Analyzing Animal Track Data. Version 4.2.4. 2023. Available online: https://CRAN.R-project.org/package=move (accessed on 6 September 2023).

31. Calenge, C. The Package "Adehabitat" for the R Software: A Tool for the Analysis of Space and Habitat Use by Animals. *Ecol. Model.* **2006**, *197*, 516–519. [CrossRef]

32. Dormann, C.F.; Elith, J.; Bacher, S.; Buchmann, C.; Carl, G.; Carré, G.; Marquéz, J.R.G.; Gruber, B.; Lafourcade, B.; Leitão, P.J.; et al. Collinearity: A Review of Methods to Deal with It and a Simulation Study Evaluating Their Performance. *Ecography* **2013**, *36*, 27–46. [CrossRef]

33. Fox, J.; Monette, G. Generalized Collinearity Diagnostics. *J. Am. Stat. Assoc.* **1992**, *87*, 178–183. [CrossRef]

34. Burnham, K.P.; Anderson, D.R. *Model Selection and Multimodel Inference: A Practical Information-Theoretic Approach*, 2nd ed.; Springer: New York, NY, USA, 2002; ISBN 978-0-387-95364-9.

35. Richards, S.A.; Whittingham, M.J.; Stephens, P.A. Model Selection and Model Averaging in Behavioural Ecology: The Utility of the IT-AIC Framework. *Behav. Ecol. Sociobiol.* **2011**, *65*, 77–89. [CrossRef]

36. R Core Team. *R: A Language and Environment for Statistical Computing*; R Foundation for Statistical Computing: Vienna, Austria, 2023.

37. Bates, D.; Mächler, M.; Bolker, B.; Walker, S. Fitting Linear Mixed-Effects Models Using Lme4. *J. Stat. Softw.* **2015**, *67*, 1–48. [CrossRef]

38. Barton, K. MuMIn: Multi-Model Inference. Version 1.47.5. 2022. Available online: https://CRAN.R-project.org/package=MuMIn (accessed on 6 September 2023).

39. Kuznetsova, A.; Brockhoff, P.B.; Christensen, R.H.B. LmerTest Package: Tests in Linear Mixed Effects Models. *J. Stat. Softw.* **2017**, *82*, 1–26. [CrossRef]

40. Wickham, H.; Chang, W.; Henry, L.; Pedersen, T.L.; Takahashi, K.; Wilke, C.; Woo, K.; Yutani, H.; Dunnington, D.; Posit, P.B.C. Ggplot2: Create Elegant Data Visualisations Using the Grammar of Graphics. Version 3.4.3. 2023. Available online: https://CRAN.R-project.org/package=ggplot2 (accessed on 6 September 2023).

41. Perry, G.; Garland, T. Lizard Home Ranges Revisited: Effects of Sex, Body Size, Diet, Habitat, and Phylogeny. *Ecology* **2002**, *83*, 1870–1885. [CrossRef]

42. Schorr, R.A.; Lambert, B.A.; Freels, E. Habitat Use and Home Range of Long-Nosed Leopart Lizards (*Gambelia wislizenii*) in Canyons of the Ancients National Monument, Colorado. *Herpetol. Conserv. Biol.* **2011**, *6*, 312–323.

43. Warrick, G.D.; Kato, T.T.; Rose, B.R. Microhabitat Use and Home Range Characteristics of Blunt-Nosed Leopard Lizards. *J. Herpetol.* **1998**, *32*, 183–191. [CrossRef]

44. Ryberg, W.A.; Garrett, T.B.; Adams, C.S.; Campbell, T.A.; Walkup, D.K.; Johnson, T.E.; Hibbitts, T.J. Life in the Thornscrub: Movement, Home Range, and Territoriality of the Reticulate Collared Lizard (*Crotaphytus reticulatus*). *J. Nat. Hist.* **2019**, *53*, 1707–1719. [CrossRef]

45. Melo, G.; Pinheiro, L.; Passos, D.; Galdino, C. Spatial Organisation of the Neotropical Lizard *Tropidurus hispidus* (Squamata: Tropiduridae). *Salamandra* **2017**, *53*, 435–438.

46. Ferner, J.W. Home-Range Size and Overlap in *Sceloporus undulatus erythrocheilus* (Reptilia: Iguanidae). *Copeia* **1974**, *1974*, 332–337. [CrossRef]

47. Chan, L.M.; Choi, D.; Raselimanana, A.P.; Rakotondravony, H.A.; Yoder, A.D. Defining Spatial and Temporal Patterns of Phylogeographic Structure in Madagascar's Iguanid Lizards (Genus *Oplurus*). *Mol. Ecol.* **2012**, *21*, 3839–3851. [CrossRef] [PubMed]

48. Randriamahazo, H.J.A.R.; Mori, A. Spatial Utilization and Social Interactions in *Oplurus Cuvieri Cuvieri* (Squamata, Opluridae) in Madagascar. *Jpn. J. Herpetol.* **1999**, *18*, 57–65. [CrossRef] [PubMed]

49. Glaw, F.; Vences, M. *A Field Guide to the Amphibians and Reptiles of Madagascar*, 3rd ed.; Vences & Glaw: Köln, Germany, 2007; ISBN 978-3-929449-03-7.

50. Row, J.R.; Blouin-Demers, G. Kernels Are Not Accurate Estimators of Home-Range Size for Herpetofauna. *Copeia* **2006**, *2006*, 797–802. [CrossRef]

51. Laver, P.N.; Kelly, M.J. A Critical Review of Home Range Studies. *J. Wildl. Manag.* **2008**, *72*, 290–298. [CrossRef]

52. Carrascal, L.M.; López, P.; Martín, J.; Salvador, A. Basking and Antipredator Behaviour in a High Altitude Lizard: Implications of Heat-Exchange Rate. *Ethology* **1992**, *92*, 143–154. [CrossRef]

53. Burke, R.L.; Ner, S.E. Seasonal and Diel Activity Patterns of Italian Wall Lizards, *Podarcis sicula campestris*, in New York. *Nena* **2005**, *12*, 349–360. [CrossRef]

54. Paulissen, M.A. Ontogenetic and Seasonal Comparisons of Daily Activity Patterns of the Six-Lined Racerunner, *Cnemidophorus sexlineatus* (Sauria: Teiidae). *Am. Midl. Nat.* **1988**, *120*, 355–361. [CrossRef]

55. Lima, S.L.; Dill, L.M. Behavioral Decisions Made under the Risk of Predation: A Review and Prospectus. *Can. J. Zool.* **1990**, *68*, 619–640. [CrossRef]

56. Ferreira, A.S.; Faria, R.G. Predation Risk Is a Function of Seasonality Rather than Habitat Complexity in a Tropical Semiarid Forest. *Sci. Rep.* **2021**, *11*, 16670. [CrossRef]

57. Shepard, D.B. Habitat but Not Body Shape Affects Predator Attack Frequency on Lizard Models in the Brazilian Cerrado. *Herpetologica* **2007**, *63*, 193–202. [CrossRef]

58. Fenner, A.L.; Godfrey, S.S.; Michael Bull, C. Using Social Networks to Deduce Whether Residents or Dispersers Spread Parasites in a Lizard Population. *J. Anim. Ecol.* **2011**, *80*, 835–843. [CrossRef] [PubMed]

59. Gracceva, G.; Bombi, P.; Luiselli, L.; Bologna, M. Do Demographic Aspects of Neighbouring Lizard Populations Differ? A Case Study with the Common Wall Lizard, *Podarcis muralis*. *Amphib.-Reptil.* **2008**, *29*, 443–448. [CrossRef]

60. Vences, M. Oplurus Cyclurus. IUCN Red List. Threat. Species 2011: e.T172831A6926304. Available online: https://doi.org/10.2305/IUCN.UK.2011-2.RLTS.T172831A6926304.en (accessed on 6 September 2023).

61. Sherwin, C.M.; Christiansen, S.B.; Duncan, I.J.; Erhard, H.W.; Lay, D.C.; Mench, J.A.; O'Connor, C.E.; Petherick, J.C. Guidelines for the Ethical Use of Animals in Applied Ethology Studies. *Appl. Anim. Behav. Sci.* **2003**, *81*, 291–305. [CrossRef]

Article

Empirical Evidence for the Rescue Effect from a Natural Microcosm

Richard M. Lehtinen [†]

Division of Reptiles and Amphibians, University of Michigan Museum of Zoology, Ann Arbor, MI 48109, USA; rlehtinen@wooster.edu

[†] Present Address: Biology Department, 931 College Mall, The College of Wooster, Wooster, OH 44691, USA.

Simple Summary: A widely assumed ecological process called the "rescue effect" holds that populations of organisms that receive immigrants from other populations should be less vulnerable to local extinction than populations that do not receive such immigrants. While sensible, the rescue effect has had little strong support from real-world populations because of the difficulty of testing this idea. Using a study system of plant-specialist frogs from the rainforests of Madagascar, this study provides evidence for the rescue effect as well as weaker evidence for a related (but opposite) process referred to as the "abandon-ship effect".

Abstract: Ecological theory predicts that populations which receive immigrants are less vulnerable to extinction than those that do not receive immigrants (the "rescue effect"). A parallel but opposite process may also exist, where emigration increases the risk of local extinction (the "abandon-ship effect"). Using a natural microcosm of plant-specialist frogs from Madagascar, empirical evidence for both processes is provided. Populations receiving immigrants were less extinction-prone than those without immigration, and those populations losing individuals through emigration were more extinction-prone than those in which no emigration occurred. The number of immigrants and emigrants was also elevated and depressed (respectively) in patches that did not go extinct. These data provide some of the first definitive empirical evidence for the rescue effect and provide suggestive initial data on the abandon-ship effect. Both of these processes may be important to understanding the dynamics of populations.

Keywords: *Guibemantis*; immigration; emigration; dispersal; population ecology; extinction risk; *Pandanus*; rescue effect; abandon-ship effect

Citation: Lehtinen, R.M. Empirical Evidence for the Rescue Effect from a Natural Microcosm. *Animals* **2023**, *13*, 1907. https://doi.org/10.3390/ani13121907

Academic Editors: Angelica Crottini, Franco Andreone, Andolalao Rakotoarison and Fandresena Rakotoarimalala

Received: 20 March 2023
Revised: 27 May 2023
Accepted: 5 June 2023
Published: 7 June 2023

1. Introduction

Small populations are widely thought to be more vulnerable to extinction from stochastic factors than larger populations [1,2]. Small populations can forestall extinction in two ways: by either successfully reproducing locally or by receiving immigrants from other populations. Either process increases the local population size, and this by itself reduces the immediate risk of local extinction. Immigration as a demographic mechanism that lowers extinction risk in small, spatially isolated populations (the "rescue effect") was proposed almost 50 years ago by Brown and Kodric-Brown [3]. Island biogeography theory [4] provided the original theoretical context for the rescue effect and its intellectual descendent, metapopulation theory [5–7], also utilizes this concept. The rescue effect mechanism is also the logical basis for theories of source–sink dynamics [8] and the emphasis on corridors and population connectivity in conservation biology [9–11].

A related but different process called the genetic rescue effect has also been proposed which was originally defined as "the increase in the probability of a population's survival due to the immigration of genes from another population" [12]. Since small, isolated populations can become inbred and lose genetic diversity via genetic drift over time, this infusion of alleles from other populations can increase subsequent offspring fitness and population growth rates [12]. The increased genetic variation from immigration can also lead to "evolutionary rescue" whereby small populations successfully adapt to a changing environment [13]. Both the genetic rescue effect and evolutionary rescue are now well-established phenomena, having been documented in varying degrees in a wide variety of organisms and circumstances [14]. However, empirical support for the original demographic rescue effect of Brown and Kodric-Brown [3] remains scarce.

If immigration can decrease the probability of extinction in small populations, the converse may also be true (i.e., emigration may increase the chance of local extinction). Since emigration necessarily reduces population size and smaller populations are more vulnerable to demographic and environmental stochasticity, emigration may have a negative effect on population persistence. Other researchers have informally recognized this parallel but opposite demographic process to the rescue effect [15–17]. This potential impact of emigration on extinction probability is referred to here as the "abandon-ship effect".

The theoretical bases for the demographic rescue and abandon-ship effects are clear and the mechanisms are plausible [3,18,19]. Nonetheless, robust empirical evidence from natural populations for these processes is minimal [17]. This relative lack of evidence does not necessarily imply that rescue and/or abandon-ship effects are absent in natural systems; rather, they may simply be empirically difficult to establish. At least three lines of evidence are necessary to document the operation of either process. First, individuals must be spatially segregated into semi-independent patches of habitat. Second, there must be at least periodic dispersal between these patches. Third, for the rescue effect, those patches that receive immigrants must have a lower extinction rate than patches that do not receive immigrants. For the abandon-ship effect, patches that lose emigrants must have a higher extinction rate than patches that did not lose emigrants.

A wide variety of studies have satisfied the first two conditions by demonstrating movement between spatially segregated subpopulations [20,21]. Some studies have also presented genetic or demographic data indicating that immigration or emigration has some impact on the focal population [22–27]. However, dispersal is often inferred rather than directly measured, as it is often very challenging to distinguish between individuals born into a given population versus those that have immigrated there [17]. Further, extinctions often play out on considerable time scales (years to decades or longer [28]), necessitating empirical research that is long-term or spatially extensive (or both). Given these challenges, many studies in natural populations focus on one or a small number of study populations. However, to satisfy the third condition, it is not enough to show that immigration is associated with population persistence; one must also simultaneously demonstrate that lack of immigration decreases the likelihood of persistence, ideally in many replicate populations.

Due to the logistical difficulties of simultaneously assessing both inter-patch dispersal and patch extinction rates at large spatial (and often temporal) scales in replicate natural populations whose fate is initially unknown, the third condition has not yet been fully satisfied. Because of these real-world empirical challenges, demographic rescue effects have primarily been explored using computational models [29] and laboratory-based experimental populations [19,30–32]. Herein, empirical evidence is presented from a unique natural microcosm system that suggests that both the rescue effect and the abandon-ship effect influence extinction rates in a real-world population. Natural microcosms have the dual advantage of being more logistically tractable than systems at larger spatial scales; yet, there is no reduced level of biological realism [33].

2. Materials and Methods

2.1. Study System Description

At least eleven rainforest frogs in the genus *Guibemantis* (Anura: Mantellidae) from Madagascar are known to complete their entire life cycle in and on *Pandanus* plants ('screw pines', Pandanaceae; [34,35]). Some species of *Pandanus* retain rainwater in their leaf axils and these plant-held water bodies are used for breeding purposes in some *Guibemantis* [36].

In these species, eggs are laid on the surface of *Pandanus* leaves and, after a period of development, the hatchlings drop into the water-filled leaf axils [37]. After several months, aquatic tadpoles metamorphose into terrestrial juveniles [38]. Both juveniles and adults also remain on *Pandanus* plants and are found in no other microhabitats. *Guibemantis wattersoni* (the focal species in the present study) has a maximum life span of less than 12 months [39]. *Guibemantis wattersoni* was studied in a fragment of littoral rainforest near the village of Sainte Luce (Manafiafy) in southeastern Madagascar (24°46′ S, 47°10′ E, elevation 5–10 m asl). Other aspects of the natural history of this system are described elsewhere [40,41]. Note that older literature refers to *G. wattersoni* as *G. bicalcaratus* or *Mantidactylus bicalcaratus* (see [42]).

2.2. Field Surveys

Because *G. wattersoni* completes its entire life cycle in the water-filled leaf axils of screw pine (*Pandanus*) plants (Figure 1), each plant is a patch of potentially suitable habitat. This system occurs on a relatively modest spatial scale; therefore, it was possible to simultaneously assess occurrence and abundance in a large number of patches as well as dispersal patterns among them. The spatial location of all *Pandanus* plants in two large networks of contiguous forest plots was mapped (total area: 18,750 m^2), resulting in 839 mapped *Pandanus* patches. However, many of these plants were seedlings and not usable by *G. wattersoni*. Only considering *Pandanus* plants > 1.0 m in size in these two plot networks, 236 *Pandanus* patches were surveyed over a three-year period (2000–2002) to gather information on occupancy and turnover (local extinction and recolonization). The average distance between *Pandanus* plants occupied by *G. wattersoni* was 7.96 ± 4.28 m (network 1) and 8.60 ± 4.65 m (network 2; data from 2001).

Figure 1. Adult *Guibemantis wattersoni* on a *Pandanus* leaf (**left**). *Pandanus* plant with three *G. wattersoni* (**right**). Photos by the author.

Three visual encounter surveys of each plant took place each rainy season between January and March (nine surveys overall for each plant throughout the study, 2124 plant surveys total). Surveys involved visually examining all leaves and leaf axils in each plant for any *G. wattersoni* individuals of any life stage. Searching continued as long as was necessary to be confident that no individuals were missed (usually less than ten minutes per plant). These small frogs are conspicuously colored, primarily diurnal and often active on the leaves of the plant [39]. This, combined with the fact that the plants are relatively small, resulted in high detection probabilities (see below). Most *Pandanus* plants were found on the forest floor; however, some ascended into the forest canopy. *Pandanus* plants whose height was greater than 1.5 m but less than 4.0 m were sampled using a stepstool constructed for this purpose. Plants higher than 4.0 m were sampled (when possible) by climbing adjacent trees using the single rope climbing technique. All field surveys were carried out by the author; therefore, inter-observer differences are not possible. The maximum number of *G. wattersoni* individuals detected in each plant in each year was used as a population size estimate in that plant for that year. Patches said to have experienced extinction were defined as those that were found to be occupied by at least one *G. wattersoni* individual in one year and were unoccupied in all surveys in the following year. Surveys were conducted between 0600 and 1700 h. Within years, each survey was separated by 14–20 days.

2.3. Mark-Recapture Study

A concurrent mark-recapture study in the same 236 patches was conducted to provide information on inter-patch emigration and immigration rates. Each rainy season, each patch was visited an additional four times within a 4-week period with each visit separated by 5–8 days (2832 visits total). These visits were in addition to, and separate from, the three annual visits to document occupancy and turnover. Individual frogs were hand-captured and marked by clipping toepads with sterilized scissors in unique combinations to allow individual recognition upon recapture [43]. Captured individuals were immediately released after marking. Since the two networks of plots were over 500 m away from one another (far beyond the dispersal ability of *G. wattersoni*), duplicate marks were used in each network to reduce the number of toepad clips needed to provide a unique identity. Clipped toepads of individuals from outside the immediate study area were clearly visible and showed no sign of regeneration after ten weeks. All captured adults and juveniles were toepad-clipped but recent metamorphs were not, as these were too small to safely capture and process. Together, these dispersal and turnover data enabled the direct comparison of patch-specific extinction rates with patch-specific dispersal patterns in the same time interval for a large number of populations. The mark-recapture study was used to document individual immigration and emigration but not to estimate population sizes, which was carried out with the field surveys (see above).

2.4. Detection Probability

Empirical estimates of local extinction or recolonization can be influenced by detection probability during field surveys [18,44]. Specifically, changes in occupancy can be inferred erroneously if field surveys miss individuals that were actually present. Therefore, it is important to demonstrate that detection probability is relatively high to have confidence in local extinction and recolonization frequency estimates [45]. To estimate the probability of detection of *G. wattersoni* from the field surveys, presence–absence data from all surveys of all *Pandanus* plants carried out in all years was assembled. Using the program Presence (version 2.13.35; [46]), we ran a simple single-season occupancy model with no covariates to provide an estimate of the detection probability of the field surveys.

2.5. Statistical Analysis

Fisher's exact test was used to assess patterns in extinction rates among patches that did and did not receive immigrants or emigrants in the previous year. For comparison, a generalized linear model with binary logistic error structure was also run. This analysis used population persistence (yes or no) as the dependent variable and the number of immigrants and the number of emigrants as main effects. Student's *t*-tests or Mann–Whitney tests were used to validate the assumption that small populations are more extinction-prone and to compare the number of immigrants and emigrants in patches in the previous year that did and did not go extinct. Statistical analyses were performed in IBM SPSS (version 28.0). Analyses only included patches that had both turnover and immigration and emigration information in each yearly time interval. *Pandanus* plants that were never occupied by *G. wattersoni* in any year were similarly excluded from the analyses.

3. Results

Guibemantis wattersoni occupied an average of 63.4% ($\pm$9.7% SD) of the available patches (*Pandanus* plants) during the study period (range: 74.6% (in 2000) to 56.8% (in 2002). On average, 22.0% ($\pm$6.1%) of these occupied patches went extinct in any given year ($n = 84$ total). Those populations of *G. wattersoni* that went extinct were significantly smaller, on average, than those populations that did not go extinct (U = 9069.5, n = 380, $p < 0.014$; Figure 2). Population size per occupied plant ranged from 1 to 12 individuals (mean 3.48 ± 2.03 SD).

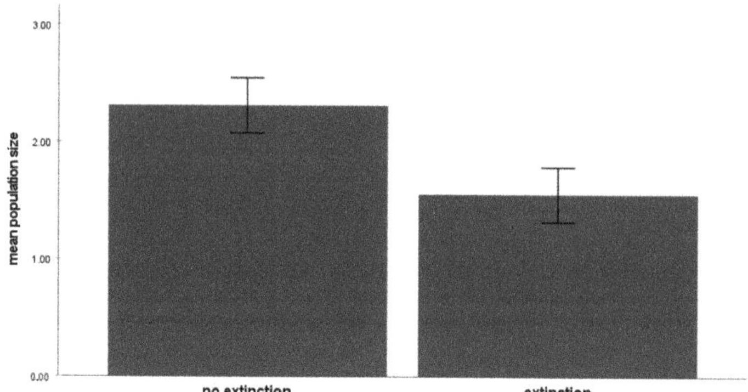

Figure 2. Patches experiencing extinction in a given year had a significantly smaller population size of *G. wattersoni* compared to patches that did not experience extinction (all years pooled, excluding unoccupied patches, Mann–Whitney test U = 9069.5, n = 380, $p < 0.014$). Error bars indicate means ± 2 SE.

A total of 567 *G. wattersoni* were marked in the mark-recapture study. Recapture rates were high within years (mean 53.8%, 305 recaptures total), but there were no inter-year recaptures. Dispersal among *Pandanus* patches was uncommon but not rare, with a mean of 13.9% (±6.8% SD) of marked individuals dispersing to a new patch per month. Mean dispersal distance for all known inter-patch movements was 15.5 m (±17.5 SD, range 0.5–123 m, n = 102).

In the 2000–2001 interval, dispersal data for *G. wattersoni* were available from 101 patches, 38 of which received at least one immigrant. Of the 38 populations which received immigrants, two went extinct in 2000–2001 (5.3%). In the remaining 63 patches (in which immigration was not detected) seven went extinct (11.1%; Fisher's exact test, $p = 0.477$; Table 1). In 2001–2002, dispersal data were available for 89 patches, 39 of which received at least one immigrant. Two of these went extinct (5.1%). In the remaining 50 patches (with no immigration), four extinctions were recorded (8.0%; Fisher's exact test, $p = 0.692$; Table 1). Pooling data for *G. wattersoni* from all years yielded dispersal and turnover information from 190 patches (77 with immigration, 113 without). The overall extinction rate in patches receiving immigrants was 5.2 ± 0.1% and the overall extinction rate in patches not receiving immigrants was 9.7 ± 2.2% (Fisher's exact test, $p = 0.288$, Table 1). The number of immigrants per patch ranged from zero to four.

Table 1. Comparison of extinction rates in patches with and without immigration and emigration for *G. wattersoni*. p-value is from Fisher's exact test.

Years	n	No Immigration	# Extinctions (%)	Immigration	# Extinctions (%)	p
2000–2001	101	63	7 (11.1)	38	2 (5.3)	0.477
2001–2002	89	50	4 (8.0)	39	2 (5.1)	0.692
all	190	113	11 (9.7)	77	4 (5.2)	0.288

Years	n	No Emigration	# Extinctions (%)	Emigration	# Extinctions (%)	p
2000–2001	101	57	5 (8.8)	44	5 (11.4)	0.911
2001–2002	89	52	1 (1.9)	37	5 (13.5)	0.078
all	190	109	6 (5.5)	81	10 (12.3)	0.115

Of patches in which no *G. wattersoni* emigration was detected in 2000–2001 (n = 57), five went extinct (8.8%). In this same interval, five extinctions (11.4%) were recorded in patches where emigration was detected (n = 44; Fisher's exact test, p = 0.744; Table 1). In 2001–2002, one extinction (1.9%) was recorded in those patches where emigration was not detected (n = 52). Among patches in which at least one emigration event was detected (n = 37), five extinctions (12.3%) were recorded (Fisher's exact test, p = 0.078; Table 1). Pooled data for *G. wattersoni* from all years yields six extinctions (5.5% ± 4.9%) where emigration was not detected (n = 109) and ten extinctions (12.3 ± 1.5%) where emigration was detected (n = 81; Fisher's exact test, p = 0.115; Table 1). The number of emigrants per patch varied from zero to three. Using data from all years, the generalized linear model did not find a significant effect of the number of immigrants or the number of emigrants on population persistence (likelihood ratio chi-square value = 9.17, df = 7, p = 0.241).

For patches that did not go extinct between 2000 and 2002, the mean number of *G. wattersoni* immigrants and emigrants per patch was very similar (0.56 ± 0.79 SD and 0.54 ± 0.81 SD, respectively, n = 175). However, patches that went extinct had significantly fewer immigrants (mean = 0.27 ± 0.46 SD; t = 2.45, df = 187, p = 0.025; Figure 3) and relatively more emigrants (mean = 0.73 ± 0.59 SD; t = −0.607, df = 187, p = 0.544; Figure 3). The single-season occupancy model confirmed that detection probability in the field surveys was high. The estimated detection probability of *G. wattersoni* in a single field survey was 0.806 (±0.011 SE; 95% C.I. 0.784–0.827).

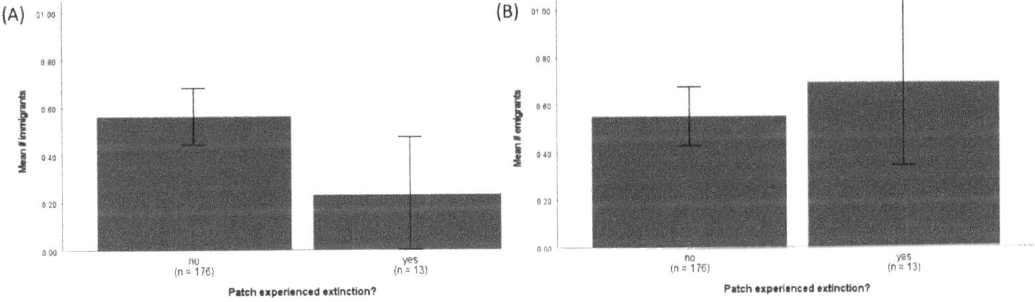

Figure 3. (**A**) Mean number (±2 SE) of immigrants to patches that went extinct ('yes') and those that did not experience extinction ('no'), pooled data from all years 2000–2002. Patches that went extinct had significantly fewer immigrants (t = 2.45, df = 187, p = 0.025). (**B**) Mean number (±2 SE) of emigrants to patches that went extinct ('yes') and those that did not experience extinction ('no'), pooled data from all years 2000–2002. Patches that went extinct had relatively more emigrants on average but this was not a significant difference (t = −0.607, df = 187, p = 0.544).

4. Discussion

Using the three criteria from above to empirically demonstrate a demographic rescue effect or abandon-ship effect, this study has shown the following: (1) individuals of *G. wattersoni* were spatially segregated into semi-independent patches of habitat; (2) there was periodic dispersal between these patches; and (3) those patches that receive immigrants did have a lower extinction rate than patches that did not receive immigrants. For the abandon-ship effect, patches that lost emigrants had a higher extinction rate than patches that did not lose emigrants. While these differences do not quite reach statistical significance, the trends are uniformly in the expected directions (Table 1). Further, the expectation that smaller populations are significantly more extinction-prone was confirmed (Figure 2) and, importantly, patches experiencing extinction had significantly fewer immigrants (Figure 3). However, patches experiencing extinction did not have significantly more emigrants than patches not experiencing extinction (Figure 3). Thus, these results are in accordance with previous theory and predictions regarding the demographic rescue effect [3] and there is some intriguing suggestion of the abandon-ship effect as well.

Previously, the best empirical data from natural populations on the rescue effect were from the Glanville Fritillary butterfly [18]. In this famous study, more connected patches (as measured using the S_i metric) experienced significantly fewer extinction events, at least in smaller populations [17]. However, even in the Glanville Fritillary butterfly dataset, the evidence for the rescue effect was only correlative, since inter-patch movement was not measured directly but rather only indirectly inferred via relationships between patch size, patch isolation and dispersal behavior [18]. In the current study, because of the smaller spatial scale of the study system, direct measurement of immigration and emigration was possible. These observations show a clear trend for patches receiving immigrants to be less extinction-prone than those that did not receive immigrants (Figure 3; Table 1). While many variables such individual-specific (e.g., age or sex) or patch-specific characteristics (e.g., patch quality) may influence dispersal decisions, the over-arching result is that the arrival of immigrants tends to reduce extinction risk, regardless of the proximate reasons for their arrival.

It is also notable that based on the estimated detection probability from the occupancy model results, the estimates of local extinction and recolonization frequency in this system seem reliable. Specifically, the chance of recording a *Pandanus* plant as unoccupied by *G. wattersoni* from a single field survey, when in fact it was occupied, was estimated using the model to be 0.194. With three surveys per plant per year, the estimated chance of mischaracterizing an occupied plant as unoccupied is less than 1% ($0.194 \times 0.194 \times 0.194 = 0.0073$). Thus, sampling error appears to be relatively unimportant in establishing the annual occupancy status of each plant and the resulting turnover estimates.

In this system, immigration reduces local extinction risk rather than increasing it through disease transmission or breaking up local adaptations (the "anti-rescue effect" [47]). The observed rescue effect is also very likely to be a demographic one (as originally envisioned by [3]) and not genetic. A genetic-rescue effect can occur when immigrants to small, genetically homogeneous populations rescue these populations from extinction by increasing subsequent genetic diversity [14,48]. The processes that result in low genetic diversity (e.g., inbreeding and genetic drift), however, can take many generations to accumulate [12]. The current study system, in contrast, is characterized by moderately low spatial isolation, substantial rates of inter-patch dispersal and rapid turnover dynamics [41], which make it very unlikely that small local populations are isolated enough or would survive long enough for severe inbreeding or genetic drift to be a problem.

Most population ecological models assume that there is either no correlation [49] or a negative relationship [8] between emigration rate and extinction risk. Alternatively, extinction probability could increase as the emigration rate increases if these departing individuals are not replaced by immigration or local recruitment. In these situations, small populations are likely to go extinct deterministically due to high emigration rates rather than as a result of demographic or environmental stochasticity. Several studies have provided some evidence that extinction is more probable as the number of individuals leaving the patch increases [16,50]; however, again, this is challenging to rigorously assess at large spatial scales in natural systems. The data presented here indicate a trend for extinction rates to be higher in patches with emigration compared to those without emigration (Table 1). However, while the number of emigrants tended to be higher in patches where extinction occurred compared to those that did not experience extinction, this was not a significant difference (Figure 3). Therefore, some of the evidence suggests that emigration by *G. wattersoni* increases the probability of local extinction; however, as of yet, this first test of the abandon-ship effect is only partially supported.

Patch occupancy in *G. wattersoni* varied over space and time during this study but was driven primarily by variation in patch height, patch size and patch quality [see [41] for full details]. The overall metapopulation of *G. wattersoni* was dynamic (approximately 22% of patches went locally extinct per year) but was fairly stable over the three years of the study. Thus, while local extinction was reasonably common, this was largely balanced by colonization. Additionally, as shown above, extinctions were made less likely by rescue from immigrants from other patches.

5. Conclusions

Many phenomena in ecology are difficult to address empirically in natural systems [51]. Nevertheless, finding ways to test theory with empirical data is critical to further our understanding of how the natural world works. The unique features of this natural microcosm from the rainforests of Madagascar permitted one such test of an important and widely assumed ecological process (the rescue effect). These data suggest that, in this system, the rescue effect does operate in the manner originally envisioned by Brown and Kodric-Brown [3] and that the converse process (referred to here as the abandon-ship effect) may also be at work. Further work in natural microcosm systems will likely continue to provide important insights into the dynamics of natural populations [33].

Funding: I gratefully acknowledge funding provided by the Horace H. Rackham Graduate School, the Department of Ecology and Evolutionary Biology, the International Institute and the Museum of Zoology at the University of Michigan.

Institutional Review Board Statement: The study was conducted according to the guidelines of the Declaration of Helsinki, and all protocols were approved in advance by The University of Michigan Committee on Use and Care of Animals (approval number 8297).

Informed Consent Statement: Not applicable.

Data Availability Statement: Data from this study are available at: https://doi.org/10.6084/m9.figshare.22304293.

Acknowledgments: I thank Greg Feinberg for help in the field in 2002 and the people of Manafiafy for patience with another crazy vazaha. Gabriella Rashelimanana, Jean-Baptiste Ramanamanjato, and QIT Madagascar Minerals, S.A. provided logistical support. Le Ministere des Eaux et Forêts issued research permits. Lastly, I thank M. Campbell for love and inspiration.

Conflicts of Interest: The author declares no conflict of interest.

References

1. Soulé, M.E. Introduction. In *Viable Populations for Conservation*; Soulé, M.E., Ed.; Cambridge University Press: Cambridge, UK, 1987; pp. 1–10.
2. Harrison, S. Local extinction in a metapopulation context: An empirical evaluation. *Biol. J. Linn. Soc.* **1991**, *42*, 73–88. [CrossRef]
3. Brown, J.H.; Kodric-Brown, A. Turnover rates in insular biogeography: Effect of immigration on extinction. *Ecology* **1977**, *58*, 445–449. [CrossRef]
4. MacArthur, R.H.; Wilson, E.O. *The Theory of Island Biogeography*; Princeton University Press: Princeton, NJ, USA, 1967.
5. Levins, R. Some demographic and genetic consequences of environmental heterogeneity for biological control. *Am. Entomol.* **1969**, *15*, 237–240. [CrossRef]
6. Hanski, I.; Simberloff, D. The metapopulation approach, its history, conceptual domain, and application to conservation. In *Metapopulation Biology*; Hanski, I.A., Gilpin, M.E., Eds.; Academic Press: San Diego, CA, USA, 1997; pp. 5–26.
7. Moilanen, A.; Smith, A.T.; Hanski, I. Long-term dynamics in a metapopulation of the American pika. *Am. Nat.* **1998**, *152*, 530–542. [CrossRef] [PubMed]
8. Pulliam, H.R. Sources, sinks, and population regulation. *Am. Nat.* **1988**, *132*, 652–661. [CrossRef]
9. Simberloff, D.; Farr, J.A.; Cox, J.; Mehlman, D.W. Movement corridors: Conservation bargains or poor investments? *Conserv. Biol.* **1992**, *6*, 493–504. [CrossRef]
10. Jangjoo, M.; Matter, S.F.; Roland, J.; Keyghobadi, N. Connectivity rescues genetic diversity after a demographic bottleneck in a butterfly population network. *Proc. Natl. Acad. Sci. USA* **2016**, *113*, 10914–10919. [CrossRef]
11. Drake, J.; Lambin, X.; Sutherland, C. The value of considering demographic contributions to connectivity: A review. *Ecography* **2022**, *2022*, e05552. [CrossRef]
12. Richards, C.M. Inbreeding depression and genetic rescue in a plant metapopulation. *Am. Nat.* **2000**, *155*, 383–394. [CrossRef]
13. Bell, G. Evolutionary rescue. *Annu. Rev. Ecol. Evol. Syst.* **2017**, *48*, 605–627. [CrossRef]
14. Bell, D.A.; Robinson, Z.L.; Funk, W.C.; Fitzpatrick, S.W.; Allendorf, F.W.; Tallmon, D.A.; Whiteley, A.R. The exciting potential and remaining uncertainties of genetic rescue. *Trends Ecol. Evol.* **2019**, *34*, 1070–1079. [CrossRef]
15. Hill, J.K.; Thomas, C.D.; Lewis, O.T. Effects of habitat patch size and isolation on dispersal by Hesperia comma butterflies: Implications for metapopulation structure. *J. Anim. Ecol.* **1996**, *65*, 725–735. [CrossRef]

16. Kuussaari, M.; Nieminen, M.; Hanski, I. An experimental study of migration in the Glanville fritillary butterfly Melitaea cinxia. *J. Anim. Ecol.* **1996**, *65*, 791–801. [CrossRef]

17. Hanski, I. Population dynamic consequences of dispersal in local populations and in metapopulations. In *Dispersal*; Clobert, J., Danchin, E., Dhondt, A.A., Nichols, J.D., Eds.; Oxford University Press: Oxford, UK, 2001; pp. 283–298.

18. Hanski, I. *Metapopulation Ecology*; Oxford University Press: Oxford, UK, 1999.

19. Hufbauer, R.A.; Szűcs, M.; Kasyon, E.; Youngberg, C.; Koontz, M.J.; Richards, C.; Tuff, T.; Melbourne, B.A. Three types of rescue can avert extinction in a changing environment. *Proc. Natl. Acad. Sci. USA* **2015**, *112*, 10557–10562. [CrossRef] [PubMed]

20. Stacey, P.B.; Taper, M.L.; Johnson, V.A. Migration within metapopulations: The impact upon local population dynamics. In *Metapopulation Biology*; Academic Press: Cambridge, MA, USA, 1997; pp. 267–291.

21. Baguette, M.; Legrand, D.; Fréville, H.; Van Dyck, H.; Ducatez, S. Evolutionary ecology of dispersal in fragmented landscape. In *Dispersal Ecology and Evolution*; Clobert, J., Baguette, M., Benton, T., Bullock, J., Eds.; Oxford University Press: Oxford, UK, 2012; pp. 381–391. [CrossRef]

22. Smith, A.T. Temporal changes in insular populations of the pika (*Ochotona princeps*). *Ecology* **1980**, *61*, 8–13. [CrossRef]

23. Olivieri, I.; Couvet, D.; Gouyon, P.H. The genetics of transient populations: Research at the metapopulation level. *Trends Ecol. Evol.* **1990**, *5*, 207–210. [CrossRef]

24. Martin, K.; Stacey, P.B.; Braun, C.E. Recruitment, dispersal, and demographic rescue in spatially-structured white-tailed ptarmigan populations. *Condor* **2000**, *102*, 503–516. [CrossRef]

25. Bowler, D.E.; Benton, T.G. Impact of dispersal on population growth: The role of inter-patch distance. *Oikos* **2009**, *118*, 403–412. [CrossRef]

26. McEachern, M.B.; Van Vuren, D.H.; Floyd, C.H.; May, B.; Eadie, J.M. Bottlenecks and rescue effects in a fluctuating population of golden-mantled ground squirrels (*Spermophilus lateralis*). *Conserv. Genet.* **2011**, *12*, 285–296. [CrossRef]

27. Millon, A.; Lambin, X.; Devillard, S.; Schaub, M. Quantifying the contribution of immigration to population dynamics: A review of methods, evidence and perspectives in birds and mammals. *Biol. Rev.* **2019**, *94*, 2049–2067. [CrossRef]

28. Schoener, T.W. Rate of species turnover decreases from lower to higher organisms: A review of the data. *Oikos* **1983**, *41*, 372–377. [CrossRef]

29. Eriksson, A.; Elías-Wolff, F.; Mehlig, B.; Manica, A. The emergence of the rescue effect from explicit within-and between-patch dynamics in a metapopulation. *Proc. R. Soc. B Biol. Sci.* **2014**, *281*, 20133127. [CrossRef] [PubMed]

30. Gonzalez, A.; Lawton, J.H.; Gilbert, F.S.; Blackburn, T.M.; Evans-Freke, I. Metapopulation dynamics, abundance, and distribution in a microecosystem. *Science* **1998**, *281*, 2045–2047. [CrossRef] [PubMed]

31. Griffen, B.D.; Drake, J.M. A review of extinction in experimental populations. *J. Anim. Ecol.* **2008**, *77*, 1274–1287. [CrossRef] [PubMed]

32. Long, Z.T.; Petchey, O.L.; Holt, R.D. The effects of immigration and environmental variability on the persistence of an inferior competitor. *Ecol. Lett.* **2007**, *10*, 574–585. [CrossRef]

33. Srivastava, D.S.; Kolasa, J.; Bengtsson, J.; Gonzalez, A.; Lawler, S.P.; Miller, T.E.; Munguia, P.; Romanuk, T.; Schneider, D.C.; Trzcinski, M.K. Are natural microcosms useful model systems for ecology? *Trends Ecol. Evol.* **2004**, *19*, 379–384. [CrossRef]

34. Lehtinen, R.M.; Lannoo, M.J.; Wassersug, R.J. Phytotelm-breeding anurans: Past, present and future research. In *Ecology and Evolution of Phytotelm-Breeding Anurans*; Lehtinen, R.M., Ed.; Miscellaneous Publications, Museum of Zoology; University of Michigan: Ann Arbor, MI, USA, 2004; Volume 193, pp. 1–9.

35. Lehtinen, R.M. Phytotelma-Breeding Frogs of the World website. Version 1.11. 2023. Available online: https://sites.google.com/site/phytotelmbreedingfrogsworld/home (accessed on 20 February 2023).

36. Lehtinen, R.M.; Vences, M.; Scherz, M.D. *Guibemantis* (Mantellidae). In *The New Natural History of Madagascar*; Goodman, S.M., Ed.; Princeton University Press: Princeton, NJ, USA, 2022; pp. 1374–1377.

37. Lehtinen, R.M. The use of screw pines (*Pandanus* spp.) by amphibians and reptiles in Madagascar. *Herpetol. Bull.* **2002**, *82*, 20–25.

38. Lehtinen, R.M. Tests for competition, cannibalism, and priority effects in two phytotelm-dwelling tadpoles from Madagascar. *Herpetologica* **2004**, *60*, 1–13. [CrossRef]

39. Lehtinen, R.M. The natural history of two plant-breeding frogs from Madagascar, *Guibemantis bicalcaratus* and *G. punctatus* (Anura: Mantellidae). *Salamandra* **2009**, *45*, 39–49.

40. Lehtinen, R.M. Parental care and reproduction in two species of *Mantidactylus* (Anura: Mantellidae). *J. Herpetol.* **2003**, *37*, 766–768. [CrossRef]

41. Lehtinen, R.M.; Carfagno, G.L.F. Habitat selection, the included niche and coexistence in plant-specialist frogs from Madagascar. *Biotropica* **2011**, *43*, 58–67. [CrossRef]

42. Lehtinen, R.M.; Glaw, F.; Vences, M. Two new plant-breeding frog species (Anura: Mantellidae, *Guibemantis*) from southeastern Madagascar. *Herpetol. J.* **2011**, *21*, 95–112.

43. Donnelly, M.A.; Guyer, C.; Juterbock, E.J.; Alford, R.A. *Techniques for Marking Amphibians. Measuring and Monitoring Biological Diversity: Standard Methods for Amphibians*; Heyer, R., Donnelly, M.A., Foster, M., Mcdiarmid, R., Eds.; Smithsonian Institution Press: Washington, DC, USA, 1994; pp. 277–284.

44. Harding, K.C.; McNamara, J.M. A unifying framework for metapopulation dynamics. *Am. Nat.* **2002**, *160*, 173–185. [CrossRef] [PubMed]

45. Saccheri, I.; Kuussaari, M.; Kankare, M.; Vikman, P.; Fortelius, W.; Hanski, I. Inbreeding and extinction in a butterfly metapopulation. *Nature* **1998**, *392*, 491–494. [CrossRef]
46. Thomas, C.D.; Hanski, I. Butterfly metapopulations. In *Metapopulation Biology*; Hanski, I.A., Gilpin, M.E., Eds.; Academic Press: Cambridge, MA, USA, 1997; pp. 359–386.
47. Crone, E.E.; Doak, D.; Pokki, J. Ecological influences on the dynamics of a field vole metapopulation. *Ecology* **2001**, *82*, 831–843. [CrossRef]
48. Steinberg, E.K.; Kareiva, P. Challenges and opportunities for empirical evaluation of "Spatial Theory". In *Spatial Ecology: The Role of Space in Population Dynamics and Interspecific Interactions*; Tilman, D., Kareiva, P., Eds.; Monographs in Population Biology; Princeton University Press: Princeton, NJ, USA, 1997; Volume 30, pp. 318–332.
49. MacKenzie, D.I.; Nichols, J.D.; Hines, J.E.; Knutson, M.G.; Franklin, A.D. Estimating site occupancy, colonization and local extinction when a species is detected imperfectly. *Ecology* **2003**, *84*, 2200–2207. [CrossRef]
50. Barata, I.M.; Griffiths, R.A.; Ridout, M.S. The power of monitoring: Optimizing survey designs to detect occupancy changes in a rare amphibian population. *Sci. Rep.* **2007**, *7*, 16491. [CrossRef] [PubMed]
51. Hines, J.E. PRESENCE. Software to Estimate Patch Occupancy and Related Parameters. USGS-PWRC. 2006. Available online: http://www.mbr-pwrc.usgs.gov/software/presence.html (accessed on 26 May 2023).

animals

MDPI

Article

Population Status and Vulnerability of *Mantidactylus pauliani* from Ankaratra Protected Area, Madagascar

Herizo Oninjatovo Radonirina [1,*], Bernard Randriamahatantsoa [2] and Nirhy H. C. Rabibisoa [1,2,*]

[1] Doctoral School of Natural Ecosystem, University of Mahajanga, Mahajanga 401, Madagascar
[2] Environmental and Life Science, Faculty of Sciences Technology and Environment, University of Mahajanga, University Campus of Ambondrona, Mahajanga 401, Madagascar; bernardzoo01@gmail.com
[*] Correspondence: radonirinaherizo@gmail.com (H.O.R.); nhcrabibisoa@gmail.com (N.H.C.R.); Tel.: +261-340471088 (H.O.R.)

Simple Summary: *Mantidactylus pauliani* is a locally endemic amphibian species restricted to mountain streams on the Ankaratra Massif, in the central highlands of Madagascar. This species has a highly restricted distribution, which makes it vulnerable to habitat destruction and, consequently, *M. pauliani* is considered one of the most threatened frog species in Madagascar. Therefore, having information concerning *M. pauliani* and its habitat is necessary for effective conservation. Our study aimed to examine the population status of *M. pauliani* by verifying its geographic distribution and elevational range, exploring habitat use, and assessing threats. *M. pauliani* occurs from 1900 m to 2378 m a.s.l. and is most abundant at an altitude between 1993 and 2166 m. Adults, juveniles, and tadpoles were associated with different levels of stream depth, speed, and width. We found that human activities are contributing to habitat losses, which is modifying its environment and threatening the species' survival. The data collected on the occurrence and habitat use of *M. pauliani* serve as an "environmental alert dashboard" to promote sustainable conservation.

Abstract: Mountain summits in Madagascar generally have species with specific habitat requirements, providing a home to a unique and locally endemic herpetofauna. Among them is *M. pauliani*, a typically aquatic and critically endangered amphibian found on the Ankaratra Massif. This species inhabits high elevations with a limited distribution range. Our study aimed to present new data on the distribution and elevational range, habitat use, and threats to *M. pauliani* and its occurrence according to habitat changes. To achieve this, annual monitoring was carried out from 2018 to 2021. Nine 100 m transects were established along streams at elevations ranging from 1762 to 2378 m a.s.l. along which we conducted visual encounter surveys. Data analysis was performed using a χ^2 test and Factor Correspondence Analysis. We found that *M. pauliani* occupies elevations between 1900 and 2378 m a.s.l. within humid forests and savannah habitats. The results showed a fluctuation in the number of animals observed and a higher occurrence at higher elevations throughout the years according to the season, stream quality, and water volume. Ongoing habitat alteration makes *M. pauliani* vulnerable to population decline, with annual bushfires likely having a negative impact on habitat.

Keywords: *Mantidactylus pauliani*; Ankaratra; landscape; habitat type; season; distribution

Citation: Oninjatovo Radonirina, H.; Randriamahatantsoa, B.; Rabibisoa, N.H.C. Population Status and Vulnerability of *Mantidactylus pauliani* from Ankaratra Protected Area, Madagascar. *Animals* **2023**, *13*, 2706. https://doi.org/10.3390/ani13172706

Academic Editors: Angelica Crottini, Franco Andreone, Andolalao Rakotoarison and Fandresena Rakotoarimalala

Received: 19 June 2023
Revised: 21 August 2023
Accepted: 23 August 2023
Published: 25 August 2023

1. Introduction

The Ankaratra forest, one of the biodiversity hotspots in Madagascar, is home to the locally endemic and Critically Endangered amphibian, *Mantidactylus pauliani* [1,2]. This species has previously been recorded at elevations above 2000 m [3–5], and specifically lives in the cold rocky streams of the Ankaratra Mountain. Since *M. pauliani* is a typically aquatic species [6], its distribution is thus restricted. Hence, it greatly relies on the streams' availability in terms of quantity and quality for breeding and sustenance [6–8].

However, in the last few decades, the forest [9] and its streams have gradually undergone changes due to various human activities, including charcoal production, timber harvesting, cattle grazing, and wildfire [5,10,11]. Anthropogenic activities increase threats to the species' habitats, leading to disruption in population connectivity [12], alteration, and degradation [9], which may induce vulnerability as amphibians are among the most highly threatened species in the world [13].

Considering the limited distribution of *M. pauliani* and its highly dependent behavior on streams, it is likely to be sensitive to habitat change [14], making it an important indicator of environmental health. However, the impact of habitat change on this locally endemic frog remains unknown because the data are still insufficient [12]. Long-term studies are also needed to produce reliable information for helping conservation activities to mitigate the vulnerability from threats that have a negative impact on the *M. pauliani* population.

The aim of this study was to present the new data on the spatial distribution and occurrence, habitat use, and threats to *M. pauliani* within a spatiotemporal approach. Based on previous research [5,12,14] and the theoretical considerations, we hypothesized that habitat degradation can change the distribution and occurrence of the *M. pauliani* population. More specifically, this study explored the relationship between this species and their occurrence by focusing on their spatio-altitudinal distribution and ecological parameters, once the knowledge on habitat use and threats has been identified. The new information gathered during this research can help us understand the ecological guilds and abundance variation of this critically endangered frog. While *M. pauliani* reproduces during both humid and dry seasons, the humid season represents the seasonality in months, and the variability is significant. These data are needed to produce an effective and sustainable conservation action plan.

2. Materials and Methods

2.1. Study Area

The study was conducted within the New Protected Area (NPA) of Ankaratra, which is one of the few remaining forests in the central highlands of Madagascar, covering about 8130 hectares (Figure 1). It extends over the slopes of the Ankaratra Massif, situated between 19°19′ and 19°24′ south latitude and 47°14′ and 47°22′ east longitude. The study area is part of the Manjakatompo Forest Station, located 84 km from the city of Antananarivo, via the RN7, and 17 km west of the city of Ambatolampy.

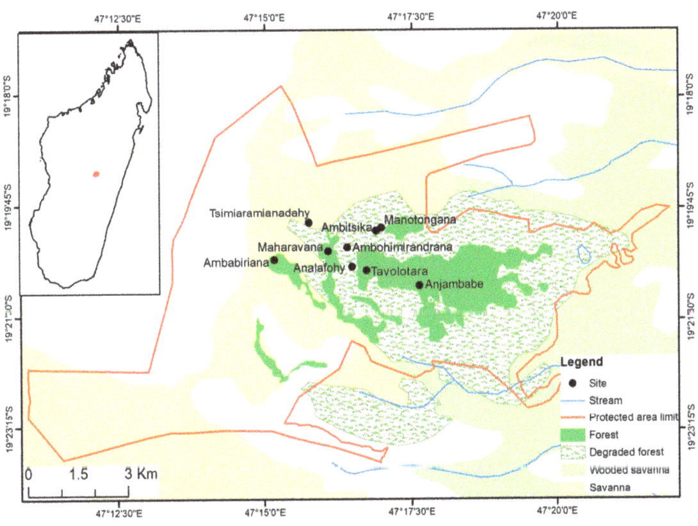

Figure 1. Map of Ankaratra showing the nine sampling sites.

2.2. Data Collection

We conducted four 15-day field trips in September 2018, March 2019, February 2020, and July 2021. These four trips made it possible to monitor variations in the population size of *M. pauliani* according to the season (dry and humid) and the different habitats. The different types of ecological guilds, the presence of permanent rock formations in the streams, and elevation levels were taken into account when selecting the study sites. The sites cover the entire mountain range (crest, valley, and slope) and differ in terms of habitat characteristics and degree of degradation. Six sites were selected in 2018 and 2019, and three additional sites were included in 2020 and 2021, namely Maharavana, Analafohy, and Ambohimirandrana (Table 1).

Table 1. Description of the sampling sites.

Transect	Elevation (m)	Description
Anjambabe	1900	Natural forest
Tavolotara	1993	Natural forest
Maharavana	2100	Natural forest
Analafohy	2126	Degraded forest
Manotongana	2167	Edge [1]
Ambaniriana	2166	Galery forest
Ambitsika	2202	Savanna
Ambohimirandrana	2244	Savanna
Tsimiaramianadahy	2378	Savanna

[1] River separating a deforested area and bare soil.

A 100 m transect was established at each site with varied widths based on the structure of the stream. We gathered data along the transects through direct observation and by examining shelters and refuges [15]. A transect consisted of a 100 m line that ran along the stream and was marked every 10 m with a flag. By surveying along transects, we were able to count individuals and categorize them by life stage and sex.

To define the stream parameters, water depth and speed were measured using the same techniques as reference [16]. To calculate the speed, a floating object (typically a cork) was timed over a one-meter distance. A straight stick held vertically was used to measure the distance between the surface and the bottom of a stream at a stable spot with buried rock. Three samples were gathered where the current speed was timed using the benchmark "flags". During each field trip, the same parameters were recorded at the same locations.

2.3. Data Analysis

Data analysis was performed using Microsoft Excel. A Chi-square test (χ^2) was used to confirm whether the elevational distribution of *M. pauliani* for each developmental stage was significantly different according to the study season. It was also performed to determine if the numbers of this species by developmental stage vary by season. In addition, a multivariate analysis was carried out using correspondence factor analysis, which is a method for analyzing data when the variables to be studied are quantitative measures. We used correspondence factor analysis to evaluate the habitat use of the individuals for each developmental stage and their characteristics according to the physical parameters of the streams. It consisted of projecting the individuals (adults, juveniles, and tadpoles) and the variables, such as water speed ranges (0.5–1 m/s, and 1–1.5 m/s, water depth ranges (0–20 cm, 20–40 cm and 40–60 cm), stream width (0–1 m, 1–2 m and 2–3 m), slopes (low slope, medium slope, and steep slope), microhabitat (still water, bank, and rock), and stream substrates (muddy, rocky, and sandy), which can influence the distribution of *M. pauliani* on a factorial plan.

We identified the threats to the species' habitat using a model adapted from reference [17]. The model was used to determine the degree of each threat by assigning a score of 1–3 points for each threat at each study site. The variables scored included their duration (permanent with a score of 3 points, temporary with 2 points, or occasional with 1 point), their intensity (strong, medium, or weak), and their importance (regional, zonal, or local). The assessment was based on the sum of these degrees of threat (Dm) according to duration, intensity, and the importance for each type of disturbance. If Dm $\leq$ 4, the pressure is minor; if 4 < Dm < 7, the pressure is medium; and if Dm $\geq$ 7 the pressure is major. Then, the pressure index (PI) was calculated to determine the overall importance of each type of disturbance. The PI is between the total number of points assigned to a site and the maximum value of the site [18]. The value of the index varies from 0 to 1. If 0.8 < PI $\leq$ 1, the threat is very high; 0.7 < PI $\leq$ 0.8, the threat is high; 0.5 < PI $\leq$ 0.7, the threat is medium; 0.3 < PI $\leq$ 0.5, the threat is low; and 0 < PI $\leq$ 0.3, the pressure is very low.

3. Results

3.1. Spatial Distribution and Elevational Range

Table 2 shows the total numbers of individuals of *M. pauliani* recorded over four years (2018, 2019, 2020, and 2021). For all developmental stages, the number of individuals encountered varied significantly by season (χ^2 = 462.89, p < 0.0001, df = 6). Adults were more abundant than juveniles and tadpoles in the dry seasons (2018, 2021), while tadpoles outnumbered adults and juveniles in the humid seasons (2019, 2020). During the dry season (2021), more juveniles than tadpoles were observed (101 vs. 48).

Table 2. Summary table of the numbers of *Mantidactylus pauliani* individuals counted by age class according to the altitude during each study season. A = Adult; J = Juveniles; T = Tadpoles; Alt = Altitude.

Alt	Dry Season (2018)			Total	Humid Season (2019)			Total	Humid Season (2020)			Total	Dry Season (2021)			Total
	A	J	T		A	J	T		A	J	T		A	J	T	
1900 m	36	16	24	76	2	7	34	43	0	0	12	12	3	0	12	15
1993 m	155	40	50	245	28	13	15	56	10	3	11	24	31	11	5	47
2100 m	-	-	-	-	-	-	-	-	9	8	48	65	26	14	0	40
2126 m	-	-	-	-	-	-	-	-	5	5	30	40	31	28	8	67
2166 m	38	39	11	88	17	6	13	36	3	2	34	39	15	39	4	58
2167 m	1	2	22	25	0	0	26	26	0	0	0	0	0	0	0	0
2202 m	36	29	9	74	2	2	44	48	2	5	61	68	11	8	5	24
2244 m	-	-	-	-	-	-	-	-	0	1	69	70	1	0	0	1
2378 m	23	6	11	40	1	1	15	17	2	0	32	34	3	1	14	18
Total	289	132	127	548	50	29	147	226	31	24	297	352	121	101	48	270

In terms of altitudinal distribution (Table 2), *M. pauliani* were recorded at an altitude between 1900 m and 2373 m. However, we observed most individuals at an altitude between 1993 and 2200 m, and the species was particularly numerous at 1993 m altitude during the years 2018 and 2019. Depending on season and altitude, the number of individuals encountered gradually decreased over time. This finding was particularly evident at the 2167 m altitude, where 1 adult, 2 juveniles, and 22 tadpoles were counted during the dry season (2018) and only 26 tadpoles during the humid season (2019). No individuals were observed here in 2020 and 2021. Generally speaking, their distribution along the mountain stream varied significantly for all developmental stages depending on the survey seasons (Adult stage: χ^2 = 222.26, p-value < 0.0001, df = 24; Juvenile stage: χ^2 = 152.52, p-value < 0.0001, df = 24; and Tadpole stage: χ^2 = 382.48, p-value < 0.0001, df = 24).

3.2. Habitat Use

For all years and seasons, we observed more individuals along the forest transect than the savannah transect, with the exception of the 2020 humid season, when more tadpoles were observed in the savannah than in the forest (Figure 2). We also noted that the number of individuals decreased during the dry season in 2018 and 2021, both in the forest (adults: 230 vs. 106) and in the savannah (adults: 59 vs. 15). The same result was also observed during the humid season in 2019 and 2020, except for tadpoles, whose numbers were high in 2020 (forest: 86 vs. 135 and savannah: 59 vs. 162; Figure 2). Overall, we observed more males than females in the forest and more females than males in the savannah. However, more males than females were observed in the savannah in the 2021 dry season (9 vs. 6; Table 3).

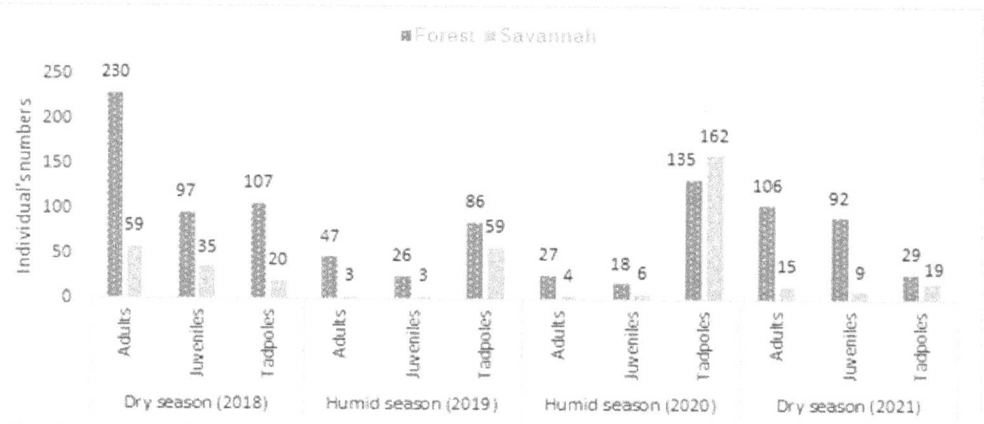

Figure 2. Numbers of individuals per age class counted according to habitat type for each season of fieldwork.

Table 3. Number of *M. pauliani* males and females according to habitat type and season.

Season	Forest		Savannah	
	Female	Male	Female	Male
Dry season (2018)	87	158	24	20
Humid season (2019)	13	34	2	1
Humid season (2020)	6	21	3	1
Dry season (2021)	35	71	6	9

We found adults, juveniles, and tadpoles to be associated with different habitat variables (Figure 3). In the Factor Correspondence Analysis, the first factorial explained 100% of the total variability, i.e., 98.2% for axis 1 and 1.8% for axis 2. Thus, both factors explain the individual distributions relative to ecological guilds. Adults were more numerous in streams that have a width between 0 and 1 m, a velocity between 0.5 and 1.5 m/s, and a depth between 0 and 20 cm. They were observed on steep slopes close to the riverbank. On the over hand, juveniles were found either under the rocky stream or hidden behind it on a medium slope. The tadpoles were found in deep, still water (40 to 60 cm depth), with a wide watercourse (2 to 3 m), and were observed in areas with sandy substrates.

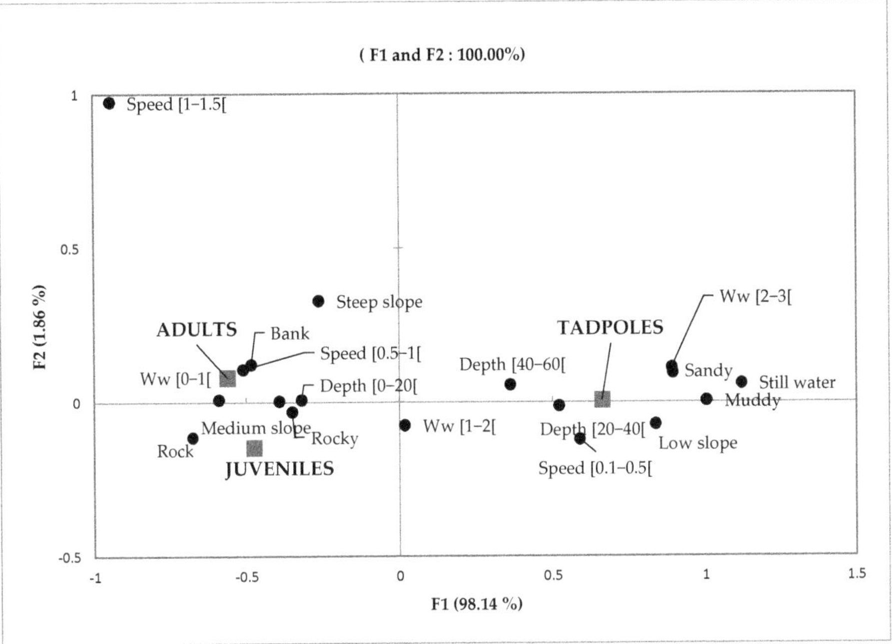

Figure 3. Projection of variables and individuals by developmental stage on the factorial plane (F1 and F2). Ww: water width.

3.3. Vulnerability Analysis

We observed six threats types, most of which are related to human activities: bush fires, charcoal production, logging, grazing fires, trampling of waterways by cattle, and the expansion of agricultural areas on slopes. These threats lead directly or indirectly to the degradation of the habitat of this species (see Figure S1).

However, the intensity of these different threat pressures varied depending on the type of formation and its proximity to the watercourse. The intensity of the threat was greatest in the open area where the pressure index was high, PI $\geq$ 0.5 (principally livestock grazing; Table 4). For the forested area, only anthropogenic activities near streams represented a high threat to the species, e.g., the edge for Manotongana, PI = 0.63 (logging and charcoal). Figure 4 demonstrates the relationship between the vulnerability of the species and changes in its habitat caused by anthropogenic pressures.

Table 4. Evaluation of pressure.

Environment Types	Sites	Threat Types	Score			Evaluation
			Length	Intensity	Importance	
	Anjambabe (1900 m)	Bushfires	1	1	1	Minor (3)
		Charcoal production	1	2	1	Minor (4)
		Logging	2	1	1	Minor (4)
		Livestock grazing	1	3	1	Medium (5)
		Evaluation total [Index of pressure PI]				16 [0.44]
	Tavolotara (1993 m)	Bushfires	2	1	1	Minor (4)
		Charcoal production	2	1	1	Minor (4)
		Logging	1	2	1	Minor (4)

Table 4. *Cont.*

Environment Types	Sites	Threat Types	Score			Evaluation
			Length	Intensity	Importance	
Forest environment		Livestock grazing	1	2	1	Minor (4)
		Evaluation total [Index of pressure PI]				16 [0.44]
	Maharavana (2100 m)	Bushfires	1	2	1	Minor (4)
		Charcoal production	1	2	1	Minor (4)
		Logging	1	1	1	Minor (3)
		Evaluation total [Index of pressure PI]				11 [0.41]
	Analafohy (2126 m)	Charcoal production	2	2	1	Medium (5)
		Logging	2	2	1	Medium (5)
		Livestock grazing	1	2	1	Minor (4)
		Expansion of agricultural land	1	1	1	Minor (3)
		Evaluation total [Index of pressure PI]				17 [0.47]
	Ambaniriana (2166 m)	Charcoal production	1	2	1	Minor (4)
		Logging	1	1	1	Minor (3)
		Livestock grazing	2	2	1	Medium (5)
		Evaluation total [Index of pressure PI]				12 [0.44]
Edge	Manotongana (2167 m)	Bushfires	2	3	1	Medium (6)
		Charcoal production	2	3	1	Medium (6)
		Logging	2	2	1	Medium (5)
		Evaluation total [Index of pressure PI]				17 [0.63]
Savannah	Ambitsika (2202 m)	Bushfires	1	2	1	Minor (4)
		Charcoal production	2	1	1	Minor (4)
		Feux de pâturage	1	2	1	Minor (4)
		Livestock grazing	2	3	1	Medium (6)
		Evaluation total [Index of pressure PI]				18 [0.5]
	Ambohimirandrana (2244 m)	Charcoal production	1	2	1	Minor (4)
		Livestock grazing	2	3	1	Medium (6)
		Evaluation total [Index of pressure PI]				10 [0.55]
	Tsimiaramianadahy (2378 m)	Bushfires	2	3	1	Medium (6)
		Livestock grazing	2	3	1	Medium (6)
		Charcoal production	1	2	1	Minor (4)
		Evaluation total [Index of pressure PI]				16 [0.59]

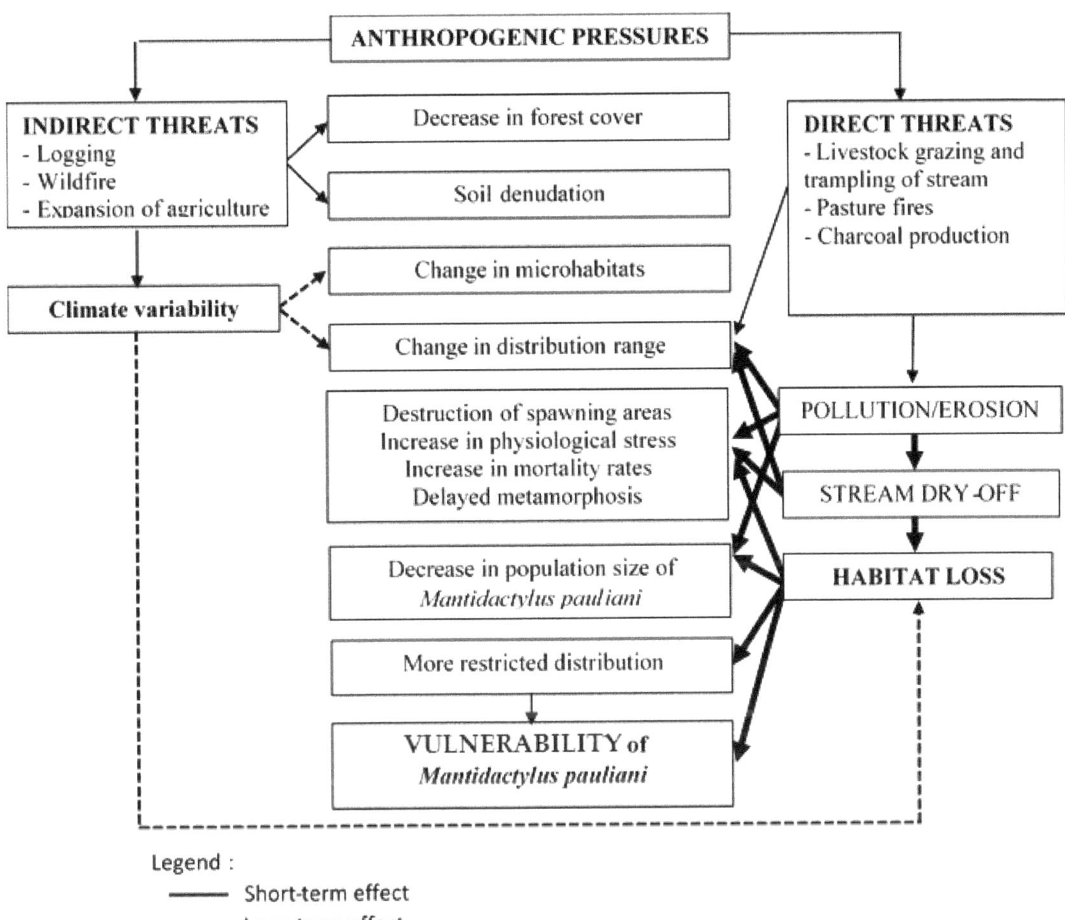

Figure 4. Diagram showing the relationship between habitat loss and the vulnerability of *M. pauliani*.

According to this diagram and threat analysis (Figure 4), the PI depends greatly on the distance relative to the stream. Indirect threats lead to decreased forest cover, soil denudation, and climate variability. In the long term, this climatic variability leads to changes in microhabitats and species distribution. Conversely, the direct threats cause erosion, which leads to pollution and drying of the streams.

4. Discussion

4.1. Spatial Distribution and Elevation Range

M. pauliani is classified as a critically endangered mountain species in decline [19]. The available data suggest a restricted range for this species at an altitude between 2000 and 2300 m [6,11,20], but our study showed that this range extends from 1900 to 2392 m. We offer two hypotheses to explain their extended distributional range. The first is that the search effort was increased because of the addition of a new transect along the slope from 1900 m to 2400 m. The second is that the current study encompassed all seasons (dry and rainy), compared with other studies that have concentrated solely on the breeding period that occurs during the humid season. Although *M. pauliani* was observed at altitudes of 1900 m and above during the study seasons, the greatest numbers of individuals were recorded in the same altitude range as in previous studies. According to reference [21], one of the biological responses to climate change is an upward shift in species distribution due to increased temperatures. Further work should therefore be carried out on this species to examine whether there has been a shift in response to climate change.

There was a difference in the number of individuals at each developmental stage for different periods. During the dry season in 2018 (pre-reproduction period), we observed more adults than juveniles and tadpoles (289 vs. 132, 127 individuals), but there were more tadpoles during the reproduction period (humid season 2019, 2020) than during the other periods (dry season 2018, 2021). This indicates that the juveniles could reach maturity (121 adults vs. 101 juveniles; dry season 2021). It was noted that the number of tadpoles decreased quickly between the humid season (297/2020) and the dry season (48/2021); this might be possibly due to the fast-flowing streams between both periods and their impact on the speed of tadpole metamorphosis [22].

The majority of the recorded individuals were found at the mid-slope (1900 m and 2200 m), where the forests are stable, and where there is a variety of refuges available in the form of rocks, fine sand, and gravel [16]. In addition, we observed more tadpoles in open habitats than in closed habitats. It is possible that strong sunlight could facilitate their hatching and development [23,24].

Regarding the habitat preferences of males and females across the survey periods, a higher number of males were found in forested areas at low elevations than in open and savannah habitats at high elevations, confirming the findings of previous studies [25,26]. Therefore, females may attract males to lay their eggs, and it was documented that Mantellinae rarely lay their eggs in water but sometimes just overhanging the water's surface [3,27].

Regarding seasonal distribution, *M. pauliani* was more abundant during the cold and dry seasons and decreased during the rainy season. This observation is similar to that observed by reference [21]. Furthermore, high-elevation species reproduce throughout the year, unlike species at medium and low elevations, which are inactive during the cold and dry seasons [28]. This explains the abundance of *M. pauliani* throughout the year. However, the authors of reference [29] explained that changes in abundance correlated with short-term and long-term habitat loss. In the case of the Manotongana transect at 2167 m altitude, the habitat was fragmented and degraded due to erosion and the eventual drying of the stream [30]. This is why this site has a very low number of individuals; only one adult male was recorded in September 2018, and no individual has been reported at this site since February 2020.

4.2. Habitat Use and Adaptation

According to reference [16], the numerous rock dimensions and shapes were favorable habitats because they balanced the speed of water. This assumption is the same as in the habitat of *M. pauliani*, where their presence and abundance correlate with the rock presence and also the water depth, degree of slope, water velocity, and stream flow.

Although forest degradation is a threat to the biodiversity of Ankaratra, this study showed that both savannah and forest represent important environments for this species to ensure its viability, a result similar to those previously put forward by the authors of reference [5]. Moreover, while the threats may result in greater water flow and floods, they are likely damaging tadpoles or washing them away in both savannah and forest, which represents a risk for their survival. However, the impact of forest degradation requires further analysis. So far, it has not had an immediate impact on the population of this species, as suggested by similar studies [31], but the long-term effects need to be analyzed. The Ankaratra Massif is vulnerable to various human activities that can have a negative impact on the survival of the species. According to the authors of reference [1], high altitudes are refuges for several species whose distribution is restricted and which are also more vulnerable. Fragmentation is one of the main threats to the disappearance of amphibians [32,33]. Logging, erosion, and livestock grazing leading to the trampling of watercourses remain the most significant threats. These threats lead to the destruction and pollution of watercourses, as in the case of Manotongana at 2167 m a.s.l., where the streams are very degraded and polluted. As a result, only one adult individual was recorded at this site in 2018, and none were recorded for three years thereafter.

We recommend strengthening monitoring through habitat patrols by forest rangers, especially in the savannah habitat which may be important for breeding. The establishment of a firebreak system around the forest is also essential to minimize the damage caused by bushfires. Environmental education, awareness-raising, and consciousness-raising within the village community must also be undertaken. This will allow them to understand the importance of the forest and make them aware that biodiversity conservation will benefit them in the long term. Finally, the livelihoods of the local population should be ameliorated through the provision of water from Ankaratra.

5. Conclusions

Overall, this study highlighted the vulnerability of *M. pauliani* to habitat degradation and provided an assessment of the population status. The species is found at altitudes between 1900 and 2378 m, and adults and juveniles were more abundant in forests than in savannahs, while tadpoles preferred the savannah habitat. Our results also showed how the number of individuals encountered in surveys varies across seasons and years. During the study, we noticed a decrease in the number of individuals encountered for each age class at the most degraded site, Manotongana. The viability of *M. pauliani* depends on the quality and quantity of its habitat on the Massif, specifically the stream, and its degradation has a direct impact on the survival of the population. To effectively manage and conserve *M. pauliani*, information is needed about their habitat preference, population number, distribution across altitudes, and the threats they face. Our study provides some of this critical baseline data. We hope decision-makers can use our results as a tool in the development of a conservation strategy with proactive measures to prevent the irreversible loss of one of the emblematic species of the Ankaratra Massif.

Supplementary Materials: The following supporting information can be downloaded at: https://www.mdpi.com/article/10.3390/ani13172706/s1, Figure S1: photos of pressure and threats in Ankaratra Massif. Photo of *Mantidactylus pauliani* and its habitat.

Author Contributions: Conceptualization, N.H.C.R. and H.O.R.; validation, N.H.C.R., B.R. and H.O.R.; writing—original draft preparation, H.O.R.; writing—review and editing, H.O.R.; supervision, N.H.C.R. and B.R. All authors have read and agreed to the published version of the manuscript.

Funding: This research received no external funding.

Institutional Review Board Statement: Not applicable.

Informed Consent Statement: Not applicable.

Data Availability Statement: The data presented in this study are available on request from the corresponding author. The data are not publicly available due to privacy or ethical restrictions.

Acknowledgments: The authors would like to thank the "Direction Générale des Forêts de Madagascar" who provided the research permits, and the community based in the Ankaratra villages who contributed to the smooth realization of the fieldwork (guidance, cooking, and camp assistance). The authors thank all of the staff of V.I.F and the former Ankaratra NPA manager for supporting them.

Conflicts of Interest: The authors declare no conflict of interest.

References

1. Andreone, F.; Cox, N.A.; Glaw, F.; Köhler, J.; Rabibisoa, N.H.C.; Randriamahazo, H.; Randrianasolo, H.; Raxworthy, C.J.; Stuart, S.N.; Vallan, D.; et al. Update of the Global Amphibian Assessment for Madagascar in Light of Species Discoveries, Nomenclature Changes, and New Field Information. *Monogr. Mus. Reg. Sci. Nat. Torino* **2008**, *XLV*, 419–438.
2. Andreone, F.; Carpenter, A.I.; Cox, N.; du Preez, L.; Freeman, K.; Furrer, S.; Garcia, G.; Glaw, F.; Glos, J.; Knox, D.; et al. The Challenge of Conserving Amphibian Megadiversity in Madagascar. *PLoS Biol.* **2008**, *6*, e118. [CrossRef]
3. Glaw, F.; Vences, M. *A Field Guide to the Amphibians and Reptiles of Madagascar*, 3rd ed.; Vences & Glaw: Cologne, Germany, 2007; ISBN 978-3-929449-03-7.

4. Rakotonoely, S.A.X. Biologie et Écologie de Deux Espèces d'Amphibiens *Boophis williamsi* (GUIBE, 1974) et *Mantidactylus pauliani* (GUIBE, 1974) Critiquement en Danger du Massif de l'Ankaratra. Master's Thesis, University of Antananarivo, Antananarivo, Madagascar, 2012.

5. Rosa, S.C.F. Is There a Future for the Amphibians of the Ankaratra Massif Reserve? Understanding the Role of Landscape Change. Master's Thesis, University of Porto, Porto, Portugal, 2017.

6. Vences, M.; Andreone, F.; Glaw, F.; Raminosoa, N.; Randrianirina, J.E.; Vieites, D.R. Amphibians and Reptiles of the Ankaratra Massif: Reproductive Diversity, Biogeography and Conservation of a Montane Fauna in Madagascar. *Ital. J. Zool.* **2002**, *69*, 263–284. [CrossRef]

7. Andreone, F.; Cadle, J.E.; Cox, N.; Glaw, F.; Nussbaum, R.A.; Raxworthy, C.J.; Stuart, S.N.; Vallan, D.; Vences, M. Species Review of Amphibian Extinction Risks in Madagascar: Conclusions from the Global Amphibian Assessment. *Conserv. Biol.* **2005**, *19*, 1790–1802. [CrossRef]

8. Ndriantsoa, S.H.; Riemann, J.C.; Raminosoa, N.; Rödel, M.-O.; Glos, J.S. Amphibian Diversity in the Matrix of a Fragmented Landscape around Ranomafana in Madagascar Depends on Matrix Quality. *Trop. Conserv. Sci.* **2017**, *10*, 1940082916686065. [CrossRef]

9. Suzzi-Simmons, A. Status of Deforestation of Madagascar. *Glob. Ecol. Conserv.* **2023**, *42*, e02389. [CrossRef]

10. Bletz, M.C.; Rosa, G.M.; Andreone, F.; Courtois, E.A.; Schmeller, D.S.; Rabibisoa, N.H.C.; Rabemananjara, F.C.E.; Raharivololoniaina, L.; Vences, M.; Weldon, C.; et al. Widespread Presence of the Pathogenic Fungus Batrachochytrium Dendrobatidis in Wild Amphibian Communities in Madagascar. *Sci. Rep.* **2015**, *5*, 8633. [CrossRef]

11. Rabemananjara, F.; Randriamahazo, H.; Rahantamalala, J.; Rahantalisoa, H.; Rakotoarisoa, J.M.; Rabibisoa, N.H.C.; Andreone, F. The Conservation Effort for Two Critically Endangered Amphibian Species of the Ankaratra Massif, Boophis Williamsi and Mantidactylus Pauliani. *FrogLog* **2012**, *103*, 29–31.

12. Hazell, D. Frog Ecology in Modified Australian Landscapes: A Review. *Wildl. Res.* **2003**, *30*, 193–205. [CrossRef]

13. Hernández-Ordóñez, O.; Martínez-Ramos, M.; Arroyo-Rodríguez, V.; González-Hernández, A.; González-Zamora, A.; Zárate, D.A.; Reynoso, V.H. Distribution and Conservation Status of Amphibian and Reptile Species in the Lacandona Rainforest, Mexico: An Update after 20 Years of Research. *Trop. Conserv. Sci.* **2014**, *7*, 1–25. [CrossRef]

14. Fu, L.; Wang, X.; Yang, S.; Li, C.; Hu, J. Morphological Variation and Its Environmental Correlates in the Taihangshan Swelled-Vented Frog across the Qinling Mountains. *Animals* **2022**, *12*, 2328. [CrossRef]

15. Raxworthy, C.J.; Nussbaum, R.A. A Rainforest Survey of Amphibians, Reptiles and Small Mammals at Montagne d'Ambre, Madagascar. *Biol. Conserv.* **1994**, *69*, 65–73. [CrossRef]

16. Moisan, J.; Pelletier, L. *Guide de Surveillance Biologique Basée sur les Macrosinvertébrées Benthiques d'Eau Douce du Québec: Cours d'Eaux Peu Profondes à Substrats Grossier*; Direction du Suivi de l'Etat de l'Environnement, Ministère du Développement Durable, de l'Environnement et des Parcs: Québec, QC, Canada, 2008; ISBN 978-2-550-53590-4.

17. ZICOMA. *Evaluation de La Faune Aviaire dans les Zones Humides entre le Parc National de Ranomafana et Celui d'Andringitra*; Projet d'Appui a la Gestion de l'Environnement, International Resources Group: Washington, DC, USA, 2000.

18. Salafsky, N.; Margoluis, R. Threat Reduction Assessment: A Practical and Cost-effective Approach to Evaluating Conservation and Development Projects. *Conserv. Biol.* **1999**, *13*, 830–841. [CrossRef]

19. IUCN. *The IUCN Red List of Threatened Species*; Version 2022-2; Red list Authorities: Gland, Switzerland, 2022.

20. Blommers-Schlösser, R.M.A. Biosystematics of the Malagasy Frogs. I. Mantellinae (Ranidae). *Beaufortia* **1979**, *29*, 1–77.

21. Rabibisoa, N.H.C.; Raxworthy, C.J.; Andreone, F. Changement Climatique et Amphibien. In *Sahonagasy Action Plan*; Andreone, F., Randriamahazo, H., Rabibisoa, N., Eds.; MRSN, CI, IUCN/SSC-ASG: Turin, Italy, 2008; pp. 43–48, ISBN 978-88-860-41-83-6.

22. Haramura, T. Microhabitat Selection by Tadpoles of Buergeria Japonica Inhabiting the Coastal Area. *J. Ethol.* **2007**, *25*, 3–7. [CrossRef]

23. Dittrich, C.; Drakulić, S.; Schellenberg, M.; Thein, J.; Rödel, M.-O. Some like It Hot? Developmental Differences in Yellow-Bellied Toad (*Bombina variegata*) Tadpoles from Geographically Close but Different Habitats. *Can. J. Zool.* **2016**, *94*, 69–77. [CrossRef]

24. Brattstrom, B.H. Thermal Control of Aggregation Behavior in Tadpoles. *Herpetologica* **1962**, *18*, 38–46.

25. Edmonds, D.; Rakotoarisoa, J.C.; Rasoanantenaina, S.; Sam, S.S.; Soamiarimampionona, J.; Tsimialomanana, E.; Rainer Dolch, Y.; Rabemananjara, F.; Rabibisoa, N.; Robsomanitrandrasana, E. Captive Husbandry, Reproduction, and Fecundity of the Golden Mantella (*Mantella aurantiaca*) at the Mitsinjo Breeding Facility in Madagascar. *Salamandra* **2015**, *51*, 315–325.

26. Eterovick, P.C.; Ferreira, A.D.M. Breeding Habitat and Microhabitat Choices by Male and Female Frogs: Are There Differences Between Sexes and Seasons? *Herpetologica* **2008**, *64*, 397–405. [CrossRef]

27. Porcel, X.; Dubos, N.; Nöel, J.; Lava, H.; Velo, J.H.; Melo, M.; Rosa, G.M.; Andreone, F.; Crottini, A. Male Parental Care in Malagasy Stream-Dwelling Frogs of the Mantidactylus Femoralis Group (Anura: Mantellidae: Ochthomantis): Egg Guarding in Ochthomantis. *Herpetol. Notes* **2022**, *15*, 55–61.

28. Raxworthy, C.J.; Nussbaum, R.A. Montane Amphibian and Reptile Communities in Madagascar. *Conserv. Biol.* **1996**, *10*, 750–756. [CrossRef]

29. Cocca, W.; Andreone, F.; Belluardo, F.; Rosa, G.M.; Randrianirina, J.E.; Glaw, F.; Crottini, A. Resolving a Taxonomic and Nomenclatural Puzzle in Mantellid Frogs: Synonymization of *Gephyromantis azzurrae* with *G. Corvus*, and Description of *Gephyromantis kintana* sp. Nov. from the Isalo Massif, Western Madagascar. *ZooKeys* **2020**, *951*, 133–157. [CrossRef]

30. Welsh, H.H.; Ollivier, L.M. Stream Amphibians as indicators of ecosystem stress:a case study from California's redwoods. *Ecol. Appl.* **1998**, *8*, 1118–1132. [CrossRef]
31. Konopik, O.; Steffan-Dewenter, I.; Grafe, T.U. Effects of Logging and Oil Palm Expansion on Stream Frog Communities on Borneo, Southeast Asia. *Biotropica* **2015**, *47*, 636–643. [CrossRef]
32. Tabarelli, M.; Gascon, C. Lessons from Fragmentation Research: Improving Management and Policy Guidelines for Biodiversity Conservation. *Conserv. Biol.* **2005**, *19*, 734–739. [CrossRef]
33. Funk, W.C.; Greene, A.E.; Corn, P.S.; Allendorf, F.W. High Dispersal in a Frog Species Suggests That It Is Vulnerable to Habitat Fragmentation. *Biol. Lett.* **2005**, *1*, 13–16. [CrossRef]

Article

Life History Traits and Longevity of the Invasive Asian Common Toad *Duttaphrynus melanostictus* (Schneider, 1799) in Madagascar

Fabio Maria Guarino [1], Franco Andreone [2], Marcello Mezzasalma [3,*], Fulvio Licata [4,5,6,*], Simona Puoti [1], Bárbara Santos [4,5], Walter Cocca [4,5,6], Jean Francois Solofoniaina Fidy [7], Serge Herilala Ndriantsoa [8], Jean Noel [7], Tsanta Fiderana Rakotonanahary [9], Rodino Fetrarijahona Harison [10], Gaetano Odierna [1] and Angelica Crottini [4,5,6]

1   Department of Biology, University of Naples Federico II, Via Cinthia 26, I-80126 Naples, Italy; fabio.guarino@unina.it (F.M.G.); si.puoti@studenti.unina.it (S.P.); gaetanodierna@gmail.com (G.O.)
2   Museo Regionale di Scienze Naturali, Via G. Giolitti, 36, I-10123 Torino, Italy
3   Department of Biology, Ecology and Earth Science (DiBEST)—University of Calabria, Via P. Bucci 4/B, I-87036 Rende, Italy
4   CIBIO, Centro de Investigação em Biodiversidade e Recursos Genéticos, InBIO Laboratório Associado, Campus de Vairão, Universidade do Porto, 4485-661 Vairão, Portugal; barbarasantosbio@gmail.com (B.S.); walter.cocca85@gmail.com (W.C.); tiliquait@yahoo.it (A.C.)
5   BIOPOLIS Program in Genomics, Biodiversity and Land Planning, CIBIO, Campus de Vairão, 4485-661 Vairão, Portugal
6   Departamento de Biologia, Faculdade de Ciências, Universidade do Porto, Rua do Campo Alegre s/n, 4169-007 Porto, Portugal
7   Madagascar Fauna and Flora Group, BP 442, Toamasina 501, Madagascar; rendrirendry@gmail.com (J.F.S.F.); karenlmfreeman@gmail.com (J.N.)
8   Département de Biologie Animale, Université d'Antananarivo, BP 906, Antananarivo 101, Madagascar; nsehel2006@gmail.com
9   Turtle Survival Alliance, BP 8511, Antananarivo 101, Madagascar; tsantafiderana@gmail.com
10 Institut Supérieur de Science, Environnement et Développement Durable (ISSEDD), Université de Toamasina, Toamasina 501, Madagascar; rijashalom501@gmail.com
*   Correspondence: marcello.mezzasalma@unical.it (M.M.); fulvio.licata@gmail.com (F.L.)

Citation: Guarino, F.M.; Andreone, F.; Mezzasalma, M.; Licata, F.; Puoti, S.; Santos, B.; Cocca, W.; Solofoniaina Fidy, J.F.; Ndriantsoa, S.H.; Noel, J.; et al. Life History Traits and Longevity of the Invasive Asian Common Toad *Duttaphrynus melanostictus* (Schneider, 1799) in Madagascar. *Animals* **2023**, *13*, 2099. https://doi.org/10.3390/ani13132099

Academic Editor: Xiang Ji

Received: 2 May 2023
Revised: 20 June 2023
Accepted: 21 June 2023
Published: 24 June 2023

Simple Summary: The Asian common toad *Duttaphrynus melanostictus* represents a serious threat to the biodiversity of Madagascar, where it was introduced in around 2010. Here, we study some traits of its life history, including its body size, age structure, and age at sexual maturity based on individuals sampled at different sites of Toamasina, on the east coast of Madagascar, in 2016, 2018, and 2019. The individuals studied here were larger, although no longer-living, with respect to individuals of the native populations in SE Asia. In the invasive population of Madagascar, the males were significantly smaller and younger than the females when data from different sites were pooled. However, when data are analyzed separately, this was true only for one site. The maximum age ascertained in this study is 3 years in males and 6 years in females. Males and females reach sexual maturity after the first and second year of age, respectively. Further studies on the life history traits of the population of *D. melanostictus* in Madagascar can be relevant for the management of this invasion as well as to better understand the biology of the species.

Abstract: We analyzed the body length, age structure, and age at sexual maturity of the invasive Asian common toad *Duttaphrynus melanostictus* from different sites in Toamasina, east Madagascar. We used skeletochronology as a proxy for age estimation, while gonads were histologically analyzed to determine the age of sexual maturity. The analysis of pooled age data from three sites investigated in 2016 showed that both sexes were larger, although not older, than those of native populations. For the individuals from Madagascar, the males were significantly smaller and younger (mean ± SD, SVL: 71.4 ± 1.6 mm; age: 1.8 ± 0.7 years) than the females (SVL: 78.42 ± 1.9 mm; age: 2.7 ± 1.3 years), when the data were pooled, but when the data were analyzed separately for each of the three sites, similar results were obtained only for one site. The oldest recorded male and female were 3 and 6 years old, respectively. Gonadal histology showed that the males and females reach sexual maturity after the first and second years of age, respectively. Further studies are needed to understand if the larger size and faster growth rates observed in the invasive population of *D. melanostictus* in Madagascar are a consequence of more favorable environmental conditions with respect to the native range (e.g., the availability of larger trophic niches, a lack of competitors, and lower predatory pressure), and we suggest to extend the monitoring of these life history traits to understand how they might influence the invasion.

Keywords: Amphibia; invasive species; skeletochronology; age estimation; gonadal analysis; age at sexual maturity

1. Introduction

Amphibians and reptiles are establishing as model organisms due to their high species richness and diversity and have been widely studied in different fields such as genetics, biogeography, and more recently invasion biology [1–4]. Due to their species richness and diversity, several areas of the world remain poorly characterized and more efforts should be invested in these places.. One of these areas is indeed Madagascar, which is characterized by an incredibly rich amphibian fauna [5].

Unluckily, Madagascar is currently the stage of an ongoing invasion determined by an introduced amphibian species: the Asian common toad *Duttaphrynus melanostictus* (Schneider, 1799) (Amphibia: Anura: Bufonidae). This species is native to Southeast Asia and was introduced to Madagascar between 2007 and 2010 [6–11]. This species is considered the third most harmful alien amphibian species, after *Rhinella marina* and *Xenopus laevis* [1], and represents a serious threat to the unique biodiversity of Madagascar, as well as to the biodiversity of other geographical areas where it was introduced [12,13]. Indeed, several biological (including the presence of parotid glands which produce bufotoxins) and ecological (including high clutch size and a generalist diet) characteristics allow this species to readily adapt to new environments, be competitive with native species, have disruptive effects on predators, and establish stable populations at faster rates [11,14]. Recent studies have revealed a drastic impact on the native Malagasy frog-eating snake *Madagascarophis colubrinus*, with an estimated monthly mortality rate due to toad poisoning of ca. 5%, which could lead to local extinction of this species if the impacts remain constant over time [14]. According to Vences et al. [15], the invasive population of *D. melanostictus* in Madagascar originated from SE Asia, namely Vietnam and/or Cambodia, likely from a single source of introduction. In this study, we investigated life history traits, including body length, age structure, and age at sexual maturity, of this invasive population using skeletochronology and histological analysis of the gonads. This information is crucial to informing management strategies and minimizing the impacts of this invasion.

2. Materials and Methods

2.1. Sampling

A total of 101 toads were sampled at 5 sites around the city of Toamasina (east coast of Madagascar), hereby called: site 1 (18.20195° S, 49.33701° E), site 2 (18. 21342° S, 49.30854° E), site 3 (18.12476° S, 49.39223° E), site 4 (18.14651° S, 49.36406° E), and site 5 (18.13951° S, 49.34605° E) (Figure 1). The toads were searched for at night and collected by hand in September 2016 (sites 1–3), and between October 2018 and May 2019 (sites 4–5). Thirty-five individuals at site 1, 26 at site 2, 20 at site 3, 9 at site 4, and 11 at site 5 were collected (see Tables S1 and S2). A detailed description of sampling sites 1–3 can be found in Santos et al. [16], while that of the sites 4 and 5 can be found in Licata et al. [17]. The climate of the study area is characterized by monthly stable temperatures (ranging from 21 °C in July and 27 °C in February) and by the occurrence of alternating dry (between May and September) and moist (between October and April) seasons [18].

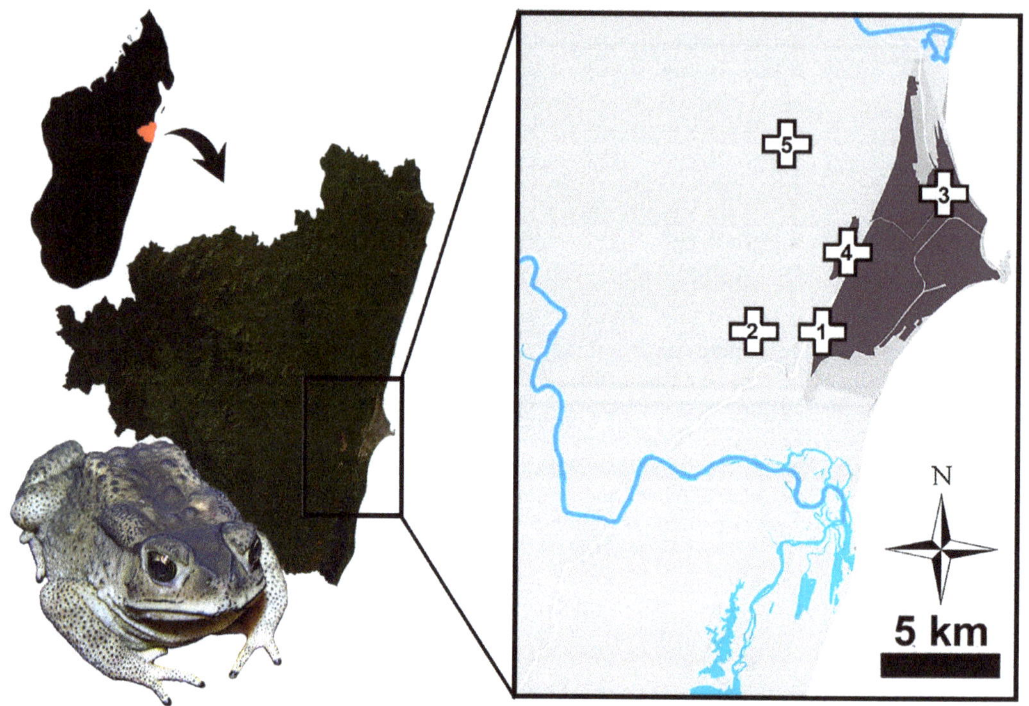

Figure 1. Geographic position of Toamasina (**left**) and the five sampling sites (**right**).

After capture, the stage of development (juvenile or adult), sex (based on the secondary sexual characteristics; see [19,20]), and snout–vent length (SVL) were collected (Tables S1 and S2). SVL measurements were collected using a caliper to the nearest 0.1 mm. The animals were anesthetized and euthanatized with 250 mg/L tricaine methane–sulfonate (MS-222, SIGMA).

2.2. Skeletochronology

For age estimation, we applied standard skeletochronological procedures [21–23] on the phalanges of 55 toads from sites 1–3 (12 males, 13 females, and 4 juveniles from site 1; 8 males and 10 females, from site 2; 2 males, 5 females, and 1 juvenile from site 3) (Table S1). From a sub-sample of randomly selected toads from sites 1–3 ($n = 26$), we also processed sections of the femur to verify the precision of the skeletochronology methods applied to the phalanges (Table S1). The phalanges and femurs were cleaned from soft tissues and preserved in 75% ethanol. They were decalcified in 5% nitric acid for 1.5–2 h (phalanges) or 3–4 h (femurs), washed under running tap water for 12 h, dehydrated through a series of graded ethanol baths, cleared with a terpene of natural origin (Bioclear, Bio Optica, Milan, Italy), and embedded with paraffin (melting point 60 °C). Diaphyseal bone cross-sections (12 μm-thick for the phalanges and 15 μm-thick for the femurs) were cut using a standard rotative microtome. The sections were stained with Mayer's hematoxylin for 30 min and then dehydrated in the open air and mounted in synthetic medium (Bio Mount HM, Bio Optica, Milan). The stained sections were examined using a Motic BA340 light microscope, equipped with a digital camera. The count of the lines of arrested growth (LAGs) was performed independently by two observers (FMG and MM) on at least ten sections per individual and without prior knowledge of the SVL and sex of the analyzed individual. According to other studies [22,24,25], we selected diaphyseal sections with the smallest medullar cavity and the widest periosteal bone, so that the total resorption of the innermost (oldest) periosteal LAG was less likely to have occurred. The number of completely resorbed LAGs was assessed based on the comparison between the phalangeal sections at the same magnification from different individuals. Age was estimated by counting the visible LAGs with the addition of any resorbed ones.

We assumed that a LAG is formed each year during the cooler and drier period (July–September), when the temperature and rainfall are lower and food availability is expected to be less abundant.

The growth rate was estimated using von Bertalanffy's model, as previously performed in several other anuran studies [22,26–28]. The general form of von Bertalanffy's growth equation used was: $SVLt = SVL \infty (1 - e^{k(t-t0)})$, where $SVLt$ is the body length at age t; $SVL\infty$ is the estimated maximum body length; e *is* the base of the natural logarithm; k is the growth coefficient; and $t0$ is the age at metamorphosis, which in amphibians represents the starting point of the growth interval. We considered 9.4 mm as the mean size at metamorphosis according to Mogali et al. [29]. Estimates of SVLmax and K with the corresponding 95% confidence interval were computed using Growth II software [30].

2.3. Sexual Maturity

To assess the age at sexual maturity, we analyzed the gonads of five specimens from sites 1–3 which were also processed for skeletochronology (two males, SVL 55.3 and 63.2 mm; three females, SVL 81.2, 95.5, and 100.9 mm) (see Table S1). Furthermore, the gonads of 20 specimens collected from sites 4–5 (six males, SVL range: 43.70–56.0 mm; fourteen females, range 58.1–79.5 mm) (see Table S2) were also analyzed to confirm the minimum body size for the onset of sexual maturity according to [31]. The gonads were fixed in neutral buffered formalin and embedded in paraffin following standard protocols [23]. They were sectioned at 7 μm thickness using a rotary microtome. The sections were stained with Mallory Trichrome (Bioptica, Milano), dehydrated through a series of graded ethanol baths, cleared with a terpene of natural origin (Bioclear, Bio Optica, Milan, Italy), and mounted in synthetic medium (Bio Mount HM, Bio Optica, Milan). The stained sections were observed under a Motic BA340 light microscope equipped with a digital camera. In the males, the presence of different germ cell stages in the seminiferous tubules was assessed according to [32], and only individuals in advanced spermatogenesis, including spermatids and sperm, were considered mature. Female sexual maturity and reproductive activity were confirmed by the presence of yolked follicles and oviductal eggs [32,33].

2.4. Statistical Analysis

The descriptive statistics included the mean ± standard deviation, minimum and maximum. For the age modal, values were also given. Since several subsets of SVL and age data differed from a normal distribution or included small sample sizes, we used the non-parametric Kruskal–Wallis test and Mann–Whitney U test to test for significant differences between the two sexes and sites. The relationship between size and age was assessed with Spearman correlation. The significance level was set at 0.05. Statistical analysis was performed using PAST software version 3.22 [34].

3. Results

3.1. Body Length

The measurements of the SVL of *D. melanostictus* individuals from sites 1–3 and the pooled data set are given in Table 1. The mean SVL of the males, as well as that of the females, did not significantly differ between the three sites (Kruskal–Wallis test, males: H = 2.78, p = 0.35; females: H = 4.4, p = 0.11). The mean SVL of the males was significantly lower than that of the females for the pooled data set (Mann–Whitney, Z = 2.11, p = 0.03). At site 3, the juveniles were the dominant cohort and included four very small-sized juveniles (25–35 mm SVL).

Table 1. SVL (in mm) of *D. melanostictus* at sampling sites 1–3 and for the pooled data set. Mean ± standard deviation (SD) and the min and max values are reported. N: number of sampled individuals; ND: phenotypically unsexed.

	Site 1	Site 2	Site 3	Pooled Sites
Males	73.4 ± 2.1	70.9 ± 2.3	67.5 ± 8.4	71.4 ± 1.6
	62.2–89.8	60.0–88.2	58.8–88.4	59.8–89.8
	$n = 15$	$n = 11$	$n = 3$	$n = 29$
Females	81.5 ± 2.2	78.7 ± 3.8	71.0 ± 5.6	78.2 ± 1.9
	61.4–89.7	57.5–104.8	57.6–109.0	57.5–104.8
	$n = 14$	$n = 13$	$n = 7$	$n = 34$
Juveniles (ND)	45.1 ± 2.0	45.7 ± 4.2	38.5 ± 3.0	41.5 ± 1.9
	36.2–49.7	41.5–49.9	25.4–49.2	25.4–49.0
	$n = 6$	$n = 2$	$n = 10$	$n = 18$

3.2. Skeletochronological Observations

Hematoxynophilic periosteal lines were always observed in the phalanx and femur diaphyseal sections and showed full correspondence in the specimens where it was possible to analyze both types of bone, indicating that skeletochronological reading can be successfully performed on either the phalanx or femur (Figure 2). The hematoxynophilic line at the edge of the medullary cavity was interpreted as a Katschenko line (KL) according to other studies [35,36]; the remaining circular periosteal lines are typical lines of arrested growth (LAGs), although their distinctiveness was very variable among the different sections and individuals (Figure 3). In the phalangeal sections of five individuals (about 9% of the whole sample), the LAGs were faint and less clear, sometimes confused with the local lamellar organization of the periosteal cortex.

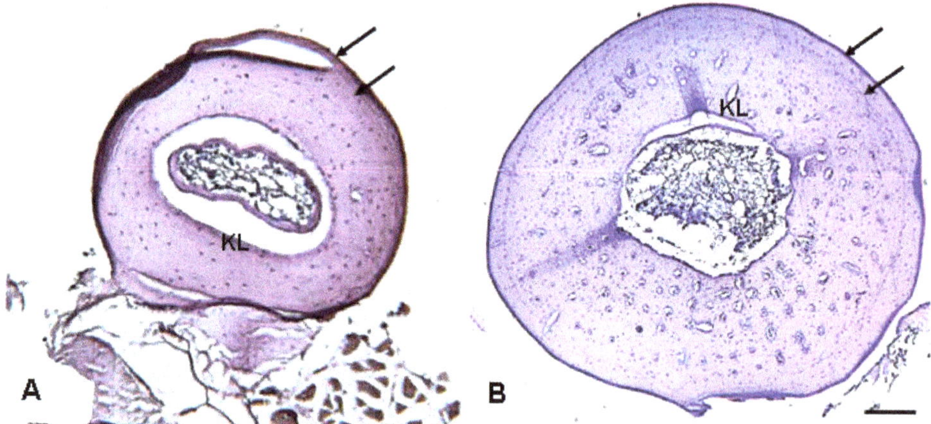

Figure 2. Representative cross-sections of the phalanx (**A**) and femur (**B**) taken from a *D. melanostictus* female, laboratory label ACZC12131, SVL: 77.9 mm. Arrows indicate the lines of arrested growth (LAGs), of which the outermost is confluent with the outer bone cortex margin. KL: Katschenko line. Scale bar: 130 µm in (**A**) and 320 µm in (**B**).

Endosteal remodeling phenomena were observed, particularly in larger (>80 mm) and likely older animals, as evidenced by the presence of conspicuous endosteal bone that had eroded the periosteal cortical layer. By comparing the diameter of the medullary cavity of the smallest individuals in the study sample (<50 mm) with those of the largest individuals (whose sections showed signs of endosteal remodeling after qualitative histological examination), it emerged that the first (innermost) LAG was sometimes (15% of the study sample) partially removed by bone remodeling.

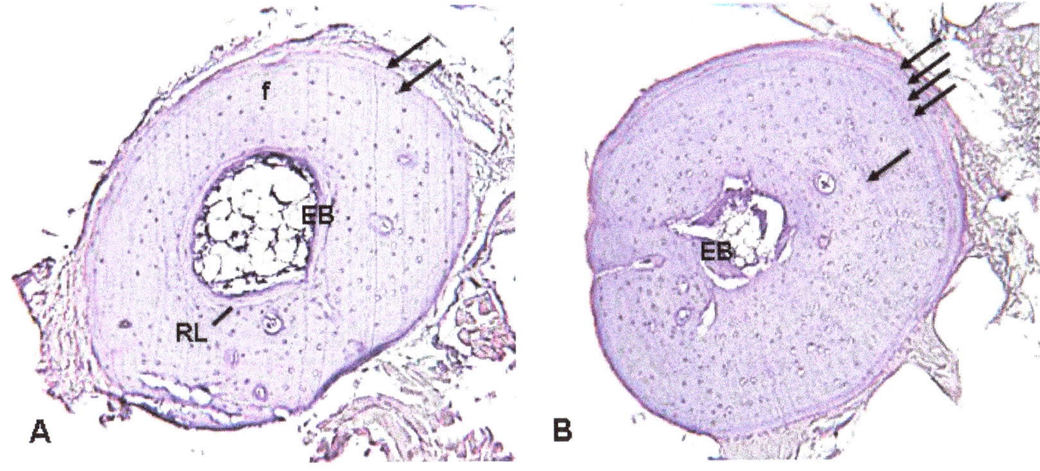

Figure 3. Representative cross-sections of the phalanx of a *D. melanostictus* (**A**) male, laboratory label ACZC12157, SVL: 70 mm; (**B**) female, laboratory label ACZC12173, SVL: 95.5 mm. Arrows indicate lines of arrested growth (LAGs). In (**B**), the faint multiple innermost circular haematoxynophilic lines were interpreted as true LAGs. EB: endosteal bone; f: false line; RL: reversal line. Scale bar: 150 μm in (**A**) and 180 μm in (**B**).

3.3. Age

Of the 55 specimens, 5 individuals showed unclear LAGs and were excluded from the analyses.

The mean and modal age (in years) of *D. melanostictus* from Sites 1–3 and for the pooled data set is given in Table 2. The mean age of the males, as well as that of the females, did not significantly differ between the three sites (Kruskal–Wallis test, males: H = 3.41, $p = 0.13$; females: H = 0.91, $p = 0.61$). The mean age of the males was significantly lower than that of the females for the pooled data set (Mann–Whitney, Z = 2.36, $p = 0.01$) and significantly lower at site 1 (Mann-Whitney, Z = 1.90, $p < 0.05$) but not for site 2 (Mann–Whitney, Z = 1.18, $p = 0.23$). At site 3, the sampling size was too low to apply statistics. For the pooled data set, the males and females exhibited unimodal (2 years) and bimodal age (2 and 3 years), respectively.

Table 2. Age (in years) of the *D. melanostictus* at the sampling sites 1–3 and for the pooled data set. The mean ± standard deviation (SD) and min and max values are reported. For the modal age of the pooled data, the frequency is also given (between brackets). N: number of sampled individuals; ND: phenotypically unsexed.

	Site 1		Site 2		Site 3		Pooled Sites	
	Mean	Mode	Mean	Mode	Mean	Mode	Mean	Mode
Males	1.7 ± 0.8 1–3 *n* = 9	3	2.1 ± 0.6 1–3 *n* = 8	2	1.0 ± 0 1 *n* = 2	-	1.8 ± 0.7 1–3 *n* = 19	2 (36.8%)
Females	2.9 ± 1.4 1–6 *n* = 12	2, 3, 4	2.7 ± 1.2 1–5 *n* = 9	2, 3	2.2 ± 1.3 1–3 *n* = 5	1	2.7 ± 1.3 1–6 *n* = 26	2, 3 (53.8%)
Juveniles ND	1.0 ± 0 1 *n* = 4	1	- - *n* = 0	-	1 1 *n* = 1	-	1 1 *n* = 5	1

The relationship between age and SVL is shown in Figure 4. Age was positively correlated with SVL in both sexes when we analyzed the pooled data set (Spearman's rank correlation, males rs = 0.74, p < 0.001; females rs = 0.73, p < 0.001), but only for the females of site 1 (Spearman's rank correlation, rs = 0.65, p = 0.02) and 2 (Spearman's rank correlation, rs = 0.84, p = 0.006) when the data were analyzed separately. The ANCOVAs showed that the difference in the linear regression line between the sexes was not significant, ($F_{1,43}$ = 0.95. p = 0.332) using age as a covariate. Von Bertalanffy's growth model using pooled data showed that for both sexes, the estimated asymptotic SVL was smaller than the maximum SVL recorded (SVL∞ ± CI, males: 75.48 ± 0.42; females: 87.52 ± 0.43). The growth coefficient significantly differed between sexes (k ± CI, males: 1.04 ± 0.02; females: 1.59 ± 0.05).

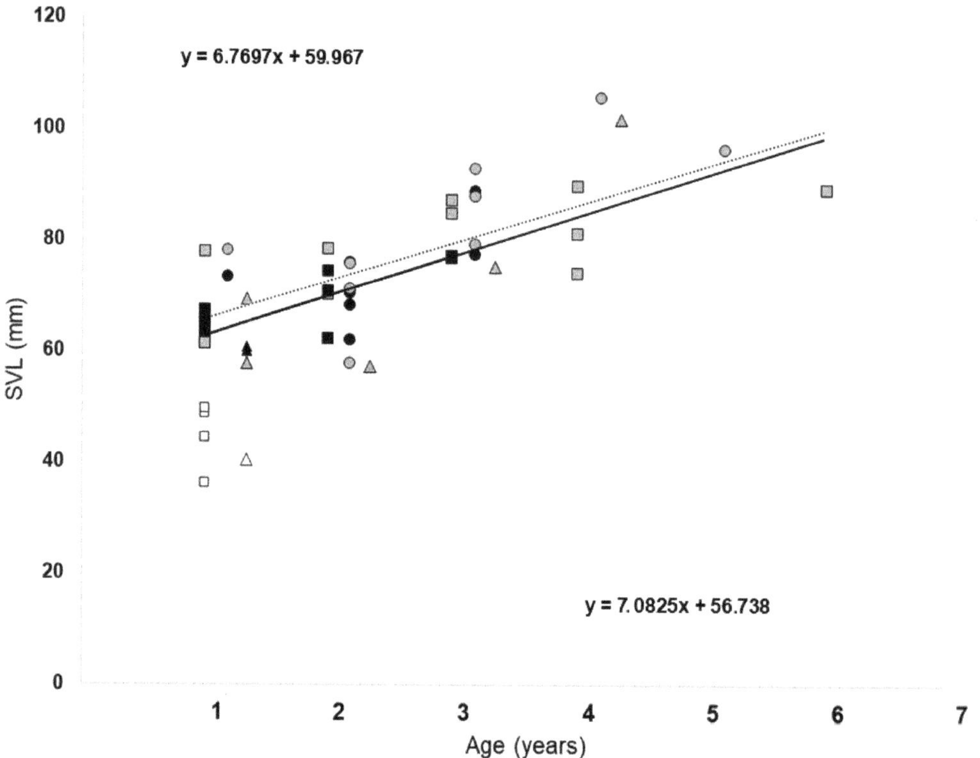

Figure 4. Relationships between age (years) and SVL in *D. melanostictus* individuals from Madagascar. Squares, circles, and triangles represent toads from sites 1, 2, and 3, respectively. The black, gray, and empty symbols refer to males, females, and juveniles, respectively. The solid line and the dotted line represent the linear regression line for males and females, respectively, for the pooled data. A linear regression equation is also given for the males (bottom) and females (top).

3.4. Assessment of Gonad Status and Individual Age

The two females (81.2 and 95.5 mm), analyzed for both skeletochronology and gonad histology (see Table S1), had an estimated age of 4 and 5 years, respectively, and their ovaries contained many postvitellogenetic and late vitellogenetic follicles. One female measuring 100.9 mm also had postvitellogenetic and late vitellogenic follicles, but it was not possible to estimate its age. The ovaries of all adult females were characterized by the presence of numerous atretic follicles (Figure 5A). The two males (55 and 63.2 mm), analyzed for both skeletochronology and gonad histology, had an estimated age of 1 year and testicles with all spermatogenetic stages, including sperm (Figure 5B). Among the specimens from sites 4–5 (see Table S2), the females ranging in SVL from 56.9 to 66.8 mm ($n = 4$) had previtellogenetic follicles, and all females with SVL ranging from 69.9 to 79.5 mm ($n = 11$) had follicles in vitellogenetic growth, thus indicating that they were sexually mature. The males ranging between 43.4 and 48.4 mm ($n = 5$) showed testicles with only spermatogonia, while one male with SVL of 56 mm had testicles with all spermatogenetic stages, including sperm (see Table S2).

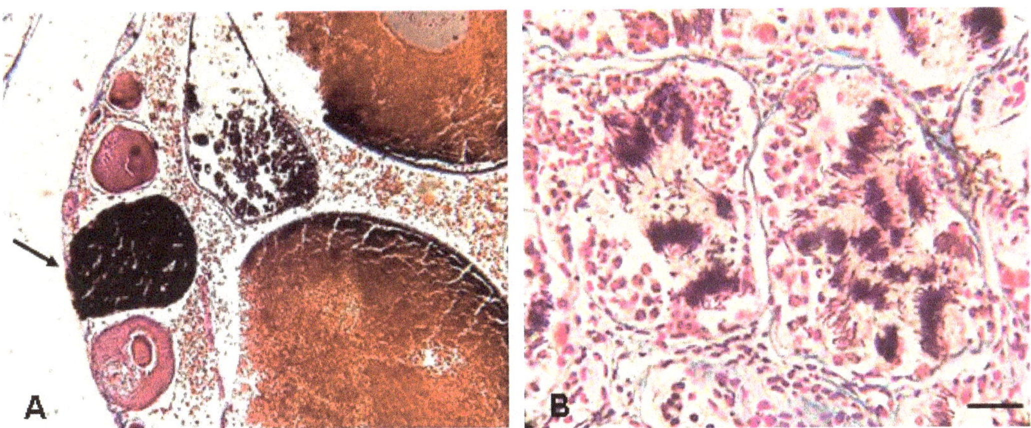

Figure 5. Histological sections of gonads of *D. melanostictus* from Madagascar stained with Mallory's trichrome. (**A**) Ovary of a female (SVL: 84.6 mm) with late vitellogenic follicles and atresia (arrow); (**B**) testis of a male (SVL: 55.2 mm) with all spermatogenetic stages. The bar corresponds to 700 µm in (**A**) and 200 µm in (**B**).

4. Discussion

Species' life history traits can be mismatched between invasive and native populations due to different environmental conditions (e.g., a lack of competitors and lower predatory pressure). On the other hand, trait mismatches can result from pre-existing differences in the native source populations and can emerge during acclimatization through genetic drift and local adaptation [37,38]. The study of life history traits, including growth rate, age, and size at maturity, of an invasive population can be useful for understanding how each trait may have contributed to successfully colonizing a new area.

In accordance with previous studies [11], males of the invasive population of the Asian common toads in Madagascar are larger in body size than the native populations, However, we did not find the same conclusion as far as the females are concerned, which were found to be slightly smaller. Indeed, the mean SVL ± SD of the adults and subadults was 71.9 ± 1.6 mm in males and 78.2 ± 1.9 mm in females in Madagascar and 60.7 ± 6.2 mm in males and 81.3 ± 9.2 in females in a native population from Vietnam [31]. Unlike what is known for native populations of Asian toad [31] as well as for other bufonids (e.g., *Bufo bufo* [39], *Sclerophrys pardalis* [40], and *Rhinella arenarum* [28]), the females were not always significantly larger than the males. This finding is in agreement with Licata et al. [11], where it has been suggested that the sexual dimorphism in the size of invasive Asian toads in Madagascar is reduced compared to that of native populations. Our data suggest that the larger size of the invasive common Asian toad of Madagascar is not dependent on greater longevity, as shown by our skeletochronological data (see below), and this might be related to local differences in food availability [11]. The prevalence of juveniles at Site 3, which at the time of sampling in 2016 was at the invasion front, suggests that the dispersal front is mainly formed by young toads [10,41].

Skeletochronology is one of the most used and reliable indirect methods for age determination in amphibians and reptiles [42–45]. In accordance with previous studies [46–49], our study shows the suitability of this method for the invasive population of this species as well, even though some difficulties were encountered in the interpretation of the results. The first (innermost) hematoxynophilic mark was often less stained than those successively layered, but it was considered here as a true LAG, based on its position with respect to the medullary cavity. The distinctiveness of growth marks is known to reflect the amplitude of the growth cycles and the severity of inactivity both at individual and population levels [50,51]. The visibility of the LAGs was a major problem, with considerable differences in the LAGs' distinctiveness both in different sections of the same individual and between the individuals. As previously proposed [27], this seems to be an individual, age-dependent phenomenon, likely related to microclimatic conditions and the length and intensity of the torpor period. This period in eastern Madagascar probably coincides with the dry season (from May to September), when more unfavorable environmental conditions (e.g., less available food) can cause some continuous period of inactivity of the individuals.

So far, a skeletochronological estimate of the age of *D. melanostictus* has been provided for populations inhabiting different areas of India [46–49]. These studies suggest that the species generally live for a maximum of four [47], five [48], or six [49] years and only reaches nine years in exceptional cases [48]. Based on the analysis of the published phalangeal sections [46], we consider the record age of 12 years (estimated for a female) dubious because we think that some of the LAGs are probably lamellae of the endosteal bone (see Figure 2B) [46]. In our study, the estimated maximum age was three years in males and six years in females, which is in general agreement with previous skeletochronological studies [47,49]. Overall, these data indicate that *D. melanostictus* is not a very long-lived species, especially if compared with other bufonids from temperate regions, such as *Bufo bufo* [39] and *Bufotes variabilis* (=*B. viridis*) [52]; although, our results are in agreement with previous studies that found that tropical or sub-tropical species are shorter-lived and mature earlier than species from temperate areas [53–55]. However, we cannot exclude that there may not have been enough time for the introduced population to achieve longer life spans due to the recent introduction of this species to Madagascar [8,9,56], and further studies are needed to confirm the maximum age of the individuals of this invasive population.

In agreement with Kumbar et al. [48], we observed a positive correlation between body length and age in both sexes, although this result was significant only with pooled data. When the data were analyzed separately for each site, the positive correlation was significant only for females from sites 1 and 2. This is probably due to the small sample size, rather than a true variability in both time and space [57]. However, in each site and for both sexes, we found very different growth rates, as confirmed by the wide overlap of SVLs among the different age classes. Furthermore, the fact that the male toads seem to be larger than their native conspecifics despite having comparable maximum age suggests that the former have higher growth rates.

By combining skeletochronological age estimation with gonad analysis, the present study shows that males and females of the *D. melanostictus* population introduced to Madagascar attain sexual maturity when they show one and two LAGs, respectively. This result suggests that sexual maturity is achieved when males are one years old and when females are two years old, in agreement with published data [49].

5. Conclusions

This study analyzes data on the body length, age structure and age at sexual maturity of the invasive Asian common toad *D. melanostictus* from different sites in Toamasina (east Madagascar). It provides useful information to better understand the life history of this species, in particular when considered as a harmful species in a biodiversity-rich context such as Madagascar, and it is also considered as a relevant threat by the "A Conservation Strategy for the Amphibians of Madagascar" project and consequent Sahonagasy Action Plans [58,59], that have been proposed over the last two decades.

Our findings show that the *D. melanostictus* males introduced in Madagascar are on average larger than the native males, although they, as well as the introduced females, do not seem to live longer (males' maximum age: 3 years; females' maximum age: 6 years).

Similar to the native populations, males and females attain sexual maturity in their first and second year after metamorphosis and exhibit a female-biased size dimorphism, although reduced when compared to native populations. Overall, the life history traits examined here largely overlap with those of native populations. Further studies are needed to understand if the larger size and faster growth rates observed in the invasive population in Madagascar are a consequence of more favorable environmental conditions, with respect to the native range (e.g., larger trophic niches, a lack of competitors, and lower predatory pressure). This aspect can have several consequences on the management of the invasive population, and we encourage the development of further studies on this matter.

Supplementary Materials: The following supporting information can be downloaded at: https://www.mdpi.com/article/10.3390/ani13132099/s1, Table S1: Sample of Malagasy *Duttaphrynus melanostiscus* used to estimate snout-vent length (SVL) e for skeletochronology. M = male, F = female; J = Juveniles (undetermined sex). **, individuals whose gonad was also available; Table S2: Specimens of Malagasy *Duttaphrynus melanostiscus* used to evaluate reproductive status. A: adult; F: female; J: juvenile; M: male.

Author Contributions: F.M.G., F.A. and A.C. conceived the study; B.S., W.C., J.F.S.F., S.H.N., J.N., T.F.R., R.F.H. and F.L., sampled the animals; F.M.G., M.M., F.L. and S.P. performed the laboratory analyses; F.M.G., M.M., F.L. and S.P. contributed to the evaluation of the results obtained; F.M.G. wrote the first version of the manuscript; F.M.G., F.A., M.M., F.L., G.O. and A.C. reviewed and edited the final version of the manuscript. All authors have read and agreed to the published version of the manuscript.

Funding: Portuguese National Funds through an FCT (Fundação para a Ciência e a Tecnologia) support research contract to AC [2020.00823.CEECIND/CP1601/CT0003].

Institutional Review Board Statement: The Ministère de l'Environnement et du Développement Durable approved all of the following permits: collection: No 226/16/MEEF/SG/DGF/DSAP/SCB.Re of 19th September 2016; transport: No 1679-16/MEEF/SG/DGF/DREEF.ATS/SREco and No 1680-; 16/MEEF/SG/DGF/DREEF.ATS/SREco of 24th September 2016; and export: No 284N-EA10/MG16 of October 5th.

Informed Consent Statement: Not applicable, this study does not involve humans.

Data Availability Statement: All the data generated are available within this manuscript and in the supplementary information.

Acknowledgments: We thank all of the villagers in and around Toamasina who helped us with finding the toads and allowed us to enter their properties to collect them.

Conflicts of Interest: The authors declare no conflict of interest.

References

1. Measey, A.; Vimercati, G.; de Villiers, F.A.; Mokhatla, M.; Davies, S.J.; Thorp, G.J.; Rebelo, A.D.; Kumschick, S. Global assessment of alien amphibian impacts in a formal framework. *Divers. Distrib.* **2016**, *22*, 970–981. [CrossRef]
2. Mezzasalma, M.; Visone, V.; Petraccioli, A.; Odierna, G.; Capriglione, T.; Guarino, F.M. Non-random accumulation of LINE1-like sequences on differentiated snake W chromosomes. *J. Zool.* **2016**, *300*, 67–75. [CrossRef]
3. Mezzasalma, M.; Andreone, F.; Odierna, G.; Guarino, F.M.; Crottini, A. Comparative cytogenetics on eight Malagasy Mantellinae (Anura, Mantellidae) and a synthesis of the karyological data on the subfamily. *Comp. Cytogen.* **2022**, *16*, 11–17. [CrossRef]
4. Sidhom, M.; Said, K.; Chatti, N.; Guarino, F.M.; Odierna, G.; Petraccioli, A.; Picariello, O.; Mezzasalma, M. Karyological characterization of the common chameleon (*Chamaeleo chamaeleon*) provides insights on the evolution and diversification of sex chromosomes in Chamaeleonidae. *Zoology* **2020**, *141*, 125738. [CrossRef]
5. Andreone, F.; Carpenter, A.; Crottini, A.; D'Cruze, N.; Dubos, N.; Edmonds, D.; Garcia, G.; Luedtke, J.; Megson, S.; Rabemananjara, F.; et al. Chapter 4. Amphibian conservation in Madagascar: Old and novel threats for a peculiar fauna. In *Status and Threats of Afrotropical Amphibians. Sub-Saharan Africa, Madagascar, Western Indian Ocean Islands*; Heatwole, H., Roedel, M., Eds.; Chimaira: Frankfurt, Germany, 2021; Volume II, pp. 147–186.
6. Andreone, F.; Rabibisoa, N.; Randrianantoandro, C.; Crottini, A.; Edmonds, D.; Kraus, F.; Lewis, J.P.; Moore, M.; Rabemananjara, F.C.E.; Rabemanantoa, J.C.; et al. Risk review is under way for invasive toad. *Nature* **2014**, *512*, 253. [CrossRef]

7.   Crottini, A.; Andreone, F.; Edmonds, D.; Hansen, C.M.; Kraus, F.; Lewis, J.P.; Moore, M.; Rabemananjara, F.; Rabemanantsoa, J.C.; Randrianantoandro, C.; et al. A new challenge for amphibian conservation in Madagascar: The invasion of *Duttaphrynus melanostictus* in Toamasina province. *FrogLog* **2014**, *22*, 46–47.
8.   Kolby, J.E.; Kraus, F.; Rabemananjara, F.; Rabesihanaka, S.; Rabibisoa, N.; Rafanomezantsoa, J. Stop Madagascar's toad invasion now. *Nature* **2014**, *509*, 563. [CrossRef]
9.   Moore, M.; Solofo Niaina Fidy, J.F.; Edmonds, D. The new toad in town: Distribution of the Asian toad, *Duttaphrynus melanostictus*, in the Toamasina area of eastern Madagascar. *Trop. Conserv. Sci.* **2015**, *8*, 440–455. [CrossRef]
10. Freeman, K.L.M.; Andreone, F.; Randriamoria, T.M.; Randrianizahana, H.; Raselimanana, A.P.; Raxworthy, C.J.; Robsomanitrandrasana, E.; Crottini, A. Bufonidae: *Duttaphrynus melanostictus*, Black-spined toad. In *The new natural history of Madagascar*; Goodman, S.M., Ed.; Princeton University Press: Princeton, NJ, USA, 2022; pp. 1404–1410. [CrossRef]
11. Licata, F.; Andreone, F.; Crottini, A.; Harison, R.F.; Ficetola, G.F. Does spatial sorting occur in the invasive Asian toad in Madagascar? Insights into the invasion unveiled by morphological analyses. *J. Zool. Syst. Evol. Res.* **2021**, *59*, 2161–2169. [CrossRef]
12. Reilly, S.B.; Wogan, G.O.; Stubbs, A.L.; Arida, E.; Iskandar, D.T.; McGuire, J.A. Toxic toad invasion of Wallacea: A biodiversity hotspot characterized by extraordinary endemism. *Glob. Change Biol.* **2017**, *23*, 5029–5031. [CrossRef]
13. Soorae, P.T.; Frankham, G.J.; Ali Mohamed, A. The first record of the Asian common toad *Duttaphrynus melanostictus* Schneider, 1799 in Abu Dhabi, United Arab Emirates. *Bioinvasions Rec.* **2020**, *9*, 434–443. [CrossRef]
14. Licata, F.; Harison, F.F.; Ficetola, F.G.; Freeman, K.; Muller, B.J.; Rodriguez Ponga, V.; Andreone, F.; Crottini, A. Toad invasion of Malagasy forests triggers severe mortality of a predatory snake. *Biol. Invasions* **2022**, *24*, 1189–1198. [CrossRef]
15. Vences, M.; Brown, J.L.; Lathrop, A.; Rosa, G.M.; Cameron, A.; Crottini, A.; Dolch, R.; Edmonds, D.; Freeman, K.L.M.; Glaw, F. Tracing a toad invasion: Lack of mitochondrial DNA variation, haplotype origins, and potential distribution of introduced *Duttaphrynus melanostictus* in Madagascar. *Amphib-reptil* **2017**, *38*, 197–207. [CrossRef]
16. Santos, B.; Bletz, M.C.; Sabino-Pinto, J.; Cocca, W.; Fidy, J.F.; Freeman, K.L.; Kuenzel, S.; Ndriantsoa, S.; Noel, J.; Rakotonanahary, T.; et al. Characterization of the microbiome of the invasive Asian toad in Madagascar across the expansion range and comparison with a native co-occurring species. *PeerJ* **2021**, *9*, e11532. [CrossRef] [PubMed]
17. Licata, F.; Mohanty, N.P.; Crottini, A.; Andreone, F.; Harison, R.F.; Randriamoria, T.M.; Freeman, K.; Muller, B.; Birkinshaw, C.; Tilahimena, A.; et al. Using public surveys to rapidly profile biological invasions in hard-to monitor areas. *Anim. Cons.*, online version; 2023. [CrossRef]
18. Merkel, A. Toamasina Climate (Madagascar). 2023. Available online: https://en.climate-data.org/africa/madagascar/toamasina/toamasina-4029/ (accessed on 18 April 2023).
19. Gordon, A. Secondary sexual characters of *Bufo melanostictus* Schneider. *Copeia* **1933**, *1933*, 204–207. [CrossRef]
20. Licata, F.; Andreone, F.; Freeman, K.; Rabesihanaka, S.; Robsomanitrandrasana, E.; Reardon, J.T.; Crottini, A. The Asian toad (*Duttaphrynus melanostictus*) in Madagascar: A report of an ongoing invasion. In *Problematic Wildlife II: New Conservation and Management Challenges in the Human-Wildlife Interactions*, 1st ed.; Angelici, F.M., Rossi, L., Eds.; Springer: Berlin, Germany, 2020; pp. 617–638.
21. Erismis, U.C.; Arikan, U.; Konuk, M.; Guarino, F.M. Age structure and growth in caucasian parsley frog *Pelodytes caucasicus* (Boulenger, 1896) from Turkey. *Russ. J. Herpetol.* **2009**, *16*, 19–26.
22. Guarino, F.M.; Crottini, A.; Mezzasalma, M.; Randrianirina, J.E.; Andreone, F. A skeletochronological estimate of age and growth in a large riparian frog from Madagascar (Anura, Mantellidae, *Mantidactylus*). *Herpetozoa* **2019**, *32*, 39–44. [CrossRef]
23. Guarino, F.M.; Di Nocera, F.; Pollaro, G.; Galiero, G.; Iaccarino, D.; Iovino, D. Skeletochronology, age at maturity and cause of mortality of loggerhead sea turtles *Caretta caretta* stranded along the beaches of Campania (south-western Italy, western Mediterranean Sea). *Herpetozoa* **2020**, *33*, 39–51. [CrossRef]
24. Driscoll, D.A. Skeletochronological assessment of age structure and population stability for two threatened frog species. *Austral. J. Ecol.* **1999**, *24*, 182–189. [CrossRef]
25. Sinsch, U.; Dehling, J.M. Tropical anurans mature early and die young: Evidence from eight Afromontane *Hyperolius* species and a meta-analysis. *PLoS ONE* **2017**, *12*, e0171666. [CrossRef]
26. Miaud, C.; Guyetant, R.; Elmberg, J. Variations in life-history traits in the common frog *Rana temporaria* (Amphibia: Anura): A literature review and new data from the French Alps. *J. Zool.* **1999**, *249*, 61–73. [CrossRef]
27. Guarino, F.M.; De Pous, P.; Crottini, A.; Mezzasalma, M.; Andreone, F. Age structure and growth in a population of *Pelobates varaldii* (Anura, Pelobatidae) from northwestern Morocco. *Amphib-reptil.* **2011**, *32*, 550–556. [CrossRef]
28. Bionda, C.L.; Kost, S.; Salas, N.E.; Lajmanovich Sinsch, U.; Martino, A.L. Age structure, growth and longevity in the common toad, *Rhinella arenarum*, from Argentina. *Acta Herpetol.* **2015**, *10*, 55–62.
29. Mogali, S.; Saidapur, S.; Shanbhag, B. Influence of desiccation threat on the metamorphic traits of the Asian common toad, *Duttaphrynus melanostictus* (Anura). *Acta Herpetol.* **2017**, *12*, 75–80.
30. Henderson, P.A.; Seaby, R.M. *Growth II. Pisces*; Conservation Ltd.: Lymington, UK, 2001.
31. Ngo, B.V.; Ngo, C.D. Reproductive activity and advertisement calls of the Asian common toad *Duttaphrynus melanostictus* (Amphibia, Anura, Bufonidae) from Bach Ma National Park, Vietnam. *Zool. St.* **2013**, *52*, 12. [CrossRef]
32. Lofts, B. Amphibians. In *Marshall's Physiology of Reproduction. I. Reproductive Cycles of Vertebrates*; Lamming, G.E., Ed.; Churchill Livingstone: Edinburgh, Scotland, 1984; Volume 1, pp. 127–205.

33. Jørgensen, C.; Barker, K.S.; Vijayakumar, S. Body size, reproduction and growth in a tropical toad, *Bufo melanostictus*, with a comparison of ovarian cycles in tropical and temperate zone anurans. *Oikos* **1986**, *46*, 379–389. [CrossRef]
34. Hammer, Ø.; Harper, D.A.T.; Ryan, P.D. PAST: Paleontological Statistics Software Package for Education and Data Analysis. *Palaeontol. Electron.* **2001**, *4*, 1–9.
35. Rozenblut, B.; Ogielska, M. Development and growth of long bones in european water frogs (Amphibia: Anura: Ranidae), with remarks on age determination. *J. Morphol.* **2005**, *265*, 304–317. [CrossRef]
36. Rahman, M.M.; Lee, Y.Y.; Park, S.M.; Ham, C.H.; Sung, H.C. Confirmation of the applicability of skeletochronology and estimating the age Structure of *Kaloula borealis* (Microhylidae: Anura) at Lake Sihwa, South Korea. *Biology* **2022**, *11*, 898. [CrossRef]
37. Ficetola, G.F.; Bonin, A.; Miaud, C. Population genetics reveals origin and number of founders in a biological invasion. *Mol. Ecol.* **2008**, *17*, 773–782. [CrossRef]
38. Tessa, G.; Del Forno, C.; Govindarajulu, P.; Tissot, N.; Miaud, C.; Andreone, F. Age and body size in four introduced populations of the American bullfrog, *Lithobates catesbeianus* (Ranidae). *Eur. Zool. J.* **2016**, *83*, 497–502. [CrossRef]
39. Hemelaar, A. Age, growth and other population characteristics of *Bufo bufo* from different latitudes and altitudes. *J. Herpetol.* **1988**, *22*, 369–388. [CrossRef]
40. Cherry, M.I.; Francillon-Vieillot, H. Body size, age and reproduction in the leopard toad, *Bufo pardalis*. *J. Zool.* **1992**, *228*, 41–45. [CrossRef]
41. Licata, F.; Ficetola, G.F.; Falaschi, M.; Muller, B.J.; Andreone, F.; Harison, R.F. Spatial ecology of the invasive Asian common toad in Madagascar and its implications for invasion dynamics. *Sci. Rep.* **2023**, *13*, 3526. [CrossRef] [PubMed]
42. Castanet, J. Age estimation and longevity in reptiles. *Gerontology* **1994**, *40*, 174–192. [CrossRef] [PubMed]
43. Castanet, J.; Francillon-Vieillot, H.; Meunier, F.J.; de Ricqles, A. Bone and individual aging. In *Bone*; Hall, B.B.K., Ed.; CRC Press: Boca Raton, FL, USA, 1993; Volume 7, pp. 245–283.
44. Smirina, E.M. Age determination and longevity in amphibians. *Gerontology* **1994**, *40*, 133–146. [CrossRef] [PubMed]
45. Peng, Z.; Zhang, L.; Lu, X. Global gaps in age data based on skeletochronology for amphibians. *Integr. Zool.* **2022**, *17*, 752–763. [CrossRef]
46. Nayak, S.; Mahapatra, P.K.; Mishra, S.; Dutta, S.K. Age determination by skeletochronology in the common Indian toad *Bufo melanostictus* Schneider, 1799 (Anura: Bufonidae). *Herpetozoa* **2007**, *19*, 111–119.
47. Andia, B.N.; Dixit, P.K.; Behera, S.; Behera, H.N. Age determination in common Indian Toad, *Bufo melanostictus*, by skeletochronology. *J. Sci. Technol.* **2010**, *6*, 42–56.
48. Kumbar, S.M.; Lad, S. Determination of age and longevity of road mortal indian common toad *Duttaphrynus melanostictus* by skeletochronology. *Rus. J. Herpetol.* **2017**, *24*, 217–222. [CrossRef]
49. Sahoo, D.D.; Kara, T.C. Determination of age, longevity and age at sexual maturity in common Asian Toad (*Duttaphrynus Melanostictus*) by skeletochronology. *Octa J. Biosci.* **2017**, *5*, 5–8.
50. Leclair, M.H.; Leclair, R., Jr.; Gallant, J. Application of skeletochronology to a population of *Pelobates cultripes* (Anura: Pelobatidae) from Portugal. *J. Herpetol.* **2005**, *39*, 199–207. [CrossRef]
51. Tessa, G.; Guarino, F.M.; Giacoma, C.; Mattioli, F.; Andreone, F. Longevity and body size in three populations of *Dyscophus antongilii* (Microhylidae, Dyscophinae), the tomato frog from north-eastern Madagascar. *Acta Herpetol.* **2007**, *2*, 139–147.
52. Altunisik, A.; Ozdemir, N. Life history traits in *Bufotes variabilis* (Pallas, 1769) from 2 different altitudes in Turkey. *Turkish J. Zool.* **2015**, *39*, 153–159. [CrossRef]
53. Guarino, F.M.; Tessa, G.; Mercurio, V.; Andreone, F. Rapid sexual maturity and short life span in the blue-legged frog and the rainbow frog from the arid Isalo Massif, southern-central Madagascar. *Zoology* **2010**, *113*, 378–384. [CrossRef] [PubMed]
54. Sinsch, U. Review: Skeletochronological assessment of demographic life-history traits in amphibians. *Herpetol. J.* **2015**, *25*, 5–13.
55. Tessa, G.; Crottini, A.; Giacoma, C.; Guarino, F.M.; Randrianirina, J.E.; Andreone, F. Comparative longevity and age at sexual maturity in twelve rainforest frogs of the genera *Boophis*, *Gephyromantis*, and *Mantidactylus* (Anura: Mantellidae) from Madagascar. *Phyllomedusa* **2017**, *16*, 13–21. [CrossRef]
56. Licata, F.; Ficetola, G.F.; Freeman, K.; Mahasoa, H.R.; Ravololonarivo, V.; Fidy, F.J.S.N. Abundance, distribution and spread of the invasive Asian toad *Duttaphrynus melanostictus* in eastern Madagascar. *Biol. Invasions* **2019**, *21*, 1615–1626. [CrossRef]
57. Halliday, T.; Verrell, P. Body size and age in amphibians and reptiles. *J. Herpetol.* **1988**, *22*, 253–265. [CrossRef]
58. Andreone, F.; Dawson, J.S.; Rabemananjara, F.C.E.; Rabibisoa, N.H.C.; Rakotonanahary, T.S. *New Sahonagasy Action Plan 2016–2020*; Museo Regionale di Scienze Naturali and Amphibian Survival Alliance: Turin, Italy, 2016; pp. 1–46.
59. Rakotoarison, A.; Ndriantsoa, S.H.; Rabemananjara, F.C.E.; Rabibisoa, N.H.C.; Rakotonanahary, T.F.; Kandriamahazo, J.A.R.H.; Andreone, F. More than 15 years of amphibian conservation in Madagascar under the flag of IUCN SSC Amphibian Specialist Group. *Naturalista sicil.* **2022**, *46*, 305–312.

Article

Malagasy Amphibian Wildlife Trade Revisited: Improving Management Knowledge of the Trade

Angus I. Carpenter [1,*] and Franco Andreone [2]

[1] Institute of Science and Environment, University of Cumbria, Ambleside Campus, Rydal Road, Ambleside, Cumbria LA22 9BB, UK

[2] Museo Regionale di Scienze Naturali, Via G. Giolitti, 36, I-10123 Torino, Italy; franco.andreone@rehione.piemonte.it

* Correspondence: carpenter.angus@gmail.com

Simple Summary: There is much debate about the wildlife trade, with arguments made both for and against the trade. If wildlife trade is to continue, both knowledge of species and their population statuses and confidence within the global mechanism that monitors and manages the wildlife trade are required. Using Madagascar's amphibian trade, this study investigates this issue. The findings of the study highlight the need to maintain awareness of changes to species descriptions and the need to cross-referencing with population status data, such as that available from the IUCN Redlist, but significantly against CITES quotas and the quality of the NDFs that support them. In this study, it was observed that Madagascar seems to have improved its management of the amphibian trade over time but that new species are constantly being described over time, which could add complications to the management of the trade.

Abstract: Madagascar is a biodiversity hotspot with a long history of trading in its wildlife, especially its hyper-diverse amphibian taxa. Due to globally raised concerns over the conservation of harvested species, CITES was introduced as a global mechanism with which to monitor and regulate the trade. Utilising data collated from the CITES Trade database, this study sought to investigate the trade and CITES' effectiveness in managing the trade with respect to Madagascar. Over a 28-year period, 20 known amphibian species were exported from Madagascar, constituting a total of nearly 271,000 individuals. Formal descriptions of Malagasy amphibian species have increased and continue to increase greatly over time. However, there was no longitudinal relationship regarding the numbers of individuals traded as new species were described. Overall, the number of individuals traded has declined over time, but where assessments were provided by the IUCN Redlist, population declines were reported in all but one species of Malagasy amphibian. *Mantella* (97.5%) continues to be the predominantly traded genus, with certain, high-conservation-concern, species continuing to be traded. Despite initial concerns over the effectiveness of CITES's actions, after concerted efforts, it appears that CITES' actions were having positive impacts on regulating the trade. However, going forward, concerns remain over the appropriateness of the quotas set and the robustness of their underpinning NDFs. Furthermore, with the increase in the number of recognised species, the potential for incorrect species labelling on the CITES permits increases and requires greater attention.

Keywords: wildlife trade; Madagascar; amphibians; CITES; IUCN Redlist; conservation

Citation: Carpenter, A.I.; Andreone, F. Malagasy Amphibian Wildlife Trade Revisited: Improving Management Knowledge of the Trade. *Animals* **2023**, *13*, 2324. https://doi.org/10.3390/ani13142324

Academic Editor: Robert K. Browne

Received: 17 June 2023
Revised: 5 July 2023
Accepted: 13 July 2023
Published: 17 July 2023

1. Introduction

Academics may debate the existence of the Anthropocene [1], but no such debate rages around the extinction crisis currently underway globally [2]. Certain taxa have been reported as being more exposed to extinction risk than others due to a variety of pressures [1,3]. One such taxon is the Amphibians, having experienced high extinction rates globally due to a range of factors [4,5], such as climate and habitat change [6], chytrids [7], and overharvesting [8], to name a few.

Madagascar has been identified as a biodiversity hotspot [9], especially due to its diverse population of amphibians, with more than 365 recorded species to date but with many more species yet to be formally described [10]. The reported conservation pressures affecting amphibians in Madagascar include habitat change [11,12], diseases [13,14], alien species [12,15,16], and the wildlife trade, with the final pressure supplying bushmeat [8] and serving the demand in the pet trade [8,17–19]. However, many of the studies investigating the levels of wildlife trade regarding amphibians destined for international pet markets were conducted over 10–15 years ago. For example, in 1994, just 1 *Mantella* species was traded, but this number jumped to 14 known species in 2002/2003, while between 1994 and 2003, in just the *Mantella* genus, 233,893 individuals were reported to have been traded [20,21]. However, using CITES import data, for the period from 1994 to 2006, a total of 162,000 individuals were reported to have been traded across 18 species, while Malagasy government data reported the trade of over 221,000 individuals across 91 species between 2000 to 2006 [20,21]. In a global review of amphibians, it was determined that between 1978 and 2007, the genus *Mantella* was the most heavily traded genus, with 193,600 individuals traded over 14 species, accounting for 40% of the global trade figure, while 999 individuals belonging to 1 *Dyscophus* species and 2239 individuals belonging to *Scaphiophryne gottlebei* were the only other Malagasy species on the list [8]. More recently, between 2007 and 2018, the number of reported exports was reported to amount to 71,050 individual amphibians, with the genus *Mantella* accounting for nearly 97% of the trade, followed by *Scaphiophryne* (3%) and *Dyscophus* (0.5%), while the top three traded species were *M. betsileo* (n = 22,737; 33%), *M. baroni* (31%), and *M. nigricans* (11%) [21]. However, whilst the values reported were appropriate at the time, the wildlife trade has since been recognised as highly dynamic in nature. For example, it has been reported that the demand in species is driven by multiple factors [22], such as what was fashionable at a given time [23,24], dynamic changes in trade networks, and both financial and actor-level participant involvement, which can all impact trading [17,19–21,25]. Furthermore, national or international legislation changes also can affect trading [18,20,26].

Legislation change may be globally applicable, such as the Convention on International Trade in Endangered Species of Wild Fauna and Flora (CITES), or have a regional influence, such as the European Union (EU). For example, within the EU, conservation legislation sets and dictates the policies member states must implement and act on. Regarding trade in wildlife, Council Regulation (EC) No 338/97 of 9 December 1996 on the protection of species of wild fauna and flora regulates trade, specifically the procedures within Article 189c and Article 4(6) [27]. These refer to 'Introduction into the Community', with Article 4(6) stating: "In consultation with the countries of origin concerned, in accordance with the procedure laid down in Article 18 and taking account of any opinion from the Scientific Review Group, the Commission may establish general restrictions, or restrictions relating to certain countries of origin, on the introduction into the Community:" [27]. Annex A of Article 4 lists species, which can change according to the information available, that were allowed or restricted entry into the EU as directed in the documentation. For example, *M. aurantiaca* was listed under Article 4(6) code 'd' (which stated "of live specimens of species for which it has been established that their introduction into the natural environment of the Community presents an ecological threat to wild species of fauna and flora indigenous to the Community") but have since been assigned to code 'b' ("on the basis of the conditions referred to in paragraph 1(e) or paragraph 2(a), of specimens of species listed in Annex B") [28]. Such changes in coding highlight the highly dynamic conditions within which the wildlife trade operates, especially in Madagascar.

Many factors have changed since the previous studies on the Malagasy amphibian trade were conducted. For example, political leadership has changed, while greater and more coordinated strategic conservation efforts have been made [29–31]. The latter actions were supported through multi-national environmental agreements (MEAs), such as CITES, with the majority of traded Malagasy species listed within CITES Appendices, with varying conditions applying to them [32,33]. While the reported numbers of the Malagasy amphibians traded has increased, so too has the number of described species, increasing from 133 species in the 1990s to 244 around 2010, 292 in 2014, and 365 in 2022 [10]. This expansion in described species affects both the number and levels of trade being reported whilst also impacting the management of the trade in various ways, such as via misidentification opportunities, taxonomic reclassifications, etc. Furthermore, the pet trade has been recognised as and reported to be a fickle trade in terms of the species in demand [22–24]. A further consideration that needs careful addressal is the open and dynamic nature of the CITES datasets, which, while often used for trade reviews, contain inherent relevant issues [25]. Therefore, a sufficient knowledge base is required when analysing and interpreting these data [25] in order to assess the efficacy of CITES management and facilitate robust evidence-based approaches to either adapting, removing, or increasing the conservation efforts applied to species.

Therefore, this study aims to provide the most up-to-date and comprehensive review of the global trade in CITES-listed amphibian species exported from Madagascar. This information will establish a platform of knowledge from which appropriate conservation actions can be developed and implemented to improve Malagasy amphibian conversation. This study seeks to identify the countries involved and the types, levels, and complexities of the international trade in amphibians. Specifically, we aim to answer the following questions: (1) What are the levels, dynamics, and trends in the trade? (2) Which species feature significantly in the trade conducted, and what is their conservation status? (3) What is the effectiveness of CITES's actions regarding the trade?

2. Materials and Methods

The Convention on International Trade in Endangered Species of Wild Flora and Fauna (CITES) was established to facilitate the monitoring of trade in wildlife-based resources with the aim of securing the species' future through sustainable trade (www.CITES.org (accessed on 14 May 2022)). Nation states, registered as party members, submit yearly trade reports that provide details of both imports and exports conducted within the year to CITES. These data were then collated and stored on a trade database, which is maintained by United Nations Environment Program—World Conservation Monitoring Centre (UNEP-WCMC) in Cambridge, UK (https://trade.cites.org/ (accessed on 25 February 2023)), on behalf of the CITES Secretariat [25].

Data on trade conducted between 1975 to 2022 were collated and downloaded from the CITES database on 25 February 2023. The search criteria and terms used in the collation of these data have been presented in Table 1 and cover all CITES-listed species of Malagasy amphibians. Due to the well-reported permutations, vagaries, and lack of congruence between the CITES 'export reported' and 'import reported' trade values, which has often been overlooked in several studies reporting on wildlife trade, only the reported import trade data set was utilised in the following analyses [8,25,32]. CITES quota data and status for each species were extracted from Species+ (https://www.speciesplus.net/species; accessed on 5 March 2023), while their IUCN Redlist statuses were extracted from the IUCN Redlist (https://www.iucnredlist.org/; accessed on 5 March 2023).

Table 1. Criteria selected prior to performing the data collation for Amphibian trade within the CITES trade database (Source: UNEP-WCMC, 2023).

Database Field	Search Input
Search date	25 February 2023
Year range	1975–2022
Exporting countries	Madagascar
Importing countries	All countries
Source	Wild (W), Ranched (R), Source unknown (U)
Purpose	Commercial (T), Bred in captivity or artificially propagated (B), Botanical Garden (G), Circus and travelling exhibitions (Q), Personal (P)

Table 1. *Cont.*

Database Field	Search Input
Trade terms	Live (LIV), Specimens (SPE), Bodies (BOD)
Taxon	Amphibia (Amphibians)

Variables were investigated using non-parametric tests, such as Spearman rho correlation to analyse significant relationships between variables and the Mann–Whitney U test to determine any significant differences between variables.

3. Results

The data collated regarding the levels of trade between 1975 and 2022 recorded the first trading event in 1994, and the last datapoint reported was in 2021, resulting in a data period covering 28 years. During this period, a total of 20 known Malagasy amphibian species, plus two unknown listings (recorded as 'genus spp.'), and a total of 270,963 individual amphibians were reported to have been exported from Madagascar (Table 2).

Upon comparing the yearly data regarding the number of species traded with the number of individuals traded, a significant, positive relationship was observed (Figure 1; $n = 27$, $r_s = 0.76481$, and $p = 0.00$). However, the number of species being traded year after year did not display any longitudinal linear increase (Figure 2). Rather, after 1996, there was a rapid rise that peaked at 15 species in 2003/4 before reducing to an average of approx. 9 species traded per year between 2005 and 2021 (Figure 2). Conversely, the yearly average number of individuals traded was 9677 over the 28-year period and 6668 between 2005 and 2021, with a peak of 33,313 individuals in 2001 (Figure 2).

All 20 known species traded (Table 3) were listed on CITES App.II, while 1 (5%) was categorised as Critically Endangered ('CR') as per the IUCN Redlist, 6 (30%) were Endangered ('EN'), 4 (20%) were Vulnerable ('VU'), 1 (5%) was Near Threatened ('NT'), and 8 (40%) were of Least Concern ('LC') (Table 3). Grouping the species into their IUCN Redlist categorises revealed that the greatest level of trade was in the Least Concern species (8 spp.; 95,902 individuals) closely followed by Endangered (6 spp.; 92,634 individuals), Vulnerable (4 spp.; 34,849 individuals), Near Threatened (1 spp.; 21,147 individuals), and Critically Endangered (1 spp.; 6,043 individuals) species. However, analysing trade proportionally, the category listings altered, with the Near-Threatened (21,147 individuals per spp.) species being the most traded, followed by Endangered (15,439 individuals per species), Least Concern (11,987 individuals per species), Vulnerable (8712 individuals per species), and Critically Endangered (6043 individuals per species) species. Furthermore, over 71% (179,763 individuals) of the trade was conducted in species with reported declining populations (Table 3).

To manage resources, CITES utilises a quota system to limit the quantity of a particular resource. In terms of managing trade, a quota is successful when the actual trade value does not exceed the quota provided for that species by the exporting country. The relationship between actual trade conducted and the quota levels set have been presented for each genus in Figure 3. At the genus level, there were just two years when the total number of exported individuals within a genus exceeded the CITES quota total; namely, 2001 for *Mantella* (when 29,567 were exported and the quota was 8000) and 2010 for *Scaphiophryne* (when 302 were exported and the CITES quota total was 250).

Table 2. The species and number of individual Malagasy amphibians reported to have been exported from Madagascar in each year over the period from 1994 to 2021 (Source: UNEP/WCMC, 2023).

Amphibian Species	1994	1995	1996	1997	1998	1999	2000	2001	2002	2003	2004	2005	2006	2007
Dyscophus antongilii					20			75						
Dyscophus guineti														
Dyscophus insularis														
Dyscophus spp.				45										
Mantella aurantiaca	100	5515	6185	10,720	13,403	7815	5676	7545	1450	2681				
Mantella baroni								10		650	313	2670	1359	1570
Mantella bernhardi					30	440	543	400	60	105	60			
Mantella betsileo				1000	435	175	872	2926	460	1490	995	3110	1599	2238
Mantella cowanii					52	150	170	434	241	500	120			
Mantella crocea					395	250	763	1223	330	125	1020	2295	346	425
Mantella expectata					100	624	105	220	660	1390	1125	1280	2475	272
Mantella haraldmeieri							180		350	410				
Mantella laevigata					100	435	415	869	2155	533	1606	1795	2910	991
Mantella madagascariensis					125	2192	1535	450	3231	3325	4873	4245	3235	329
Mantella milotympanum									710	1780	850	1575	304	400
Mantella nigricans											200	315	150	382
Mantella pulchra					784	905	270	1658	1870	2585	2205	3455	1269	868
Mantella spp.					330	820	260	6779	9738	545	1366	255	200	
Mantella viridis					125	690	385	1434	2945	1110	2065	955	1260	295
Scaphiophryne gottlebei										980	776	270	216	465
Scaphiophryne marmorata														
Scaphiophryne spinosa														

Amphibian Species	2008	2009	2010	2011	2012	2013	2014	2015	2016	2017	2018	2019	2020	2021
Dyscophus antongilii														
Dyscophus guineti										170	437	202	30	182
Dyscophus insularis										62	286	215	60	108
Dyscophus spp.														
Mantella aurantiaca		135	1490	396	230	341	170	13	298	191	176	119	38	58
Mantella baroni	2237	1872	2100	415	2302	3005	2880	1142	2003	1584	1642	1811	80	160
Mantella bernhardi			88	49	14	82	12							
Mantella betsileo	1340	1818	1845	3396	2366	2239	2187	1331	2423	1248	1472	1551	22	392
Mantella cowanii														
Mantella crocea	410	436												
Mantella expectata	278	145	219		45	147	11							
Mantella haraldmeieri														
Mantella laevigata	665	973	808	813										
Mantella madagascariensis	212	192	203	102	85	53	81	50	73	105	20	20	8	9
Mantella milotympanum	267	157												
Mantella nigricans	192	272	144	1421	721	556	853	709	1048	721	716	956	204	282
Mantella pulchra	1197	1480	1116	241	297	193	184	78	193	122	68	97		12
Mantella spp.												50		
Mantella viridis	269	224	299											
Scaphiophryne gottlebei	171	377	302	191	163	171	48							
Scaphiophryne marmorata										22	80	93		
Scaphiophryne spinosa												170	220	20

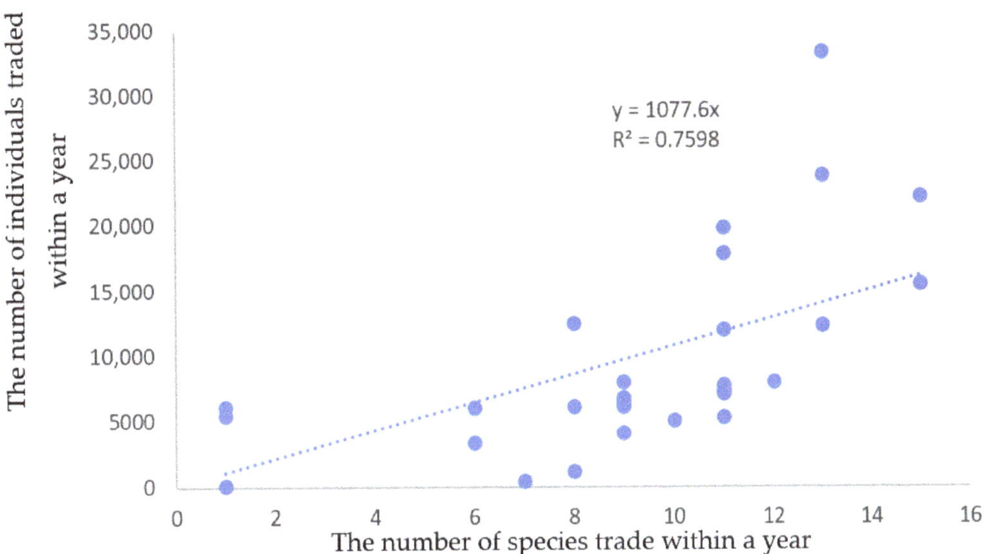

Figure 1. The relationship between the number of amphibian species traded and the total number of individuals traded on a yearly basis from Madagascar between 1994 and 2021 (Source: UNEP/WCMC, 2023).

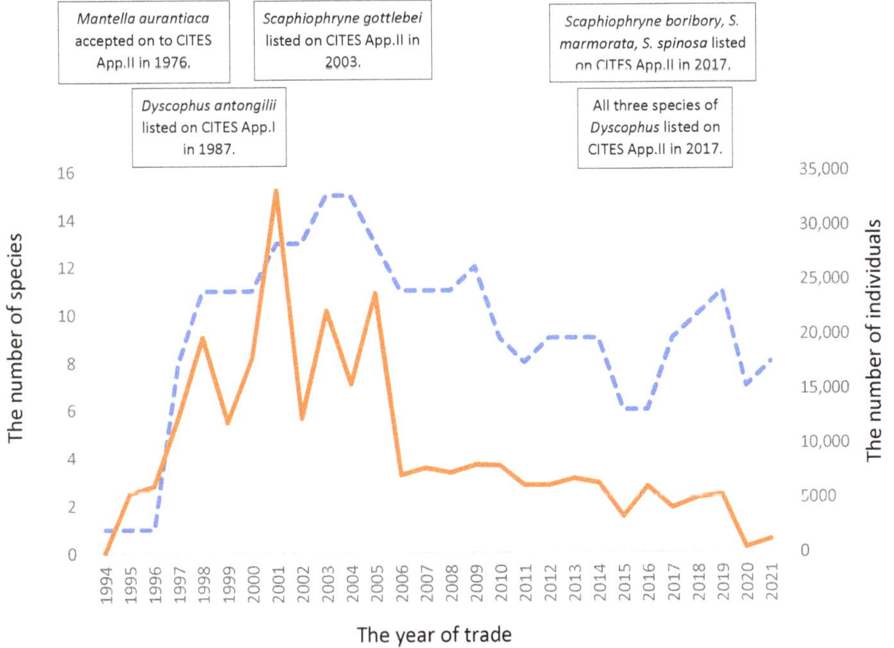

Figure 2. The non-linear increase in the number of species (blue dashed line and primary 'y' axis) exported from Madagascar each year and the total number of individuals (orange line and 2nd 'y' axis) reported to have been exported between 1994 and 2021 (Source: UNEP/WCMC, 2023).

However, at the species level and for each year, the reported number of amphibians exported from Madagascar exceeded the CITES quota level on many more occasions (Figures 4 and 5). As *Mantella* accounted for nearly 98% of the reported total export number, this genus was focused on to explore the relationship between the reported number exported and the CITES quotas per species for each year for which comparative data existed (Figures 4 and 5). The total number of occasions when the reported export number exceeded its CITES quota within a set year was 130 times over the whole period for *Mantella* species (Figure 4). Nearly 90% of these events were recorded leading up to 2005, with 16 species traded in numbers higher than their CITES quotas in 2004 (Figure 4). There were no recorded events in 2006 and 2007, while just two were recorded in 2008 with 2009 and four and five species were reported to have had their quotas exceeded in 2010; subsequently, these events reduce to nearly zero for the remainder of the period (Figures 4 and 5). Where the reported number exported for a species equals the CITES quota, it is shown as zero in Figure 5, while numbers below zero correspond to species traded at levels below their CITES quotas for the corresponding year, and vice versa. The scale of departure from the CITES quota was highest in 1998 for *M. aurantiaca* (Figure 5). After an initial period of a large number of above-quota events, post-2005–2006 trade levels rarely exceeded their CITES quota but were still highly fluctuating, albeit in a negative relationship (Figure 5). Much greater consistency in the relationships between these datasets was observed post-2014 (Figure 5).

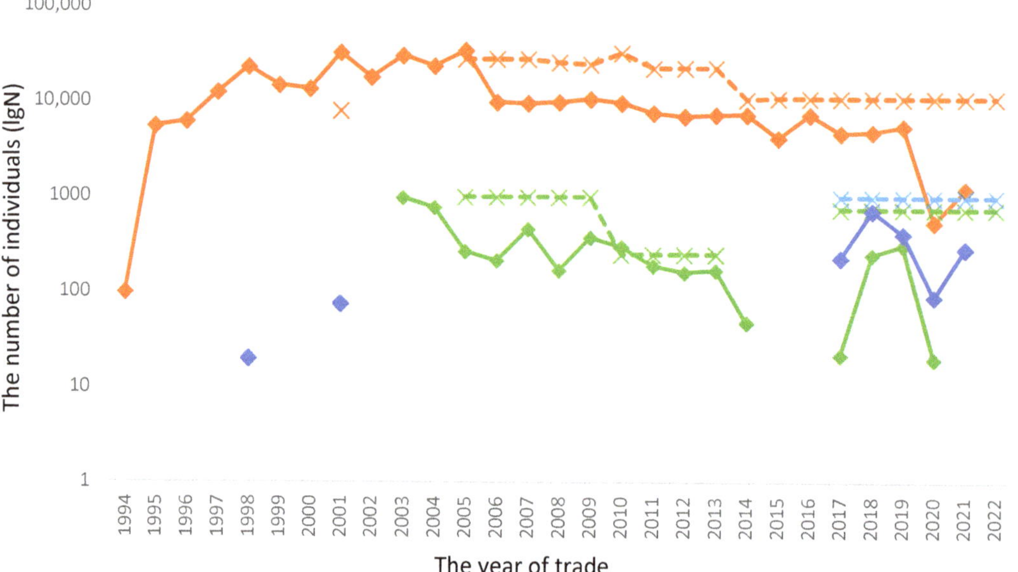

Figure 3. The relationship between the actual number of individuals exported from Madagascar for each of the three amphibian genera (*Mantella*, *Scaphiophryne*, and *Dyscophus*) and the CITES quotas implemented each year as reported between 1994 and 2021 (Source: UNEP/WCMC, 2023). Brown represents *Mantella*; green represents *Scaphiophryne*, and blue represents *Dyscophus*, while solid lines with diamonds denote reported export numbers, and dashed lines with open crosses denote the CITES quota levels.

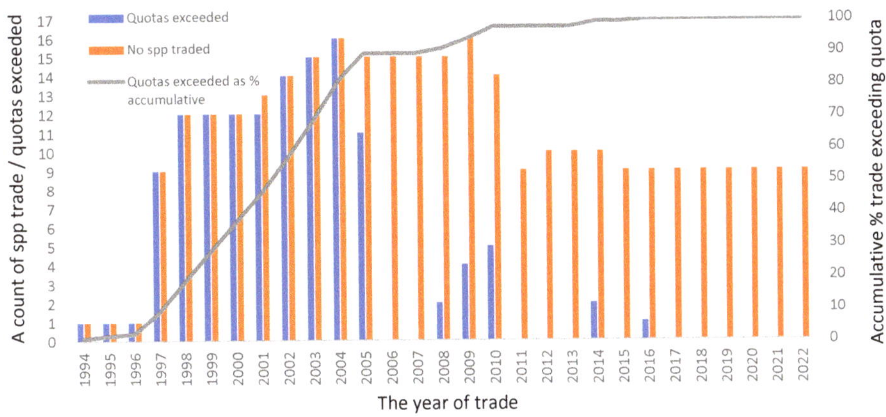

Figure 4. The events when reported trade numbers exceeded CITES quotas within a year as reported between 1994 and 2021 (Source: UNEP/WCMC, 2023).

Table 3. The Malagasy amphibian species recorded in the trade data, which are presented in order of the total number of individuals recorded in the trade and the species status according to CITES and the IUCN Red List.

Amphibian Species	Total Traded	% of Trade	CITES Listing	Population Trend	IUCN Redlist Status
Mantella aurantiaca	64,745	23.89	II	↓	EN
Mantella betsileo	38,930	14.37	II	↔	LC
Mantella baroni	29,805	11.00	II	?	LC
Mantella madagascariensis	24,753	9.14	II	↓	VU
Mantella pulchra	21,147	7.80	II	↓	NT
Mantella spp.	20,343	7.51		/	/
Mantella laevigata	15,068	5.56	II	↓	LC
Mantella viridis	12,056	4.45	II	↓	EN
Mantella nigricans	9842	3.63	II	↓	LC
Mantella expectata	9096	3.36	II	↓	EN
Mantella crocea	8018	2.96	II	↓	VU
Mantella milotympanum	6043	2.23	II	↓	CR
Scaphiophryne gottlebei	4130	1.52	II	↓	EN
Mantella bernhardi	1883	0.69	II	↓	VU
Mantella cowanii	1667	0.62	II	?	EN
Dyscophus guineti	1021	0.38	II	↓	LC
Mantella haraldmeieri	940	0.35	II	↓	EN
Dyscophus insularis	731	0.27	II	↓	LC
Scaphiophryne spinosa	410	0.15	II	?	LC
Scaphiophryne marmorata	195	0.07	II	↓	VU
Dyscophus antongilii	95	0.04	II	↓	LC
Dyscophus spp.	45	0.02		/	/

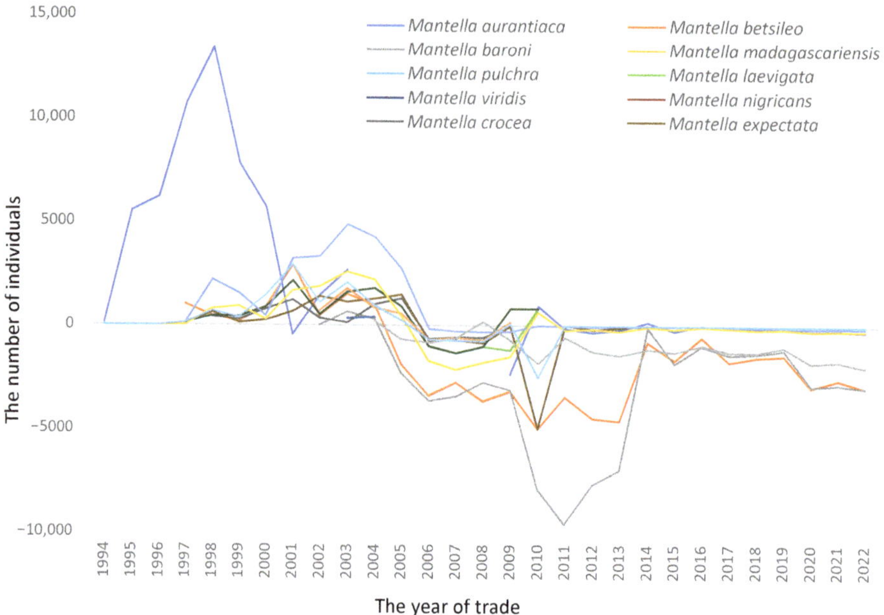

Figure 5. The graph shows when CITES quotas were exceeded (numbers presented > zero) or not met (numbers < zero), with the zero-reference line indicating when the reported number exported was equal to a given species' CITES quota for data reported between 1994 and 2021 (Source: UNEP/WCMC, 2023).

4. Discussion

This study aimed to investigate the trade in CITES-listed Malagasy amphibians and how effective the management of this trade was. This study has shown that the amphibians exported in the highest numbers from Madagascar to supply the international trade demand were still predominantly *Mantella* species. Of the 270,963 individual amphibians exported, over 97.5% were *Mantella* species, highlighting their continued demand in the trade, followed by *Scaphiophryne* spp. (1.7%) and *Dyscophus* spp. (nearly 0.7%). However, whilst an increasing number of *Mantella* species were recorded within the trade, especially in the post-2003/04/05 periods (with 15, 16, and 15 species exported, respectively), there was no linear increase in the yearly number of individuals reported to have been exported (Figures 1 and 4). This suggests that the international trade in *Mantella* had a demand ceiling, which could be met by a few or many *Mantella* species. From a population-harvesting-impact perspective, this information suggests there was no longer a favourite *Mantella* species. Rather, the harvesting impact could be spread across many *Mantella* species and even away from species that might have experienced high levels of exploitation or other perturbations to their populations via adaptive management processes.

A similar trading pattern was observed within the *Dyscophus* genus. Prior to 2017, *Dyscophus antongilii* was listed on CITES App. I, while *D. insularis* and *D. guineti* were not listed at all. At the CITES meeting in South Africa in 2017, *D. antongilii* was downlisted from App. I to App. II, while both *D. insularis* and *D. guineti* were uplisted to CITES App. II. However, after these regulatory changes, only trade in *D. insularis* and *D. guineti* was reported, thus indicating that the pre-2017 export trade fears that had previously listed *D. antongilii* as belonging to CITES App. I had altered. It is possible that this species was no longer 'fashionable', that the demand had already been satiated, or that the international demand window had been usurped by ex situ, captive breeding supplying the trade, with each scenario highlighting how extremely dynamic the wildlife trade can be for these species. Therefore, this could have wider consequences for any attempts to seek local community benefits from engaging in the sustainable harvesting of species for poverty alleviation and conservation benefits.

The advances in the ex situ captive breeding of Malagasy amphibians, such as reported in other studies [8], highlight the fact that the window of opportunity for native country supply is limited, which has financial implications for conservation within these source countries. For example, in 2023, various species of captive-bred *Mantella* could be purchased for the retail price of USD 87 (GBP 70; www.reptiles.swelluk.com (accessed 11 May 2023)). The average exchange rate for the last 5 years has been 1.25, resulting in the US $87 retail price. A reported 12,046 individuals were exported from Madagascar, equating to a financial income of USD 1,048,002. It has been stated that 3% of the retail value reaches local communities [17,18], resulting in a potential loss of over USD 31k in local income in Madagascar if the native supply of such species is lost to ex situ breeders. This also excludes the income generated for intermediaries, who were also Malagasy [17,18], in addition to the Government taxes accrued and the wider contributions to businesses and jobs along the wider supply chain, such as transport.

Madagascar's management of the trade as a potential sustainable resource was noticeable due to its apparent lack of any control on the levels of trade in amphibian species or its inability to implement CITES quotas early on (Figures 4 and 5). This led to greater international attention and scrutiny in relation to Madagascar by CITES, such as the performance of country reviews. [26]. However, trading levels exceeding CITES quota levels appears to have almost ceased following 2010 (Figure 4), with just 14 events between 2010 and 2022 across a variety of *Mantella* spp. Whether this was due to increased effectiveness in the management of the trade by Madagascar, captive breeding advances in the non-native countries supplying international demands, or a combination of these and other factors requires further study. One area for future study is the appropriateness and robustness of the datasets provided by Madagascar's government representatives to CITES that are used to calculate CITES Non-Detrimental Findings (NDFs) and serve as their basis. Upon viewing the quota values available, there seems to be a great deal of commonality in the values used despite the highly variable factors influencing each species. Thus, much greater attention needs to be paid toward increasing the robustness of the datasets that are used to calculate NDFs.

Furthermore, regarding the management of the trade and conservation statuses, as indicated by the 'IUCN Red List of Threatened Species' categorisation, there appears to be a lack of alignment or synchronisation between the two sectors. For example, one (5%) species was Critically Endangered ('CR'), six (30%) were Endangered ('EN'), and four (20%) were Vulnerable ('VU') under the IUCN Red List categorisation, yet they appeared to be highly traded (Table 3). However, at a finer scale, there were variabilities; for example, *Mantella milotympanum*, an IUCN Red List CR species, has not been recorded to have been exported since 2010. Conversely, *Mantella aurantiaca* (EN) has been reportedly exported every year from Madagascar over the 28-year period, except for 2004 to 2008, when no trade was recorded. Almost all the amphibian species were stated to have declining populations under the IUCN Red List; however, CITES quotas were set at 250, 500, or 3000, apparently without regard to the population trend. This highlights further possible research areas for the future, such as information sharing and synergies between CITES Non-Detrimental Findings (NDFs) and IUCN Red List categorisation. It is still unclear whether the most up-to-date information was being shared and utilised within each of the relevant working groups. Furthermore, it would be interesting to determine what degree of researcher collaboration transpired between the two organisations' working groups to ensure alignment in actions that is collaborative rather than appearing independent.

Thus, this study highlights areas where the management of wildlife trade practice can be improved, such as the CITES management of the trade. However, this study also highlights several areas for future research that could initiate further improvements to the management of Malagasy wildlife resources. Furthermore, there must be greater communication and alignment between international institutions to improve the management and knock-on sensitivities of these wildlife resources to attain conservation benefits.

5. Conclusions

Malagasy amphibian biodiversity has increased continuously over time and continues to increase. The addition of new species into the trade brings into question the suitability of the data that support their inclusion. For example, the NDFs produced within a country by the CITES Management Authority (MA) to determine the quotas submitted to CITES must be investigated with regard to their robustness and reliability. However, CITES capacity building efforts could be, in part, the reason for the apparent improvement of the amphibian trade in Madagascar. Whilst the conservation and poverty alleviation benefits will continue to be debated, it is imperative that the areas of both species knowledge and management effectiveness are constantly being reviewed and improved. Studies such as this one allow for those involved in the management of wildlife trade and conservation on Madagascar to identify areas where potential easy victories can be achieved with active changes.

Author Contributions: Conceptualization, A.I.C. and F.A.; methodology, A.I.C.; formal analysis, A.I.C.; investigation, A.I.C.; data curation, A.I.C.; writing—original draft preparation, A.I.C.; writing—review and editing, A.I.C. and F.A.; project administration, A.I.C. All authors have read and agreed to the published version of the manuscript.

Funding: This research received no external funding.

Institutional Review Board Statement: This study did not require ethical approval.

Informed Consent Statement: Not applicable.

Data Availability Statement: CITES species trade data were openly and freely available from CITES Trade Database.

Acknowledgments: AIC thanks the University of Cumbria for the administrative and technical support and the donations in kind towards producing this study.

Conflicts of Interest: The authors declare no conflict of interest.

References

1. Cowie, R.H.; Bouchet, P.; Fontaine, B. The Sixth Mass Extinction: Fact, fiction or speculation? *Biol. Rev.* **2022**, *97*, 640–663. [CrossRef] [PubMed]
2. IPBES. *Global Assessment Report on Biodiversity and Ecosystem Services of the Intergovernmental Science-Policy Platform on Biodiversity and Ecosystem Services*; IPBES Secretariat: Bonn, Germany, 2019; Available online: https://ipbes.net/global-assessment (accessed on 15 January 2023).
3. Jeliazkov, A.; Gavish, Y.; Marsh, C.J.; Geschke, J.; Brummitt, N.; Rocchini, D.; Haase, P.; Kunin, W.E.; Henle, K. Sampling and modelling rare species: Conceptual guidelines for the neglected majority. *Glob. Chang. Biol.* **2022**, *28*, 3754–3777. [CrossRef] [PubMed]
4. *Amphibian Conservation Action Plan*; Gascon, C.; Collins, J.P.; Moore, R.D.; Church, D.R.; McKay, J.E.; Mendelson, J.R., III. (Eds.) IUCN/SSC Amphibian Specialist Group: Gland, Switzerland; Cambridge, UK, 2007; 64p, ISBN 978-2-8317-1008-2.
5. Andreone, F.; Loarie, S.R.; Pala, R.; Luiselli, L.M.; Carpenter, A.I. Trade and exploitation of amphibians and reptiles: A short overview of the conservation impacts. *Atti dell'Accademia delle Scienze di Torino. Classe di Scienze Fisiche, Matematiche e Naturali* **2012**, *146*, 85–93.
6. Mims, M.C.; Drake, J.C.; Lawler, J.J.; Olden, J.D. Simulating the response of a threatened amphibian to climate-induced reductions in breeding habitat. *Landsc Ecol.* **2023**, *38*, 1051–1068. [CrossRef]
7. Wetsch, O.; Strasburg, M.; McQuigg, J.; Boone, M.D. Is overwintering mortality driving enigmatic declines? Evaluating the impacts of trematodes and the amphibian chytrid fungus on an anuran from hatching through overwintering. *PLoS ONE* **2022**, *17*, e0262561. [CrossRef] [PubMed]
8. Carpenter, A.I.; Andreone, F.; Moore, R.D.; Griffiths, R.A. A review of the global trade in amphibians: The types of trade, levels and dynamics in CITES listed species. *Oryx* **2014**, *48*, 565–574. [CrossRef]
9. *The New Natural History of Madagascar*; Goodman, S.M. (Ed.) Princeton University Press: Princeton, NJ, USA, 2022; ISBN 9780691222622.

10. Glaw, F.; Crottini, A.; Rakotoarison, A.; Scherz, M.D.; Vences, M. Diversity and exploration of the Malagasy amphibian fauna. In *The New Natural History of Madagascar*; Goodman, S.M., Ed.; Princeton University Press: Princeton, NJ, USA, 2022; Volume 2, pp. 1305–1322, ISBN 9780691222622.

11. Riemann, J.C.; Crottini, A.; Lehtinen, R.M.; Ndriantsoa, S.H.; Rödel, M.O.; Vallan, D.; Glos, J. Consequences of forest fragmentation and habitat alterations for amphibians. In *The New Natural History of Madagascar*; Goodman, S.M., Ed.; Princeton University Press: Princeton, NJ, USA, 2022; Volume 2, pp. 1336–1641. ISBN 9780691222622.

12. Irwin, M.T.; Wright, P.C.; Birkinshaw, C.; Fisher, B.F.; Gardner, C.J.; Glos, J.; Goodman, S.M.; Loiselle, P.; Rabeson, P.; Raharison, J.-L.; et al. Patterns of species change in anthropogenically disturbed forests of Madagascar. *Biol. Conserv.* **2010**, *143*, 2351–2362. [CrossRef]

13. Bletz, M.; Crottini, A.; Weldon, C.; Vences, M.; Rabemananjara, F.C.E.; Fisher, M.; Du Preez, L. Amphibian diseases and parasites. In *The New Natural History of Madagascar*; Goodman, S.M., Ed.; Princeton University Press: Princeton, NJ, USA, 2022; Volume 2, pp. 1342–1349, ISBN 9780691222622.

14. Bletz, M.; Rosa, G.; Andreone, F.; Courtois, E.A.; Schmeller, D.S.; Rabibisoa, N.H.C.; Rabemananjara, F.C.E.; Raharivololoniaina, L.; Vences, M.; Weldon, C.; et al. Widespread presence of the pathogenic fungus *Batrachochytrium dendrobatidis* in wild amphibian communities in Madagascar. *Sci. Rep.* **2015**, *5*, 8633. [CrossRef] [PubMed]

15. Fulvio, L.; Ficetola, G.F.; Freeman, K.; Mahasoa, R.H.; Ravololonarivo, V.; Solofo Niaina Fidy, J.F.; Koto-Jean, A.B.; Nahavitatsara, E.R.; Andreone, F.; Crottini, A. Abundance, distribution and spread of the invasive Asian toad *Duttaphrynus melanostictus* in eastern Madagascar. *Biol. Invasions* **2019**, *21*, 1615–1626.

16. Andreone, F.; Carpenter, A.I.; Crottini, A.; D'Cruze, N.; Dubos, N.; Edmonds, D.; Garcia, G.; Luedtke, J.; Megson, S.; Rabemananjara, F.C.E.; et al. Amphibian conservation in Madagascar: Old and novel threats to a peculiar fauna. In *Status and Threats of Afrotropical Amphibians—Sub-Saharan Africa, Madagascar, Western Indian Ocean Islands. Amphibian Biology*; Heatwole, H., Rödel, M.O., Eds.; Chimaira: Frankfurt am Main, Germany, 2021; Volume 11, pp. 147–187, ISSN 1613-2327.

17. Carpenter, A.I.; Robson, O. Madagascan amphibians as a wildlife resource and their potential as a conservation tool: Species and numbers exported, revenue generation and bio-economic models to explore conservation benefits. In *A Strategy for the Amphibians of Madagascar*; Andreone, F., Ed.; Monografie del Museo Regionale di Scienze Natural di Torino: Torino, Italy, 2008; pp. 357–376.

18. Rabemananjara, F.C.E.; Raminosoa, N.R.; Ramilijaona, O.R.; Rakotondravony, D.; Andreone, F.; Bora, P.; Carpenter, A.I.; Glaw, F.; Razafindrabe, T.; Vallan, D.; et al. Malagasy poison frogs in the pet trade: A survey of levels of exploitation of species in the genus *Mantella*. In *A Strategy for the Amphibians of Madagascar*; Andreone, F., Ed.; Monografie del Museo Regionale di Scienze Natural di Torino: Torino, Italy, 2008; pp. 277–300.

19. Rabemananjara, F.C.E.; Raminosoa, N.R.; Ramilijaona, O.R.; Andreone, F.; Bora, P.; Carpenter, A.I.; Glaw, F.; Razafindrabe, T.; Vallan, D.; Vieites, D.R.; et al. Malagasy poison frogs in the pet trade: A survey of levels of exploitation of species in the genus *Mantella*. *Amphib. Reptile Conserv.* **2007**, *5*, 3–16.

20. Carpenter, A.I.; Dublin, H.; Lau, M.; Syed, G.; McKay, J.E.; Moore, R.D. Over-harvesting. In *Amphibian Conservation Action Plan*; Gascon, C., Collins, J.P., Moore, R.D., Church, D.R., McKay, J.E., Mendelson III, J.R., Eds.; Proceedings: IUCN/SSC Amphibian Conservation Summit 2005; IUCN/SSC Amphibian Specialist Group: Gland, Switzerland, 2007; Chapter 5, pp. 26–31.

21. Carpenter, A.I.; Rabemananjara, F.C.E.; Behra, O.; Raselimanana, A.P.; Ratsoavina, F.M.; Wollenberg-Valero, K.C.; D'Cruze, N. Valourization of reptiles and amphibians from the wildlife trade and ecotourism. In *The New Natural History of Madagascar*; Goodman, S.M., Ed.; Princeton University Press: Princeton, NJ, USA, 2022; pp. 1452–1455.

22. Carpenter, A.I.; Slade, J. A Review of the Trade in Toucans (Ramphastidae): Levels of Trade in Species, Source and Sink Countries, Effects from Governance Actions and Conservation Concerns. *Conservation* **2023**, *3*, 153–174. [CrossRef]

23. Braunschneider, T. The Lady and the Lapdog: Mixed Ethnicity in Constantinople, Fashionable Pets in Britain. In *Humans and Other Animals in Eighteenth-Century British Culture*; Palmeri, F., Ed.; Routledge: London, UK, 2006; Chapter 2, ISBN 9781315252896.

24. Valdez, J.W. Using Google Trends to Determine Current, Past, and Future Trends in the Reptile Pet Trade. *Animals* **2021**, *11*, 676. [CrossRef] [PubMed]

25. Carpenter, A.I. The Ecology and Exploitation of Chameleons in Madagascar. Ph.D. Thesis, University of East Anglia, Norwich, UK, 2003.

26. Carpenter, A.I.; Robson, O.; Rowcliffe, M.; Watkinson, A.R. The impacts of international and national governance on a traded resource: A case study of Madagascar and its chameleon trade. *Biol. Conserv.* **2005**, *123*, 279–287. [CrossRef]

27. EU. Council Regulation (EC) No 338/97 of 9 December 1996 on the Protection of Species of Wild Fauna and Flora by Regulating Trade Therein; Official Journal of the European Union. 1997, 61/1. Available online: https://eur-lex.europa.eu/legal-content/EN/TXT/?uri=OJ:L:1997:061:TOC (accessed on 11 May 2023).

28. EU. Commission Implementing Regulation (Eu) 2019/1587 of 24 September 2019, Official Journal of the European Union. 2019, 248/5, EUR-Lex-31997R0338-EN-EUR-Lex (europa.eu). Available online: https://eur-lex.europa.eu/legal-content/EN/TXT/?uri=CELEX:32019R1587 (accessed on 11 May 2023).

29. Andreone, F.; Carpenter, A.I.; Cox, N.; du Preez, L.; Freeman, K.; Furrer, S.; Garcia, G.; Glaw, F.; Glos, J.; Knox, D.; et al. The challenge of conserving amphibian megadiversity in Madagascar. *PLoS Biol.* **2008**, *6*, e118. [CrossRef] [PubMed]

30. Andreone, F.; Carpenter, A.I.; Copsey, J.; Crottinid, A.; Garciac, G.; Jenkins, R.; Köhlerg, J.; Rabibisoah, N.; Randriamahazo, H.; Raxworthy, C.J. Saving the diverse Malagasy amphibian fauna: Where are we four years after implementation of the Sahonagasy Action Plan? *Alytes* **2012**, *29*, 43–58.

31. Crottini, A.; Ndriantsoa, S.H.; Rakotoarison, A.; Rakotonanahary, T.F.; García, G.; Rabibisoa, N.H.C.; Andreone, F. Amphibian conservation through the implementation of 'A conservation strategy for the amphibians of Madagascar–ASCAM'. In *The New Natural History of Madagascar*; Goodman, S.M., Ed.; Princeton University Press: Princeton, NJ, USA, 2022; Volume 2, pp. 1326–1330, ISBN 9780691222622.
32. CITES. Conservation of amphibians (Amphibia spp.); Nineteenth Meeting of the Conference of the Parties, Panama City, Panama, 14–25 November, 2022, CoP19 Doc. 60. Available online: https://cites.org/sites/default/files/documents/E-CoP19-60.pdf (accessed on 25 February 2023).
33. van den Burg, M.P.; Weissgold, B.J. Illegal trade of morphologically distinct populations prior to taxonomic assessment and elevation, with recommendations for future prevention. *J. Nat. Conserv.* **2020**, *57*, 125887. [CrossRef]

Article

Chromosome Diversity and Evolution of the Endemic Malagasy Velvet Geckos of the Genus *Blaesodactylus* (Reptilia, Gekkonidae)

Marcello Mezzasalma [1,*], Elvira Brunelli [1,*], Gaetano Odierna [2] and Fabio Maria Guarino [2]

[1] Department of Biology, Ecology and Earth Science, University of Calabria, Via P. Bucci 4/B, 87036 Rende, Italy
[2] Department of Biology, University of Naples Federico II, Via Cinthia 26, 80126 Naples, Italy;
      gaetanodierna@gmail.com (G.O.); fabio.guarino@unina.it (F.M.G.)
*   Correspondence: authors: marcello.mezzasalma@unical.it (M.M.); elvira.brunelli@unical.it (E.B.)

Simple Summary: We implemented a molecular and phylogenetic analysis and a comparative karyological investigation with standard and chromosome banding methods on different taxa of the endemic Malagasy velvet geckos of the genus *Blaesodactylus*. We present the description of the karyotype of three different taxa and a characterization of the chromosomal diversity in the genus. We show the occurrence of karyological variability in the genus *Blaesodactylus* in terms of chromosome number (2n = 40–42), morphology, chromosome position of loci of NORs, and distribution pattern of heterochromatin. Considering our results together with the available information on evolutionary related gecko species, we hypothesize that the karyotype evolution in *Blaesodactylus* mostly involved a decrease in the total number of chromosomes and the formation of biarmed elements. We also highlight that similar pathways of chromosomal rearrangements have been previously observed in other geckos, possibly representing a convergent karyotype evolution.

Citation: Mezzasalma, M.; Brunelli, E.; Odierna, G.; Guarino, F.M. Chromosome Diversity and Evolution of the Endemic Malagasy Velvet Geckos of the Genus *Blaesodactylus* (Reptilia, Gekkonidae). *Animals* 2023, 13, 2068. https://doi.org/10.3390/ani13132068

Academic Editor: Andrew Shedlock

Received: 24 May 2023
Revised: 15 June 2023
Accepted: 20 June 2023
Published: 22 June 2023

Abstract: We performed a molecular and phylogenetic analysis and a comparative cytogenetic study with standard karyotyping, silver staining (Ag-NOR) and sequential C-banding + Giemsa, + fluorochromes on several *Blaesodactylus* samples. The phylogenetic inference retrieved two main clades, the first comprises *B. victori*, *B. microtuberculatus* and *B. boivini*, while the second includes *B. sakalava*, *B. antongilensis* and *B. ambonihazo*. The available samples of *B. sakalava* form two different clades (here named *B. sakalava* clade A and clade B), which probably deserve a taxonomic re-evaluation. We found a karyological variability in *Blaesodactylus* in terms of chromosome number (2n = 40–42), morphology, location of NORs, and heterochromatin distribution pattern. *Blaesodactylus antongilensis* and *B. sakalava* clade A and B showed a karyotype of 2n = 40 mostly telocentric chromosomes. Pairs 1 and 6 were metacentric in *B. sakalava* clade A and B, while pair 1 was composed of subtelocentric/submetacentric elements in *B. antongilensis*. In contrast, *B. boivini* displayed a karyotype with 2n = 42 only telocentric chromosomes. NORs were on the first chromosome pair in *B. boivini*, and on the second pair in *B. antongilensis*. Adding our data to those available from the literature on evolutionarily related species, we highlight that the chromosome diversification in the genus probably proceeded towards a progressive reduction in the chromosome number and the formation of metacentric elements.

Keywords: cytogenetics; evolution; karyotype; Madagascar; Squamata

1. Introduction

Madagascar is well-known for its extraordinary biodiversity and remarkable degree of endemism and represents a unique model region for evolutionary studies [1–3]. The terrestrial reptile fauna of the island includes more than 450 endemic species of squamates belonging to six families of snakes (Boidae, Elapidae, Psammophiidae, Pseudoxyrhophiidae, Typhlopidae, Xenotyphlopidae) and six families of lizard (Agamidae, Chamaeleonidae, Gekkonidae, Gerrhosauridae, Opluridae, and Scincidae) [4,5]. Among them, the Malagasy

Gekkonidae include eleven different genera (*Blaesodactylus, Ebenavia, Geckolepis, Gehyra, Hemidactylus, Lygodactylus, Matoatoa, Paragehyra, Paroedura, Phelsuma* and *Uroplatus*) and more than 140 currently described species [5]. Nevertheless, although there has been significant progress achieved in the last few decades, the diversity of Malagasy reptiles is still relatively poorly known, with several new species discovered every year [4,5]. Even if recent research began to better describe the taxonomy and the evolutionary relationships of several different groups, only a small fraction of Malagasy reptile species has been analyzed with cytogenetic methods, despite increasing evidence that their diversity is reflected at the karyotypic level (see e.g., [6–9]). In fact, karyotype changes may either precede or follow molecular evolution, directly promoting cladogenesis or deriving from phylogenetic divergence [10,11]. In either circumstance, different karyological characteristics (e.g., different ploidy, total haploid number of chromosomes, chromosome morphology, presence or absence of differentiated sex chromosomes and occurrence and localization of particular DNA sequences) represent discrete cytogenetic markers which are helpful to describe different evolutionary trends or apomorphisms (see e.g., [12–14]).

Furthermore, squamates reptiles represent emergent model organisms in evolutionary cytogenetics as they are characterized by a high variability in chromosome number (from 2n = 16 to 2n = 62) and morphology and by the evolution of simple (XY, ZW) and multiple sex chromosome systems (X_1X_2Y and Z_1Z_2W) with either male or female heterogamety (see e.g., [15]). Two different general karyotype organizations have been described in squamates as "asymmetrical" (with macro- and microchromosomes) and "symmetrical" (with chromosomes that gradually decrease in length). Asymmetrical karyotypes are common in Iguania, many Scincomorpha and Platynota, while symmetrical karyotypes are often found in Lacertidae and Gekkota (see e.g., [15]).

The genus *Blaesodactylus* currently includes six described species (*B. ambonihazo, B. antongilensis, B. boivini, B. victori, B. microtuberculatus* and *B. sakalava*), which have been recently studied with morphological and molecular methods [16–19]. In contrast, chromosome analyses have been performed so far only on samples ascribed to *B. boivini* based on morphological characters [20], leaving the karyological variability of the genus completely unexplored. Concerning *B. boivini*, Chrostek et al. [20] analyzed one male and one female sample of the species and described a karyotype composed of 2n = 42, with only telocentric chromosomes which gradually decrease in length. Blocks of heterochromatin were identified on the centromeres of the six largest chromosome pairs and both sexes showed the same chromosome complement, without any evident heteromorphic sex chromosome pair. NORs were identified on the centromeric region of the first chromosome pair and telomeric repeats were localized at the centromeres of all chromosome pairs as well as at interstitial positions of several pairs [20].

In this paper we performed a preliminary molecular and phylogenetic analysis, using a segment of the mitochondrial ND4, and a comparative cytogenetic study with standard karyotyping, Ag-NOR staining and sequential C-banding on different *Blaesodactylus* samples from distinct Malagasy areas. We provide the first karyotype description of different taxa of the genus and a characterization of their chromosomal diversity. We show that chromosome variability in terms of total chromosome number, number of uni- and biarmed chromosomes and a different localization of chromosomal markers (NORs loci and heterochromatic regions) characterizes the studied taxa of the genus *Blaesodactylus*.

Finally, adding our newly generated karyotype data to those already available from the literature on evolutionarily closely related species and genera, we advance a hypothesis on the chromosome diversification in different genera of the family Gekkonidae.

2. Material and Methods

2.1. Sampling

We studied ten specimens of three species of Malagasy geckos of the genus *Blaesodactylus*. The taxonomic attribution after the molecular analysis (see below), field number, sex, and origin of all the samples analyzed in this study are provided in Table 1. All the specimens used in the molecular and cytogenetic analyses (Table 1) were collected between 1997–2002 for other research purposes and no animal was sampled during the realization of this study. After capture, animals were injected with a 0.5 mg/mL colchicine solution (0.1 mL/10 g body weight). Tissue samples (intestine, spleen and gonads) were incubated for 30 min in hypotonic solution (KCl 0.075 M + sodium citrate 0.5%, 1:1), fixed and conserved in Carnoy's buffer solution (methanol and acetic acid, 3:1). The fixed material was then preserved at 4 °C and transferred to the laboratory of University of Naples Federico II where it was processed as described below.

Table 1. Taxonomic attribution, origin, sex, field number of the study specimens of *Blaesodactylus*.

Species	Specimen	Locality	Sex
B. antongilensis	GA 483	Masobe Forest, Betampona	female
B. antongilensis	GA 1049	Masobe Forest, Betampona	female
B. boivini	FGMV 3010	Montagne des Français	female
B. sakalava clade A	FGMV 2029	Ifaty	female
B. sakalava clade A	FGMV 2030	Ifaty	male
B. sakalava clade A	FGMV 2032	Ifaty	male
B. sakalava clade B	GA 80	Isalo	male
B. sakalava clade B	GA 376	Marofandilia	juvenile
B. sakalava clade B	GA 377	Marofandilia	male
B. sakalava clade B	GA 378	Marofandilia	male

2.2. Molecular and Phylogenetic Analysis

A molecular and phylogenetic analysis was performed to ascertain the taxonomic status and the phylogenetic position of all the samples studied and to associate DNA sequences to the newly described karyotypes. The molecular analysis was performed using a fragment of the mitochondrial NADH dehydrogenase subunit 4 (ND4), which has been previously used in phylogenetic inferences in the genus *Blaesodactylus* [16,17,19,21]. Total genomic DNA was extracted following Sambrook et al. [22] and PCR amplification of the chosen ND4 fragment was performed according to Bauer et al. [17]. Amplicons were sequenced on an automated sequencer ABI 377 (Applied Biosystems, Foster City, CA, USA) using BigDye Terminator 3.1 (ABI). Chromatograms were checked and manually edited using Chromas Lite 2.6.6 and BioEdit 7.2.6.1 [23] and compared with available homologous traits deposited in GenBank. All the newly determined sequences were deposited in GenBank (accession numbers: OR113357-OR113363). A phylogenetic analysis using the newly determined ND4 sequences along with the homologous traits presented by Ineich et al. [19] (accession numbers: KX101035-KX101047, including those of the samples FGMV 2029, FGMV 2030 and FGMV 2032, here used in the cytogenetic analysis) was performed using MrBayes 3.2 [24]. The best fitting substitution model was selected using jModeltest 2.1.7 under the corrected Akaike information criterion (AICc) [25]. We run two independent Monte Carlo Markov Chains (MCMC) for 8,000,000 generations, sampling the chains every 1000 generations and discarding the first 25% of the trees sampled as burnin. We used as the outgroup the homologous ND4 trait of *Geckolepis maculata* (accession number JQ974269).

2.3. Cytogenetic Analysis

Chromosomes were obtained from preserved tissue samples and cell suspensions using the air-drying method, as described in Mezzasalma et al. [9]. The chromosome analysis was performed with standard karyotyping (5% Giemsa solution at pH 7 for 10 min) and different chromosome staining and banding methods. C-banding was performed following Sumner [26], and sequential C-banding + CMA_3 + DAPI according to Mezza salma et al. [27]. Active nucleolus organizing regions (NORs) were identified following the Ag-NOR staining method described by Howell and Black [28]. Metaphase plates were scored and recorded with an optical and an epifluorescence microscope (Axioscope Zeiss) equipped with an image analysis system. Karyotype reconstruction and calculation of chromosome relative length (RL = length of a chromosome/total karyotype length) (see Table 2) were performed after scoring at least ten metaphase plates per sample. Chromosomes were classified following Levan et al. [29] in metacentric (m), submetacentric (sm), subtelocentric (st) and telocentric (t).

Table 2. Chromosome morphometric parameters of the study species. RL% = Relative Length (length a chromosome/total chromosome length $\times$ 100). m = metacentric; sm = submetacentric; t = telocentric.

Species Chrom.	*B. antongilensis* RL%	*B. boivini* RL%	*B. sakalava* cl. A RL%	*B. sakalava* cl.B RL%
1	9.6 ± 0.7 25.0 ± 4.0 (st/sm)	9.7 ± 0.6 (t)	9.8 ± 1.0 39.3 ± 4.5 (m)	9.1 ± 0.8 43.1 ± 3.4 (m)
2	8.7 ± 1.1 (t)	8.2 ± 0.4 (t)	8.2 ± 0.9 (t)	8.1 ± 0.9 (t)
3	7.4 ± 0.4 (t)	7.4 ± 1.2 (t)	8.1 ± 0.7 (t)	7.8 ± 0.6 (t)
4	7.3 ± 0.5 (t)	6.6 ± 0.9 (t)	7.1 ± 0.6 (t)	6.6± 0.8 (t)
5	7.0 ± 0.7 (t)	5.8 ± 0.7 (t)	6.8.1 ± 0.5 (t)	6.4 ± 0.6 (t)
6	6.7 ± 0.6 38.4 ± 3.5 (m)	5.8 ± 1.0 (t)	6.2 ± 0.8 47.0 ± 3.0 (m)	6.2 ± 0.6 46.3 ± 3.1 (m)
7	6.2 ± 0.9 (t)	5.7 ± 0.5 (t)	5.4 ± 0.6 (t)	6.1 ± 0.8 (t)
8	5.6 ± 0.8 (t)	5.6 ± 0.4 (t)	5.3 ± 0.7 (t)	5.7 ± 0.6 (t)
9	5.0 ± 0.6 (t)	5.6 ± 0.6 (t)	5.3 ±0.8 (t)	5.4 ± 0.5 (t)
10	4.9 ± 0.5 (t)	5.6 ± 0.6 (t)	5.2 ± 0,6 (t)	4.9 ± 0.4 (t)
11	4.8 ± 0.8 (t)	4.6 ± 0.5 (t)	4.2 ± 0.6 (t)	4.5 ± 0.6 (t)
12	4.6 ± 0.4 (t)	3.9 ± 0.5 (t)	4.0 ± 0.4 (t)	4.1 ± 0.7 (t)
13	4.0 ± 0.6 (t)	3.3 ± 0.4 (t)	3.8 ± 0.5 (t)	3.8 ± 0.7 (t)
14	3.3 ± 1.0 (t)	3.3 ± 0.8 (t)	3.7 ± 0.7 (t)	3.6 ± 0.9 (t)
15	3.2 ± 0.8 (t)	3.2 ± 0.6 (t)	3.7 ± 0.7 (t)	3.4 ± 0.5 (t)
16	2.8 ± 0.9 (t)	3.2 ± 0.7 (t)	3.5 ± 0.6 (t)	3.1 ± 0.8 (t)
17	2.6 ± 0.7 (t)	2.9 ± 0.5 (t)	3.3 ± 0.6 (t)	3.0 ± 0.5 (t)
18	2.3 ± 0.6 (t)	2.8 ± 0.4 (t)	2.8 ± 0.7 (t)	2.8 ± 0.4 (t)
19	2.0 ± 0.5 (t)	2.4 ± 0.4 (t)	2.7 ± 0.8 (t)	2.7 ± 0.6 (t)
20	2.0 ± 0.8 (t)	2.3 ± 0.3 (t)	2.6 ± 0.5 (t)	2.6 ± 0.7 (t)
21		1.9 ± 0.5 (t)		

3. Results

3.1. Molecular and Phylogenetic Analysis

The phylogenetic analysis with the selected ND4 segment retrieved seven species-level lineages in *Blaesodactylus* (Figure 1). Six of these lineages correspond to the currently described species of the genus (*B. victorii*, *B. microtuberculatus*, *B. boivini*, *B. sakalava*, *B. ambonihazo* and *B. antongilensis*) (Figure 1). In addition, the available sequences of *B. sakalava* were divided into two distinct molecular clades here named *B. sakalava* clade A and *B. sakalava* clade B (Figure 1).

The phylogenetic relationships reported in our tree show that the seven species-level lineages are comprised of two different main clades: the first includes *B. victorii*, *B. microtuberculatus* and *B. boivini*, while the second comprises *B. sakalava* clade A and B, *B. ambonihazo* and *B. antongilensis* (Figure 1). In the first clade, *B. victorii* is the sister group of a clade composed of *B. microtuberculatus* and *B. boivini* (Figure 1). In the second clade, the specimens of *B. sakalava* were grouped into two different clades, together representing the sister group of the clade composed by *B. ambonihazo* and *B. antongilensis* (Figure 1).

3.2. Cytogenetic Analysis

Blaesodactylus sakalava and *B. antongilensis* had a karyotype of 2n = 40, mostly composed of telocentric chromosomes (Figure 2; Table 2). However, the elements of pairs 1 and 6, were metacentric in either *B. sakalava* clade A or *B. sakalava* clade B. In contrast, in *B. antongilensis*, the centromeric index (CI) of pair 1 (25.0%) corresponded to the transition value between a subtelocentric and a submetacentric element (Figure 2; Table 2) [29] .

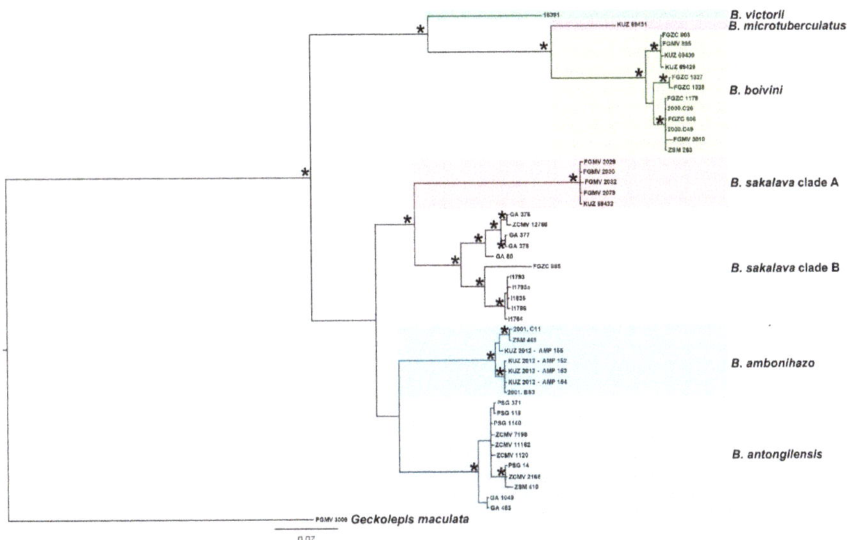

Figure 1. Phylogenetic tree with Bayesian inference using MrBayes 3.2. Asterisks at nodes represent posterior support values >0.97.

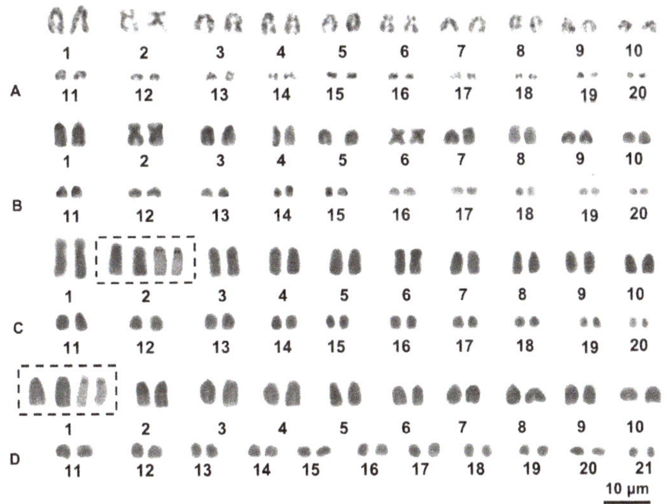

Figure 2. Giemsa-stained karyotypes of *B. sakalava* clade B (**A**), *B. sakalava* clade A (**B**), *B. antongilensis*, (**C**), *B. boivini* (**D**). The boxes show the NOR bearing pair, stained with Giemsa (left) and Ag-NOR staining (right).

Blaesodactylus boivini showed a karyotype of 2n = 42, composed of all telocentric chromosomes which gradually decrease in length (Figure 2D, Table 2).

Given the quantity and quality of metaphase plates, Ag-NOR staining and sequential C-banding + fluorochromes were successfully performed only on samples of *B. boivini* and *B. antongilensis*. Loci of NORs were close to the centromeres of the chromosome of pair 2 in

B. antongilensis and to the centromeres of the elements of pair 1 in *B. boivini* (FGMV 3010) (Figure 2C,D).

In both *B. boivini* and *B. antongilensis*, heterochromatic blocks were mostly localized on the centromeres of several chromosome pairs and were clearly evident after C-banding + Giemsa (Figure 3). In *B. boivini*, the main heterochromatic blocks were positive with either CMA_3 or DAPI, while in *B. antongilensis* they were more visible with CMA_3 (Figure 3).

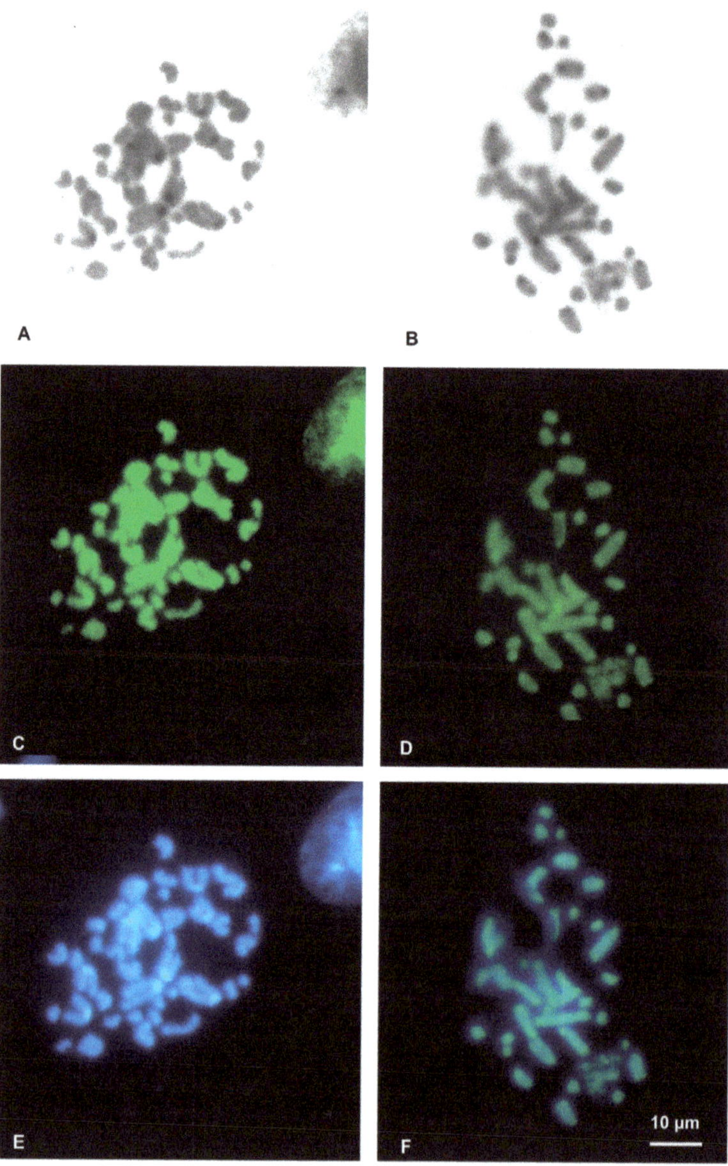

Figure 3. Metaphase plates of *B. boivini* (**A,C,E**) and *B. antongilensis* (**B,D,F**) sequentially stained with C-banding + Giemsa (**A,B**), +CMA_3 (**C,D**) and +DAPI (**E,F**).

4. Discussion

4.1. Molecular Analysis

The molecular and phylogenetic analysis using the selected ND4 segment allowed us to unambiguously assign all the studied samples to different *Blaesodactylus* genetic lineages as reported in Table 1 (see Figure 1). Our phylogenetic analysis using Bayesian inference showed similar evolutionary relationships and posterior support values to those previously reported by Ineich et al. [19]. All the currently recognized *Blaesodactylus* species show a relatively deep intraspecific mitochondrial genetic divergence (13–16% of ND4 uncorrected p-distance). We also highlight that the two distinct genetic clades found in *B. sakalava* (here named *B. sakalava* clade A and B) probably correspond to species-level lineages as their genetic distance is similar to that found between other *Blaesodactylus* sister species (see Figure 1) (see also [19]). This is not surprising, considering that different new *Blaesodactylus* species have been described in the last years and recent studies highlighted that the species diversity of the genus is probably still underestimated [18,19]. We also highlight that the DNA sequences of two different species of *Blaesodactylus* (*B. microtuberculatus* and *B. victorii*) are currently available from a single specimen, respectively, from the Ankarana National Park (north Madagascar) and the Tsingy limestone outcrops in the Namoroka National Park (northwestern Madagascar) (see [18,19]). In addition, most of the species-level lineages in *Blaesodactylus* appear to be allopatric, but *B. microtuberculatus* is sympatric with *B. boivini* and *B. victori* is sympatric with *B. sakalava* [18,19]. Further molecular analyses using a combination of mitochondrial and nuclear markers and a wide sampling across different Malagasy ecoregions are probably needed to better assess the molecular and species diversity in the genus *Blaesodactylus*.

4.2. Cytogenetic Analysis

In this study we performed the first comparative cytogenetic study on *Blaesodactylus*, providing the karyotype description of three different clades of this endemic Malagasy genus (*B. sakalava* clade A and B and *B. antongilensis*) and re-description of the karyotype of *B. boivini* (see also [20]). This contribution represents the first step in describing the karyological variability occurring in the genus *Blaesodactylus*, as well as a new contribution to reconstruct chromosomal evolutionary dynamics in a larger clade of the family Gekkonidae, which also includes the genera *Homopholis* and *Geckolepis* (Figure 1) (see also [19]).

Our chromosome analysis showed the occurrence of karyological variability between different studied species in terms of chromosome number (2n = 40–42), morphology, chromosome localization of loci of NORs, and pattern of heterochromatin. In particular, *B. sakalava* clade A and B and *B. antongilensis* showed a similar karyotype structure (with 2n = 40 and the occurrence of biarmed elements), with differences concerning the morphology of pairs 1 and 6. On the other hand, a different chromosome number (2n = 42) with only telocentric chromosomes was shown by *B. boivini*, as recently reported also by Chrostek et al. [20].

To advance a comparative hypothesis on the karyotype diversification in *Blaesodactylus* we superimposed all the currently available haploid karyograms of the genus, as well as those of the closely related genera *Homopholis* and *Geckolepis* [7,30,31], on their phylogenetic relationships ([32], this study). In particular, chromosome data are currently available for *G. typica* [7], *H. fasciata* and *H. walbergii* [20,30,31] (Figure 4). Notably, *G. typica* displays a karyotype structure similar to those of *B. sakalava* and *B. antongilensis* (2n = 40 and the occurrence of biarmed elements), while *H. fasciata* and *H. walbergii* present a lower chromosome number (2n = 36) [7,20,30,31] (Figure 4).

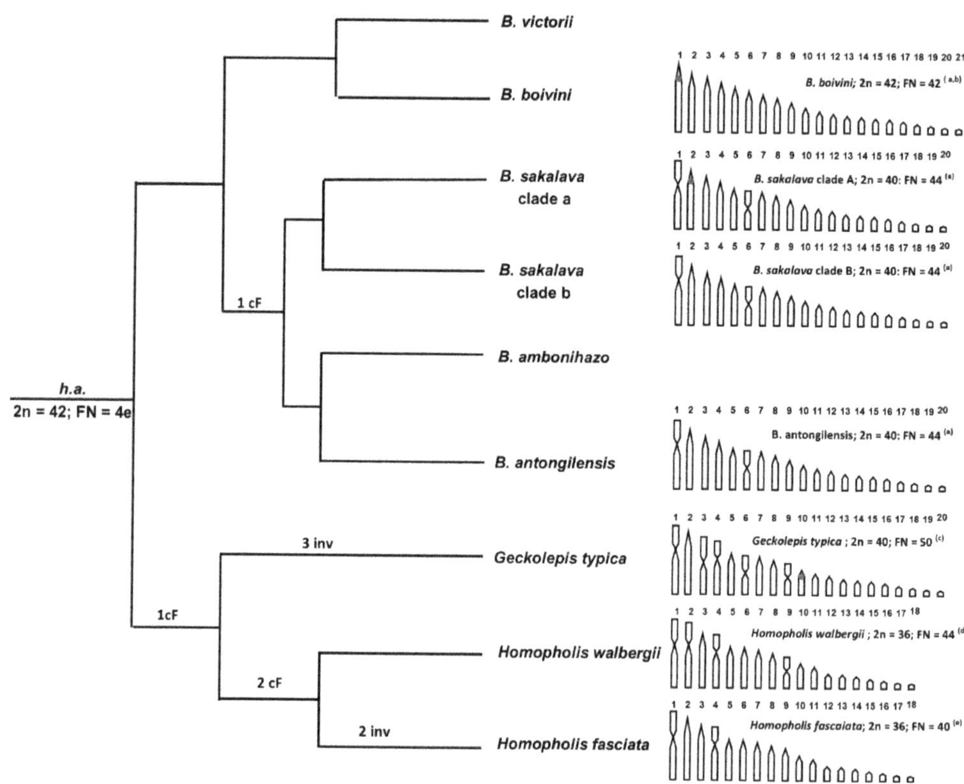

Figure 4. Phylogenetic relationships of *Blaesodactylus, Homopholis* and *Geckolepis* with the available haploid karyograms. h.a. = hypothesised ancestral condition, cF = centric fusion, inv = inversion. a = present paper; b = [20]; c = [7]; d = [30]; e = [31]. Phylogenetic relationships were redrawn from ([19,32], present study).

Among the family Gekkonidae, chromosome evolution has been hypothesized to have occurred either by an increase (as for example in the genus *Hemidactylus*) (see [33]) or by a reduction in the total chromosome number (e.g., in Diplodactylinae) [34,35].

In the case of *Blaesodactylus*, the most parsimonious chromosomal evolutionary scenario (including the lowest possible number of chromosome changes in order to generate all the different observed karyotypes) is to consider a karyotype composed of 2n = 42 and all telocentric chromosomes as the ancestral condition in the genus (Figure 4). In addition, a higher total chromosome number and a higher number of telocentric elements are generally considered ancestral karyotype characters in squamates (see e.g., [15]).

Our results show that a karyotype similar to the putative ancestral condition in *Blaesodactylus* (of 2n = 42) has probably been conserved in *B. boivini*, while different centric fusions likely occurred in *B. sakalava* clade A and B, reducing the total chromosome number to 2n = 40 and shaping the first and sixth chromosome pairs as metacentric. In *B. antongilensis*, a pericentromeric inversion (or, alternatively, a centromere repositioning) [36] likely shaped the first chromosome pair as a submeta-/subtelocentric chromosome (Figure 4). Similarly, in the clade including *Geckolepis* and *Homopholis* two centric fusions probably occurred in their common ancestor, producing a reduction in the total chromosome number (from 2n = 42 to 2n = 40) and the formation of two metacentric pairs (Figure 4). *Geckolepis typica* conserved a karyotype composed of 2n = 40 chromosomes, but three inversions produced an increase in biarmed elements (pairs 3, 5 and 8) (Figure 4). Two additional centric fusions occurred in *Homopholis* reducing the chromosome number to 2n = 36 and producing two additional biarmed chromosome pairs (pairs 2 and 8) (2n = 36; FN = 44). During the chromosomal diversification of this genus, *H. walbergii* conserved the primitive *Homopholis* karyotype, while two inversions occurred in *H. fasciata* (2n = 36; FN = 40) (Figure 4). Other than the hypothesized chromosome inversion, it is also possible that centromere repositioning has been involved in generating the chromosomal variability currently observable in *Geckolepis* and *Homopholis*.

Similar examples of a reduction in the total chromosome number and of a progressive formation of biarmed elements have been observed in different gecko genera, including *Uroplatus*, *Lygodactylus*, *Matoatoa*, *Paroedura* and *Christinus* [9,37–40]. In particular, all these different gecko genera show a karyotype composed of 2n = 34–42 mostly telocentric chromosomes and are characterized by the progressive appearance of biarmed elements by means of chromosome fusions in karyotypes with a reduced chromosome number and/or the translocation of small NOR-bearing chromosomes on larger chromosomes.

The repeated observation in different gecko groups of similar, independent instances of reduction in the total chromosome number by means of chromosome fusions and inversions suggests the possible occurrence of a convergent karyotype evolution.

Concerning the chromosome localization of NORs, they are close to the centromeres of the chromosomes of the first pair in *B. boivini* (see also [20]). Probably, this configuration is also conserved in *B. sakalava* clade A. In fact, this clade shows NORs on the second (telocentric) pair, but its first (metacentric) pair likely derived from the centric fusion of smaller telocentric chromosomes. In *G. typica*, NORs are close to the centromeres of the chromosomes of pair 10, but considering the lack of information on *Homopholis*, as well as on other *Geckolepis* species, additional data are needed to advance any hypothesis on NOR diversification.

Analysis with C-banding did not reveal the presence of differentiated sex chromosomes in either *B. boivini* or *B. sakalava* clade A. Nevertheless, we cannot exclude the possible presence of mostly pseudoautosomal sex chromosomes at an early stage of differentiation, which are known to occur in different groups of geckos (e.g., [9,15,39,41,42]). Alternatively, it is also possible that the sex is determined by environmental factors in the genus *Blaesodactylus* (e.g., temperature-dependent sex determination) as it has been previously documented in several gecko species (see [15]). Future research should employ a combination of molecular and cytogenetic methods to uncover the mechanism of sex determination in this group of geckos. Overall, C-banding + Giemsa produced similar results in *B. boivini* and *B. antongilensis*, evidencing a relatively low content of heterochromatin which is mostly localized at the centromeric regions of several chromosome pairs and particularly evident on the largest pairs (see Figure 3). This result was expected considering that a low content of heterochromatin generally characterizes the genome of squamate reptiles and is asymmetrically abundant on fully differentiated sex chromosome pairs (e.g., [15,43]). The differences between *B. boivini* and *B. antontongilensis* concerning the results obtained with C-banding + CMA$_3$ and C-banding + DAPI (which highlight CG- and AT-rich regions, respectively) suggest that a different nucleotide composition might characterize the repeated DNA content of the two species.

5. Conclusions

We performed a molecular analysis and a comparative cytogenetic study on four taxa of *Blaesodactylus*, presenting the first karyotype description of three of them. We show the occurrence of karyological variability in the genus concerning the total chromosome number (2n = 40–42), chromosome morphology, the karyotype localization of loci of NORs, and pattern of heterochromatin.

Our findings allowed us to hypothesize that the chromosomal diversification in the genus involved a reduction in the chromosome number and the progressive formation of biarmed chromosomes by means of fusions and inversions (or centromere repositioning). Following this hypothesis, the ancestral karyotype condition in *Blaesodactylus* is 2n = 42, similar to that shown by *B. boivini*, while apomorphic conditions of 2n = 40 with the formation of different biarmed chromosome pairs were observed in *B. sakalava* and *B. antongilensis*. Comparing our data to those available from the literature, we suggest that comparable chromosome rearrangements might also have characterized the karyotype evolution of the closely related *Geckolepis* and *Homopholis*. We also highlight that a similar pattern of chromosome diversification has already been observed in other gecko genera (e.g., *Paroedura*, *Lygodactylus*, *Matoatoa*, and *Uroplatus*), possibly representing independent events of convergent karyotype evolution.

This contribution confirms that chromosome changes often characterize the evolution and diversification of independent lineages of Malagasy squamates and that comparative cytogenetics provides a useful set of tools to describe evolutionary dynamics.

Author Contributions: Conceptualization, M.M.; methodology, M.M. and G.O.; validation, M.M. and G.O.; data curation, M.M., G.O., F.M.G. and E.B.; writing—original draft preparation, M.M.; writing—review and editing, M.M., G.O., F.M.G. and E.B.; visualization, M.M., G.O., F.M.G. and E.B. All authors have read and agreed to the published version of the manuscript.

Funding: This research received no external funding.

Institutional Review Board Statement: For this study we used samples already collected for other previously published studies with the approval of institutional committees and no further sampling was performed.

Informed Consent Statement: Not applicable, as this research did not involve humans.

Data Availability Statement: Newly generated cytogenetic data are available within this manuscript. DNA sequences are available on GenBank (accession numbers: OR113357-OR113363).

Acknowledgments: We are grateful to Malagasy authorities for granting research and export permits. We thank Gennaro Aprea and Franco Andreone for providing us with the tissue samples.

Conflicts of Interest: The authors declare no conflict of interest.

References

1. Myers, N.; Mittermeier, R.A.; Mittermeier, C.G.; da Fonseca, G.A.B.; Kent, J. Biodiversity hot-spots for conservation priorities. *Nature* **2000**, *403*, 853–858. [CrossRef] [PubMed]
2. Ganzhorn, J.U.; Lowry II, P.P.; Schatz, G.E.; Somme, S. The biodiversity of Madagascar: One of the world's hottest hotspots on its way out. *Oryx* **2001**, *35*, 346–348. [CrossRef]
3. Vences, M.; Wollenberg, K.C.; Vieites, D.R.; Lees, D.C. Madagascar as a model region of species diversification. *Trends Ecol. Evol.* **2009**, *24*, 456–465. [CrossRef] [PubMed]
4. Glaw, F.; Vences, M. *A Field Guide to the Amphibians and Reptiles of Madagascar*, 3rd ed.; Verlags GbR: Köln, Germany, 2007; p. 496.
5. Uetz, P.; Freed, P.; Hošek, J. The Reptile Database. 2022. Available online: http://www.reptile-database.org (accessed on 22 March 2023).
6. Koubová, M.; Pokorná, M.J.; Rovatsos, M.; Farkacová, K.; Altmanová, M.; Kratochvíl, L. Sex determination in Madagascar geckosof the genus *Paroedura* (Squamata: Gekkonidae): Are differentiated sex chromosomes indeed so evolutionary stable? *Chromosom. Res.* **2014**, *22*, 441–452. [CrossRef]
7. Mezzasalma, M.; Andreone, F.; Glaw, F.; Odierna, G.; Petraccioli, A.; Guarino, F.M. Chromosome aneupolyploidy in an endemic Malagasy gecko (Gekkonidae: *Geckolepis*). *Salamandra* **2018**, *54*, 56–62.
8. Mezzasalma, M.; Andreone, F.; Glaw, F.; Guarino, F.M.; Odierna, G.; Petraccioli, A.; Picariello, O. Changes in heterochromatin content and ancient chromosome fusion in the endemic Malagasy boid snakes *Sanzinia* and *Acrantophis* (Squamata: Serpentes). *Salamandra* **2019**, *55*, 140–144.
9. Mezzasalma, M.; Brunelli, E.; Odierna, G.; Guarino, F.M. First Insights on the Karyotype Diversification of the Endemic Malagasy Leaf-Toed Geckos (Squamata: Gekkonidae: *Uroplatus*). *Animals* **2022**, *12*, 2054. [CrossRef]
10. Ayala, F.J.; Coluzzi, M. Chromosome speciation: Humans, *Drosophila*, and mosquitoes. *Proc. Natl. Acad. Sci. USA* **2005**, *102*, 6535–6542. [CrossRef]
11. Faria, R.; Navarro, A. Chromosomal speciation revisited: Rearranging theory with pieces of evidence. *Trends Ecol. Evol.* **2010**, *25*, 660–669. [CrossRef]
12. Olmo, E. Trends in the evolution of reptilian chromosomes. *Integr. Comp. Biol.* **2008**, *48*, 486–493. [CrossRef]
13. Petraccioli, A.; Guarino, F.M.; Kupriyanova, L.; Mezzasalma, M.; Odierna, G.; Picariello, O.; Capriglione, T. Isolation and characterization of interspersed repeated sequences in the common lizard, *Zootoca vivipara*, and their conservation in squamata. *Cytogenet. Genome Res.* **2019**, *157*, 65–76. [CrossRef] [PubMed]

14. Sidhom, M.; Said, K.; Chatti, N.; Guarino, F.M.; Odierna, G.; Petraccioli, A.; Picariello, O.; Mezzasalma, M. Karyological characterization of the common chameleon (*Chamaeleo chamaeleon*) provides insights on the evolution and diversification of sex chromosomes in Chamaeleonidae. *Zoology* **2020**, *141*, 125738. [CrossRef] [PubMed]

15. Mezzasalma, M.; Guarino, F.M.; Odierna, G. Lizards as Model Organisms of Sex Chromosome Evolution: What We Really Know from a Systematic Distribution of Available Data? *Genes* **2021**, *12*, 1341. [CrossRef]

16. Greembaum, E.; Bauer, A.M.; Jackmann, T.R. *Homopholis* and *Blaesodactylus* (Squamata: Gekkonidae) revisited new insights from a molecular phylogeny. *Afr. J. Herpetol.* **2007**, *56*, 101–114. [CrossRef]

17. Bauer, A.M.; Glaw, F.; Gehring, P.-S.; Vences, M. New species of *Blaesodactylus* (Squamata: Gekkonidae) from Ankarafantsika National Park in north-western Madagascar. *Zootaxa* **2011**, *2942*, 57–68. [CrossRef]

18. Jono, T.; Bauer, A.M.; Brennan, I.; Mori, A. New species of *Blaesodactylus* (Squamata: Gekkonidae) from Tsingy karstic outcrops in Ankarana National Park, northern Madagascar. *Zootaxa* **2015**, *3980*, 406–416. [CrossRef]

19. Ineich, I.; Glaw, F.; Vences, M. A new species of *Blaesodactylus* (Squamata: Gekkonidae) from Tsingy limestone outcrops in Namoroka National Park, north-western Madagascar. *Zootaxa* **2016**, *4109*, 523–541. [CrossRef]

20. Chrostek, G.; Domaradzka, A.; Yurchenko, A.; Kratochvíl, L.; Mazzoleni, S.; Rovatsos, M. Cytogenetic Analysis of Seven Species of Gekkonid and Phyllodactylid Geckos. *Genes* **2023**, *14*, 178. [CrossRef]

21. Jacobsen, N.H.; Kuhn, A.L.; Kuhn, A.L.; Jackman, T.R.; Bauer, A.M. A phylogenetic analysis of the southern African gecko genus *Afroedura* Loveridge (Squamata: Gekkonidae), with the description of nine new species from Limpopo and Mpumalanga provinces of South Africa. *Zootaxa* **2014**, *3846*, 451–501. [CrossRef]

22. Sambrook, J.; Fritsch, E.F.; Maniatis, T. *Molecular Cloning: A Laboratory Manual*, 2nd ed.; Cold Spring Harbor Lab Press: New York, NY, USA, 1989.

23. Hall, T.A. BioEdit: A user-friendly biological sequence alignment editor and analysis program for Windows 95/98/NT. *Nucleic Acids Symp. Ser.* **1999**, *41*, 95–98.

24. Ronquist, F.; Teslenko, M.; van der Mark, P.; Ayres, D.L.; Darling, A.; Höhna, S.; Larget, B.; Liu, L.; Suchard, M.A.; Huelsenbeck, J.P. MrBayes 3.2: Efficient Bayesian Phylogenetic Inference and Model Choice Across a Large Model Space. *Syst. Biol.* **2012**, *61*, 539–542. [CrossRef] [PubMed]

25. Darriba, D.; Taboada, G.; Doallo, R.; Posada, D. jModelTest 2: More models, new heuristics and parallel computing. *Nat. Methods* **2012**, *9*, 772. [CrossRef]

26. Sumner, A.T. A simple technique for demonstrating centromeric heterochromatin. *Exp. Cell Res.* **1972**, *75*, 304–306. [CrossRef] [PubMed]

27. Mezzasalma, M.; Brunelli, E.; Odierna, G.; Guarino, F.M. Comparative cytogenetics of *Hemorrhois hippocrepis* and *Malpolon monspessulanus* highlights divergent karyotypes in Colubridae and Psammophiidae (Squamata: Serpentes). *Eur. Zool. J.* **2023**, *90*, 201–210. [CrossRef]

28. Howell, W.M.; Black, D.A. Controlled silver staining of nucleolus organizer regions with a protective colloidal developer: A1-step method. *Experientia* **1980**, *36*, 1014–1015. [CrossRef] [PubMed]

29. Levan, A.; Fredga, K.; Sandberg, A.A. Nomenclature for centromeric position on chromosomes. *Hereditas* **1964**, *52*, 201–220. [CrossRef]

30. De Smet, W.H.O. Description of the orcien stained karyotypes of 27 lizard species (Lacertilia: Reptilia) belonging to the families Iguanidae, Agamidae Chameleontidae and Gekkonidae (Ascalabota). *Acta Zool. Pathol. Antverpiensia* **1981**, *76*, 35–72.

31. Rovatsos, M.; Kratochvíl, L.; Altmanová, M.; Johnson Pokorná, M. Interstitial Telomeric Motifs in Squamate Reptiles: When the Exceptions Outnumber the Rule. *PLoS ONE* **2015**, *10*, e0134985. [CrossRef]

32. Pyron, R.A.; Burbrink, F.T.; Wiens, J.J. A phylogeny and revised classification of Squamata, including 4161 species of lizards and snakes. *BMC Evol. Biol.* **2013**, *13*, 93. [CrossRef]

33. Trifonov, V.A.; Giovannotti, M.; O'Brien, P.C.; Wallduck, M.; Lovell, F.; Rens, W.; Parise-Maltempi, P.P.; Caputo, V.; Ferguson-Smith, M.A. Chromosomal evolution in Gekkonidae. I. Chromosome painting between *Gekko* and *Hemidactylus* species reveals phylogenetic relationships within the group. *Chromosome Res.* **2011**, *19*, 843–855. [CrossRef]

34. King, M. Chromosomal evolution in the Diplodactylinae (Gekkonidae: Reptila). I. Evolutionary relationships and patterns of change. *Aust. J. Zool.* **1987**, *35*, 507–531. [CrossRef]

35. Oliver, P.; Hutchinson, M.; Hutchinson, R. Karyotypic Variation in the Australian Gecko *Diplodactylus tessellatus*, with the Description of a New Karyotypic Complement for Diplodactyline Geckos. *J. Herpetol.* **2007**, *41*, 540–543. [CrossRef]

36. Amor, D.J.; Bentley, K.; Ryan, J.; Perry, J.; Wong, L.; Slater, H.; Choo, K.H. Human centromere repositioning "in progress". *Proc. Natl. Acad. Sci. USA* **2004**, *101*, 6542–6547. [CrossRef] [PubMed]

37. King, M.; Rofe, R. Karyotypic variation in the Australian Gekko *Phyllodactylus marmoratus* (Gray) (Gekkonidae: Reptilia). *Chromosoma* **1976**, *54*, 75–87. [CrossRef] [PubMed]

38. Aprea, G.; Andreone, F.; Fulgione, D.; Petraccioli, A.; Odierna, G. Chromosomal rearrangements occurred repeatedly and independently during species diversification in Malagasy geckos, genus *Paroedura*. *Afr. Zool.* **2013**, *48*, 96–108. [CrossRef]

39. Mezzasalma, M.; Andreone, F.; Aprea, G.; Glaw, F.; Odierna, G.; Guarino, F.M. Molecular phylogeny, biogeography and chromosome evolution of Malagasy dwarf geckos of the genus *Lygodactylus* (Squamata, Gekkonidae). *Zool. Scr.* **2017**, *46*, 42–54. [CrossRef]

40. Mezzasalma, M.; Guarino, F.; Loader, S.; Odierna, G.; Streicher, J.; Cooper, N. First karyological analysis of the endemic Malagasy phantom gecko *Matoatoa brevipes* (Squamata: Gekkonidae). *Acta Herpetol.* **2020**, *15*, 137–141.
41. Gamble, T.A. Review of sex determining mechanisms in geckos (Gekkota: Squamata). *Sex Dev.* **2010**, *4*, 88–103. [CrossRef]
42. Pokorná, M.; Rábová, M.; Ráb, P.; Ferguson-Smith, M.A.; Rens, W.; Kratochvíl, L. Differentiation of sex chromosomes and karyotypic evolution in the eye-lid geckos (Squamata: Gekkota: Eublepharidae), a group with different modes of sex determination. *Chromosome Res.* **2010**, *18*, 809–820. [CrossRef]
43. Mezzasalma, M.; Visone, V.; Petraccioli, A.; Odierna, G.; Capriglione, T.; Guarino, F.M. Non-random accumulation of LINE1-like sequences on differentiated snake W chromosomes. *J. Zool.* **2016**, *300*, 67–75. [CrossRef]

MDPI AG
Grosspeteranlage 5
4052 Basel
Switzerland
Tel.: +41 61 683 77 34

Animals Editorial Office
E-mail: animals@mdpi.com
www.mdpi.com/journal/animals